How Surfaces Intersect in Space

SERIES ON KNOTS AND EVERYTHING

Editor-in-charge: Louis H. Kauffman

Published:

Vol. 1: Knots and Physics
L. H. Kauffman

Vol. 2: How Surfaces Intersect in Space
J. S. Carter

Vol. 3: Quantum Topology
edited by L. H. Kauffman & R. A. Baadhio

Vol. 4: Gauge Fields, Knots and Gravity
J. Baez & J. P. Muniain

Vol. 6: Knots and Applications
edited by L. H. Kauffman

Vol. 7: Random Knotting and Linking
edited by K. C. Millett & D. W. Sumners

Forthcoming:

Vol. 5: Gems, Computers and Attractors for 3-Manifolds
S. Lins

Vol. 8: Symmetric Bends: How to Join Two Lengths of Cord
R. E. Miles

Vol. 9: Combinatorial Physics
T. Bastin & C. W. Kilmister

Vol. 10: Nonstandard Logics and Nonstandard Metrics in Physics
W. M. Honig

K&E Series on Knots and Everything — Vol. 2

How Surfaces Intersect in Space

An introduction to topology

Second Edition

J Scott Carter

Department of Mathematics and Statistics
University of South Alabama, USA

World Scientific
Singapore • New Jersey • London • Hong Kong

Published by

World Scientific Publishing Co. Pte. Ltd.

P O Box 128, Farrer Road, Singapore 9128

USA office: Suite 1B, 1060 Main Street, River Edge, NJ 07661

UK office: 57 Shelton Street, Covent Garden, London WC2H 9HE

First edition: 1993

HOW SURFACES INTERSECT IN SPACE: AN INTRODUCTION TO TOPOLOGY (Second Edition)

ISBN 981-02-2082-0
ISBN 981-02-2066-9 (pbk)

Printed in Singapore by Uto-Print

Dedicated to:

Huong, Albert, and Alexander

Contents

Preface

Milnor's book "Topology from a Differential Viewpoint" is a model for clarity, conciseness, and rigor. The current text might be subtitled "Topology from Scott Carter's Viewpoint," and critics will, no doubt, say it is a model for none of the above. However, I make no apology for presenting things from my point of view. I have been working with geometric topology for about 12 years now, and I have a pretty clear picture of what axioms I use and need. Furthermore, within the text I have paid appropriate homage to Thom, Milnor, Smale, Thurston, Freedman, and Jones. We should be reading topology from their points of view since their perspective of the subject has shaped the way the rest of us approach it.

In regard to rigor, there is an axiomatic approach to this book. First, golden fleece is axiomatized as an infinitely malleable, infinitesimally thin material that fuses seamlessly. The mathematical reader will recognize that this conceptual surface lies at the heart of the idea of a manifold. The Jordan Curve Theorem is the principal postulate of the first chapter. Furthermore, the quotient topology is axiomatized as glue or sewing. Finally, and most importantly, the general position assumption has been made throughout even though only an inkling of the proof is indicated.

From my point of view, a single post-calculus course should be devoted to the proofs of the general position hypotheses. Meanwhile, people should not postpone their mathematical enrichment for the sake of seeing those proofs. In short, general position is a completely reasonable assumption, that should be made unabashedly. And the need for the proof of these hypotheses only comes *a posteriori* when you find that general position allows you to prove a huge bunch of stuff.

The motivation for writing this book came from the physicists who explain their theories without letting partial differential equations get in the way. They have consistently been willing to step down from their ivory towers and explain to the rest of us how their magic works. Mathematicians, have for the most part, not been willing to do so. Maybe our stuff doesn't sell as well as physics does, but a sales pitch once in a while would help the mathematics enterprise.

People's reluctance to read about math comes from our Aristotelian upbringing. There is nothing wrong with starting a discussion with the definition of *all* of the subsequent terms to be used, except that the text that results is tedious and boring. Rather than assuming that a reader either knows what a Lie group is or doesn't know, it would be better for both the reader and the writer if the latter started writing about the set of (2×2) matrices that have determinant 1. What are these matrices? Why are they important? What properties does this set have? In short, let's start from examples for a change.

I did not write this book for mathematicians, but I think many of them (you) will benefit from the treatment I have given. I have touched on most of the major topics in topology, but I often stopped short of the punch. The references given at the end of each chapter certainly fill in the details. Discussions that required much algebra were omitted, and that is why the punches were pulled. A lay-person's book about algebra that goes beyond Hermann Weyl's "Symmetry" is sorely needed.

This book started from lectures I have given over the years and comprises a course that I imagine myself teaching. The lectures have been given to various math clubs and undergraduate organizations. The most recent one of these was presented at the Alabama Academy of Mathematics and Science. The audiences at these talks have been fairly forgiving, and I hope that this book will fill the gaps that I left open.

I ask only one indulgence of the lay reader. Please, don't ask why we study topology. After you read a while, you will see the intrinsic value and beauty of it. Topology is not necessarily useful. The exploration of topological spaces is the exploration of the last great frontier — the human mind. These spaces have been created by humans for the purpose of understanding the world in which we live. But ultimately they lead to an understanding of our mind, for It can only be understood in terms of Its creations. Topological space, then, are at once a form of art and a form of science, and as such they reflect our deepest intellect.

The figures were drawn on plain white paper in india ink using a set of rapiodgraph pens. Most figures were preliminarily sketched with non-photo blue pencil. A ruler, a template, and a flexible ruler were used to keep the lines from being shakey where ever this was possible. Finally, all too much liquid paper was used to cover the mistakes.

⋆

Many people deserve special thanks for their help. Lou Kauffman had enough confidence in my abilities to put this in the *Knots and Everything* Series. Arunas Liulevicius was the first person ever to suggest that I put my pictures in book form. Ronnie Lee guided me through my Ph.D. dissertation — glimpses of which are seen in the last chapter. I thank all those people who have written letters of recommendation for me over the years, and the colleagues with whom I have worked at The University of Texas, Lake Forest College, Wayne State University, and The University of South Alabama. Several people have read over parts of the manuscript with helpful suggestions: Deirdre Reynolds, Richard Hitt, José Barrionuevo, Dan Flath, and Claire Carter had kind and well thought out criticisms. John Hughes gave me a key suggestion that was used in turning the 2-sphere inside-out. Nguyen Thom gave me a lecture on shading and light sources, and his criticisms of my drawings have always been appreciated. I especially want to thank Masahico Saito for his work with me, as well as his patiently holding back on our joint research while I finished the book. P. E. Zap of AMOEBA Enterprises typed the manuscript and is responsible for all of the mistakes.

To Albert, Alexander, and Huong, I am sorry if we missed some time together while this was being written.

List of Figures

How Surfaces Intersect in Space

Chapter 1

Surface and Space

1.1 Prologue

What is Space? Physicists and architects might tell you that space is defined by the things that occupy it. Virtual particles and antiparticles are continually being created and annihilated in the vacuum. The geometry of large scale space is determined by those massive bodies that occupy it. The ceiling, floor, and walls — rather their configuration and their relationships — define the space of the room in which you are seated. In music, the time between notes defines rhythm; the intervals between notes define melody. Even silence is defined by the sounds that fill it. Listen.

Mathematicians understand sets and spaces in terms of their constituents, or more precisely, in terms of the relations among the aggregates of elements that form subspaces. The relations are sometimes analytic, sometimes algebraic, and sometimes geometric depending on the nature of the space being studied.

In this book, we will study a variety of 2-, 3-, and 4-dimensional spaces. Our study will always begin from simple examples and build intuition from there. We will examine what happens in the mundane space in which we are seated and abstract concepts from this world to other possible worlds. These other worlds are indeed possible although they will only be imagined.

(The word "only" is misleading in this regard: Imaginations are unlimited, and if you and I agree on all of the rigors of an imagined concept and if we can carefully explain that concept, then its veracity or reality is undeniable. For example, Hamlet is a real character in our collective imagination even if he never existed in our real history.)

In all of these spaces, we will consider the intersections among flexible 2-dimensional subspaces. The patterns of intersection give topological information about the ambient space. In this way properties of space are determined by the relations among the constituents. Thus when we build spaces, we will supply them with floors, walls, and ceilings, but as one might imagine in a topology text, these will be convoluted to great degree.

⋆⋆⋆

This chapter will begin with the study of 2-dimensional objects called (compact) surfaces. Here is why.

First, I want you to gain intuition about higher dimensions (including dimension 3). Now what does intuition mean? If you are in a situation that resembles a past experience, you can try to predict what will happen in the present situation. For example, some of you do not like math very much. When you get to a formula or picture, you may reject the entire book as too difficult because your intuition leads you to believe that studying math is an unpleasant experience. But on the other hand, if I can convince you that a rather difficult theorem is easy to understand, you may stick with it for a while. Even more to the point, if you can understand how certain things happen on surfaces, there is a hope that you will be able to see how they happen in space. That is, you will be able to draw on your experiences in the early chapters to understand (and predict) things in the later chapters.

Second, we will classify surfaces. That is, we will associate certain numbers or invariants to each surface; if the numbers are the same, the surfaces are the same. These invariants are easy to compute no matter how strangely the surface is given to you. Even though no such classification can exist for higher dimensional objects, the classification of surfaces serves as a model for all further topological theorems.

Third, we will indeed examine how surfaces intersect in space as the book's title promises. Surfaces can be formed from a fabric that fits together seamlessly, yet they can intersect in a manifold variety of configurations. And these intersections can be decomposed into a few fundamental pieces. As we discuss the notions of higher dimensions, we will not truly define space, but we will give a plethora of examples of it.

Thus surfaces exemplify the tools of the trade.

1.2 Surfaces

Imagine the plane depicted in Figure 1.1 to be infinitely malleable and infinitesimally thin. Thus, it can be hammered into any thin shape we desire; it has no thickness, and it can be shrunk and stretched at will. The surface does not tear, nor does it puncture. On occasion, the surface may be cut. Imagine that the supply of such a surface is unlimited.

Figure 1.1: The plane

The standard metaphor for such a surface is an elastic or rubber sheet. I prefer to think of it as a fabric woven of a pure metal, a metal that gold only imitates. Perhaps Jason's golden fleece could be woven into this fabric.

Once the fabric is shaped, it retains its shape. Its size is of no concern: Sometimes we need acres, and sometimes we need square inches. When the edges of this fabric touch they fuse seamlessly, and so perfect one-size-fits-all garments can be made. And when we cut it, the fabric remembers which threads are to be reattached.

By a **surface** we mean any object that can be made from such a fabric. However, this definition contains an important caution: Since we are imagining a fabric that does not exist in this space, we can imagine surfaces that cannot be embedded in our ordinary 3-dimensional world. Also, a given surface can be deformed into an infinite variety of other surfaces. These will have differing topographic properties, but the topological properties will be the same. For example, we consider the surface of a triangle and the surface of a square to be the same since either can be made from a single piece of our fabric. They have the same topology; their topography is different.

The terms **topology and topography** distinguish two properties of surfaces. The *topographic* structure is a geometric property of the surface. It is the bumpiness of the surface. The bumpiness determines length and angles; these are the stuff of geometry. Topological properties are properties that are preserved under continuous transformations that are continuously invertible. We take continuity to be a primitive concept for now, but we will discuss its technical properties as we continue.

The book you are reading is about topology! Any concerns about topographical properties are secondary to the discussion. Thus the geometry (distances, angles, curvature, *etc.*) of the surface is mentioned only in passing. Remember that. If you want to know more about geometry at an introductory level, read Jeff Weeks's book [74].

The Jordan Curve Theorem

To further illustrate the concepts of topology, simple closed curves in the plane are examined. A **simple closed curve** is the continuous image of a circle that does not intersect itself. The standard circle formed by, say, the edge of a coin is a simple closed curve. The edge of a potato chip is another example, and if the bag has been dropped, an individual chip will likely have a ragged edge: This is no matter. A

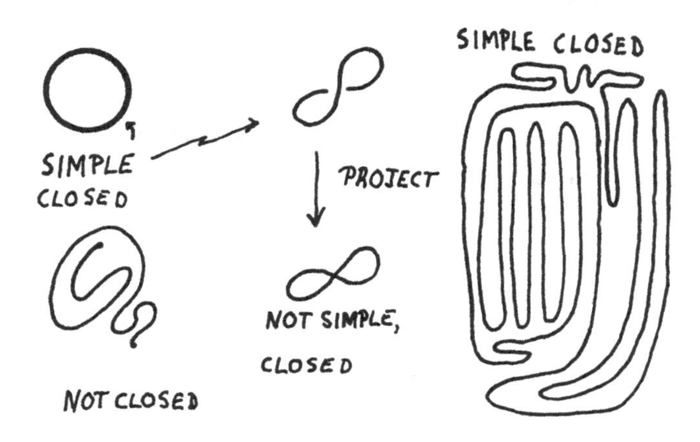

Figure 1.2: Curves: simple, closed, and otherwise

figure 8 in the plane is not simple, but it is closed. The term **simple** means that the curve does not intersect itself. The term **closed** means that the curve has no ends. Rubber bands form simple closed curves. The shadows of twisted rubber bands are not simple. See Figure 1.2.

Theorem 1 (Jordan Curve Theorem). *A simple closed curve that lies in the plane separates the plane into two pieces: an inside and an outside. The inside is always topologically a disk.*

This simple fact is remarkably difficult to prove. One reason for the difficulty is the generality of the statement. Simple closed curves can be extraordinarily complicated. Indeed, they can be fractal. Another reason for the difficulty is that people tend to assume the result within the guts of the proof. The part of the theorem that says the inside is a disk is called the **2-dimensional Schoenflies Theorem**. We will not prove the Jordan Curve Theorem here, but we will use it as our principal postulate. That is, we will prove that the Classification Theorem of Surfaces follows from the Jordan Curve Theorem.

The Compact Planar Surfaces

What kind of surfaces can be made from golden fleece? At least every surface with which you are familiar and quite a few more. The first family of surfaces that we will discuss fits into the plane. They have no infinitesimal holes, and they have finite area. These are the **compact planar** surfaces.

The Disk

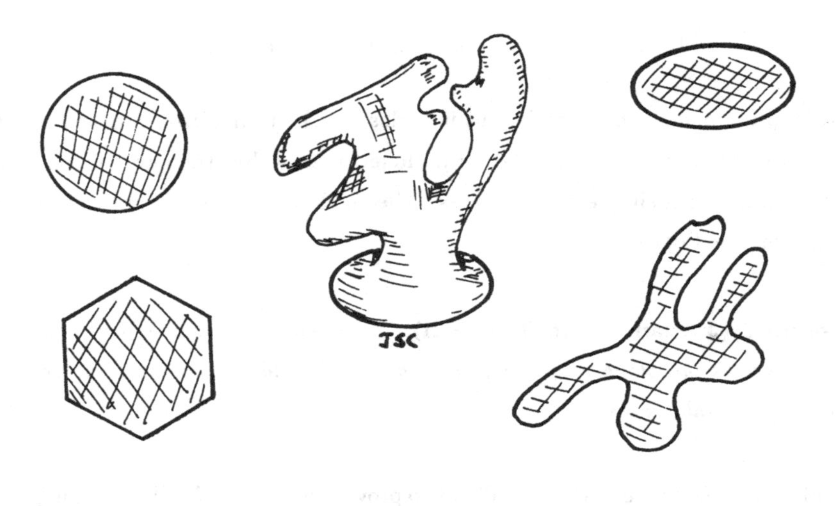

Figure 1.3: A variety of disks

In common usage the words "disk" and "circle" are synonyms. Not here. The word "disk" always refers to the 2-dimensional object that is bounded by a circle. The word "circle" always means the 1-dimensional boundary of some surface. The circle is 1-dimensional because we can locate a position on a fixed circle by angle alone.

The most elementary of all surfaces is the disk. Our fabric allows disks of any shape and size. All the surfaces depicted in Figure 1.3 are disks. The disk has a boundary that we think of as a circle even though the disk itself needn't be round. Roundness is a topographical, rather than a topological, property.

There are surfaces that have no boundaries: the surfaces that bound solids, for example; there are surfaces that have one boundary circle, and there are surfaces that have many boundary circles. But the disk has an important property that is shared by no other surfaces. Namely, when a disk is cut along an arc that connects any two points on its boundary, it becomes disconnected. To reiterate:

Theorem 2 Characterization of the Disk. *Any arc between boundary points separates the disk into two pieces each of which is a disk. The disk is the only surface with this property.*

An **arc** is a simple, *i.e.* no self-intersections, continuous image of a line segment. A simple closed curve, for example, can be made from two arcs joined together at their ends.

The Characterization of the Disk is a theorem that distinguishes a disk from other surfaces by means of a separation property. That is, you can recognize a surface that is perhaps given in some extremely abstract way as being a disk if you can prove the surface has the property that is stated in Theorem 2.

The proof that Theorem 2 characterizes the disk goes like this: If two filled triangles are glued to each other along edges, the resulting space is a space that has the same *topology* as either of the original triangles. (In fact, an explicitly continuous function, in which each point in the triangle maps to one and only one point in the quadrilateral, can be given by formulas, and compact spaces have the same topology if there is such a function.) Consequently, any space that is made from gluing two disks together along proper segments of their boundary is still a disk: For you can rewrite the gluing process in terms of triangles. This shows that any space with separation property stated in Theorem 2 is a disk.

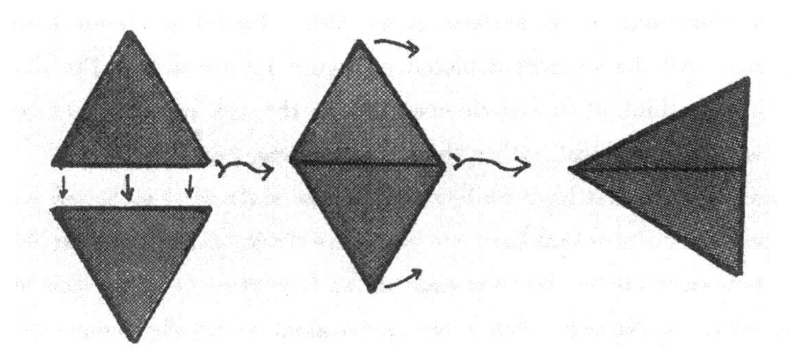

Figure 1.4: Gluing triangles to get triangles

Now we sketch the proof that the disk has the property stated in Theorem 2. Consider an arc that connects a pair of points on the boundary of the disk but otherwise lies inside the disk. Then apply the Jordan Curve theorem to the simple closed curve consisting of the right arc of the boundary together with the interior arc as in Figure 1.5. This simple loop separates the disk into two pieces. The piece that is inside the loop is a disk. The piece outside the right loop is a disk because the Jordan Curve Theorem also applies to the loop that is formed from the left segment of the boundary and the inside arc.

As long as you believe the Jordan Curve Theorem, then you should believe Theorem 2. The Jordan Curve Theorem is a very believable result. We shouldn't worry that it is hard to prove.

The most important idea within the Jordan Curve Theorem and within the proof of Theorem 2 is the notion of a separating arc or a separating closed curve. Compare this with the analogous situation in one lower dimension: The line can be separated by removing a point; the circle is separated only after two points are removed. Thus the line and the circle are different. Surfaces will be distinguished by the number of non-intersecting arcs that can be removed from them without separating them. That

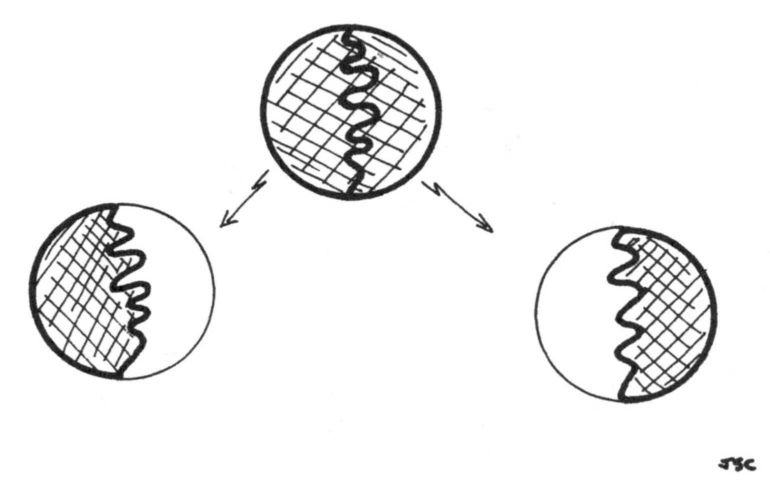

Figure 1.5: Any arc separates the disk

separation datum and two other data will completely classify surfaces. Let us look at a more general case.

Other Planar Surfaces

Consider a piece of swiss cheese, a gasket, a pair of pants, or a T-shirt made of golden fleece. More generally, consider an article of clothing suitable for a multiped as in Figure 1.6. Each such surface can be flattened out into the plane. The surfaces can be distinguished by the number of holes or boundary circles that they have. Furthermore, for the planar surfaces, the number of circles on the boundary is always one more than the number of cuts that can be made without separating the surface. Explaining and verifying this last statement is the current goal.

A **cut-line** is an arc in a surface that connects two points on the boundary. The Disk Characterization Theorem (Theorem 2) says that every cut-line in the disk separates the disk. A cut-line **does not separate** if after cutting along the line, the surface remains connected. An annular gasket has one such cut-line, a pair of pants has two, and a shirt has three. Even if a given surface is hammered into an

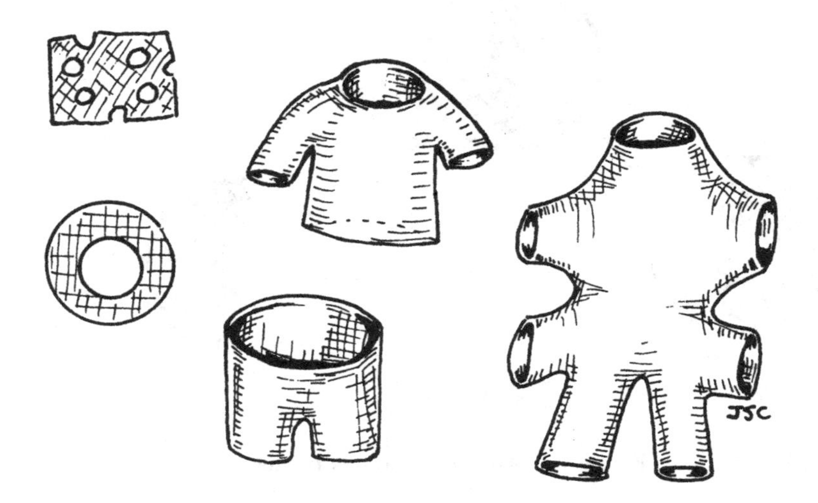

Figure 1.6: Some planar surfaces

abstract shape, then a given non-separating cut-line remains non-separating under the deformation.

The combination of the terms "non-separating" and "cut-line" seems oximoronic. Instead, let us say that a cut-line that does not separate is a **substantial arc**. A substantial arc represents some underlying property of the surface. For example, the disk has no substantial arcs, and this fact characterizes the disk among surfaces.

Theorem 3 Characterizing Planar Surfaces. *In the planar surfaces the number of substantial arcs is one less than the number of circles on the boundary. Either the number of boundary circles or the number of substantial arcs can be used to distinguish the planar surfaces.*

Even under the plastic deformations that are allowed for golden fleece, neither the number of substantial arcs nor the number of boundary components changes. Let me elaborate.

A deformation of the surface that involves hammering, shaping, dilating, or contracting is called a **homeomorphism** — a change in topography in which the un-

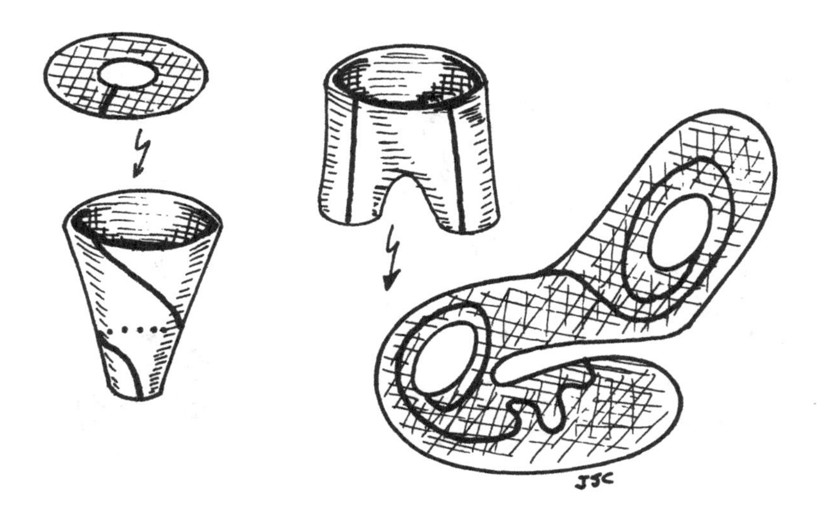

Figure 1.7: Deforming surfaces and their substantial arcs

derlying topology remains unchanged. More specifically, a homeomorphism is a continuous transformation that can be undone in a continuous fashion. Many people study calculus to understand the technical notions of continuity and differentiability. I am not assuming that you have learned calculus, but I am assuming that you know what a continuous deformation is. When you pull toffee, you can do so continuously without tearing the toffee. If the toffee breaks, this is discontinuous.

But we must be careful. An annulus (rubber gasket) can be continuously deformed into a disk: Heat the inner boundary and pull it together. When an annulus is mapped onto a disk, different points map to the same point. The annulus has two boundary circles, and the disk has one boundary circle. When the inner boundary of the annulus is melted, all of its points are mapped to a single point. So even though this transformation is continuous, it is not one-to-one — the center point of the disk has a whole circle mapped to it. Therefore, it is not a homeomorphism. Similarly,

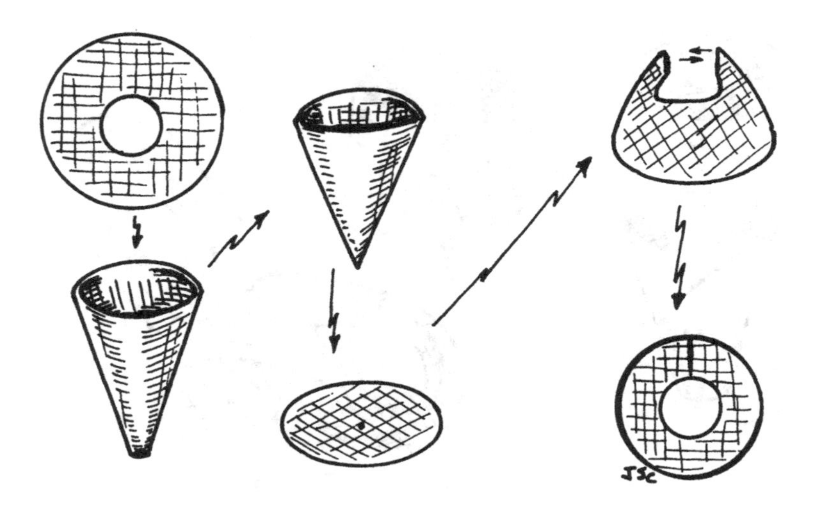

Figure 1.8: Continuous maps between disks and annuli that are not homeomorphisms

when two segments of the boundary of the disk are fused together so that the result is an annulus, points on the boundary of the disk wind up inside the annulus.

The continuous image of a connected set is connected. Furthermore, if a set X maps surjectively to a space Y and a subset B separates Y, then the set of points in X that map to B separates X. (A map is **surjective** if each point in the range comes from at least one point in the domain.) The second fact can be proven to follow from the first. The first sentence of this paragraph is an **intuitive property of continuous functions**.

Substantial arcs in a surface are topologically invariant. That is, if one surface can be transformed continuously and invertibly to another, then the substantial arcs on the one are transformed into substantial arcs on the other. The arcs are connected, and so their continuous images are connected. The continuous images are also substantial: If the image of an arc separated the range, then its preimage separated the domain. For example, the annulus has a substantial arc, and the disk has none. So the disk and the annulus are different because every arc in the annulus maps to an insubstantial arc in the disk.

There is a subtle point to be made here. Technically speaking, **a continuous function** is a function such that points that are arbitrarily close in the range come from points that are sufficiently close in the domain. But that technical definition does not explicitly say a word about connectivity and separation, and there are discontinuous functions that take connected sets to connected sets.

In college level math courses, we formulate and prove theorems such as the Intermediate Value Theorem which can be paraphrased to say, "If you are on the north side of a river at 1:00 and on the south side of the river at 2:00, then some time between 1 and 2 you crossed the river." The proof of this theorem and the more general result that says, "the continuous image of a connected set is connected," is generally presented in an advanced calculus course.

So when your calculus teacher said (or will say when you take the course), "A continuous function is one for which you can draw the graph without lifting your pencil," she or he was telling you that the technical notion of continuity obeys the intuitive ideas of what is continuous. In this book, we will play the topology game in the intuitive framework while keeping in mind that a great amount of work has been devoted to constructing a rigorous framework in which the intuitive notions hold. Moreover, the subtle functions that preserve connected sets but are not continuous will not be considered here.

My point is that surfaces, even though they be made of golden fleece, are ordinary objects. And it is clear that if you deform a surface by hammering, stretching, or shrinking, then the arcs and circles that didn't separate before the deformation, won't separate after the deformation. Similarly, it is clear from the intuitive property of continuity that two surfaces that are homeomorphic have the same number of boundary circles.

⋆⋆⋆

Back to the matter at hand: Separation properties are preserved under continuous deformation. Thus the number of non-intersecting cut-lines that can be fit into a surface so that their union does not separate the surface is a topological invariant: Two surfaces that have different numbers of non-intersecting substantial arcs are different. The term "the number of non-intersecting substantial arcs" is so cumbersome let us call this the **rank**, and let us remember that this is the rank *of a surface with boundary*.

The rank of a surface with boundary is an invariant of the surface, but it is not complete: There are surfaces of the same rank that are different. Consider the pair of pants and the nice basket shaped thingy in Figure 1.9. Both have rank 2, but the basket shaped thingy has only one boundary component.

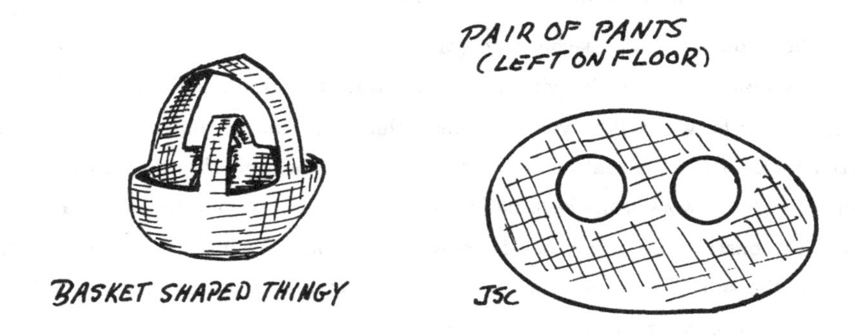

Figure 1.9: Two surfaces with the same rank

But the rank is a complete invariant for the **planar** surfaces with boundary, and I will in the next paragraphs prove this by mathematical induction. Since the rank is a complete invariant of planar surfaces and since the basket shaped thingy is not the same as a pair of pants, then it is not planar. The basket shaped thingy is not a pair of pants because it has one boundary circle, the pair of pants has three boundary circles, and the number of boundary circles is preserved under homeomorphism (which means topologically the same). So in proving the completeness of rank as an invariant of planar surfaces, I have been able to show that some surfaces are not planar. Let us proceed to the proof.

A disk has one boundary component and rank 0. So the statement: "the rank of a planar surface is one less than the number of boundary components" is true when there is only one boundary component. This is because Theorem 2 says that the disk is the only surface of rank 0: Rank is the number of substantial arcs from one point on the boundary to another, the disk has only separating arcs, and a substantial arc is non-separating.

Now suppose that we have a planar surface with k boundary components. We want to show that the rank is $k-1$.

(The letter k here indicates an arbitrary number. When presented with the letter k as an arbitrary number, I generally *pretend* $k = 138$, or $k = 5$. That is, we're suppose to imagine an arbitrary number, and 138 seems arbitrary to me. It is also reasonably large: larger than, say, most numbers with which I would care to calculate. But since I might actually have to do some calculation to check things out, 5 is usually small enough to complete the calculation in a reasonable amount of time, yet large enough so that apparent patterns that occur with smaller numbers will disappear. So let k be arbitrary, but for your own imagination's sake make it a specific value.)

The first thing we observe is that a substantial arc has to connect different boundary circles. See Figure 1.10. For if not, then we can embed the planar surface in the plane and take the cut-line and a segment on the boundary to find a simple closed curve in the plane. This separates the plane into two pieces, and consequently part of our planar surface is inside one of these. So a substantial arc has its ends on different pieces of the boundary.

Now if the planar surface is cut along this substantial arc, the two pieces of the boundary that the line connects are going to be conjoined into one. See Figure 1.11. Thus the case of k circles on the boundary reduces to the case of $k-1$ circles on the boundary. Let us run through that step again. If we cut along a substantial arc, we reduce the rank by 1 and we reduce the number of circles on the boundary by 1. Now

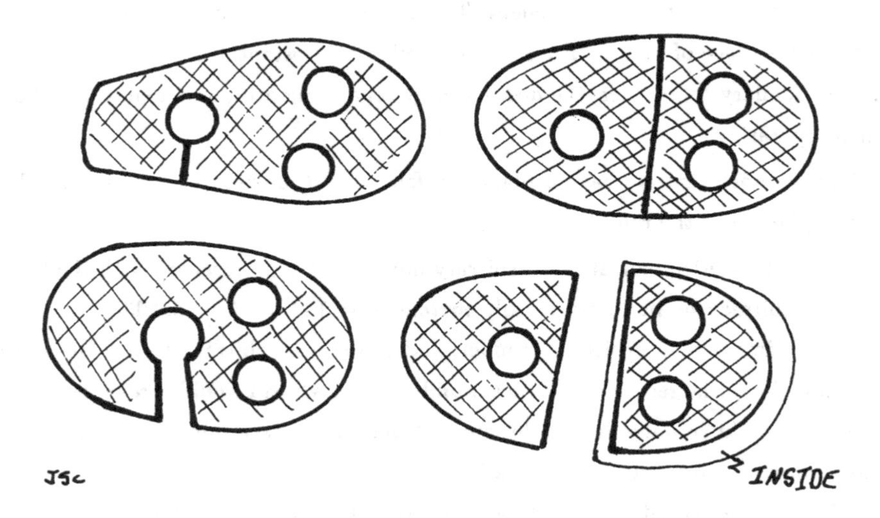

Figure 1.10: A substantial arc on a planar surface runs between different boundary circles

if k really is 138, then we have made 1 cut to get a surface with 137 boundary circles. Another cut will reduce it to a surface with 136 boundary circles. And we continue until we get to the disk. The number of cuts that were needed were 137.

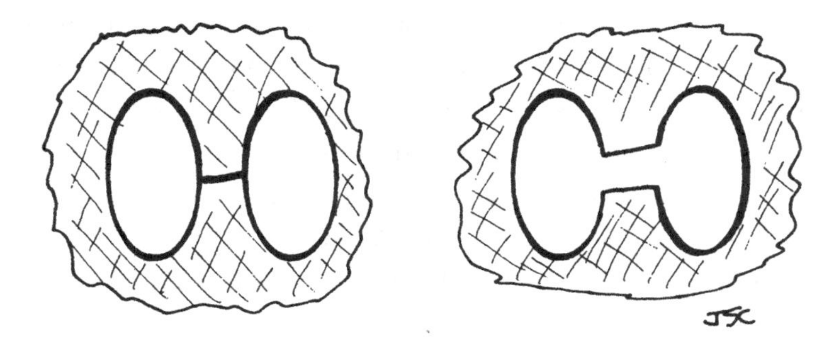

Figure 1.11: Cutting along a substantial arc connects boundary circles

We assumed that if there were $k-1$ circles on the boundary, then the rank would be $k-2$. And we showed in the case of k circles on the boundary that cutting along one cut-line would reduce the rank by 1 and the number of boundary components by 1. Thus the assertion for $k-1$ would prove the assertion for k.

⋆⋆⋆

Mathematical induction teaches us to sing "99 bottles of beer on the wall:" The last verse of "99 bottles of beer on the wall," starts with the phrase, "1 bottle of beer on the wall ... ." We know that the kth verse of "99 bottles of beer on the wall" ends in "Take one down and pass it around, $k-1$ bottles of beer on the wall." Now by induction we know the whole song. Even better we know how to sing, "138 bottles of beer on the wall, or "4534 bottles of beer on the wall." And for an arbitrary number of bottles of beer on the wall, the song will end.

In that language, the bottles of beer are the number of circles on the boundary, and the act of taking one down and passing it around is the act of cutting the planar surface along a substantial arc. This process is finished once we get to a disk.

⋆⋆⋆

So far, we have examined the separation properties of the disk and the planar surfaces: The disk is the only surface of rank 0; the rank of the planar surfaces is one less than the number of boundary components. And in this sense we have begun the classification, for we are writing down surfaces in an order that reflects how complicated the surfaces are. The ordering so far goes: disk, annulus, pair of pants, shirt, pants with a hole in each knee, dog shirt, and so forth. But in the process we depicted a surface that is not in this sequence — the surface I called the basket shaped thingy. We turn now to identifying that surface and to classifying all the surfaces in its family.

Other Orientable Surfaces with Boundary

The basket shaped thingy is actually a torus with a hole cut from it. Figure 1.12 indicates a homeomorphism. If we take lots of these basket shaped thingies and glue them together along segments of their boundaries, we get a surface that has one boundary component, but its rank is twice the number of baskets with which we started.

Let's see why the punctured torus (formerly known as the basket shaped thingy) has rank 2. First, we observe that a cut-line necessarily joins two points on the same boundary circle since there is only one such circle. Next, we observe that there is a substantial arc that cuts the surface into an annulus. Topologically this is an annulus, the flaps can be hammered into the main portion of the surface. The annulus has rank 1, and so the rank of the punctured torus is 2 because we have found 2 substantial arcs and there can be no more. See Figure 1.14.

Let's look at what happened with the first substantial arc. Recall the case of the planar surfaces: A substantial arc connected two boundary circles, and by cutting along it the two boundary circles were joined into one. In the case of the punctured torus, the substantial arc cut one boundary circle into two. These phenomena are apparently different. But seen in the correct fashion, one is the upside down image

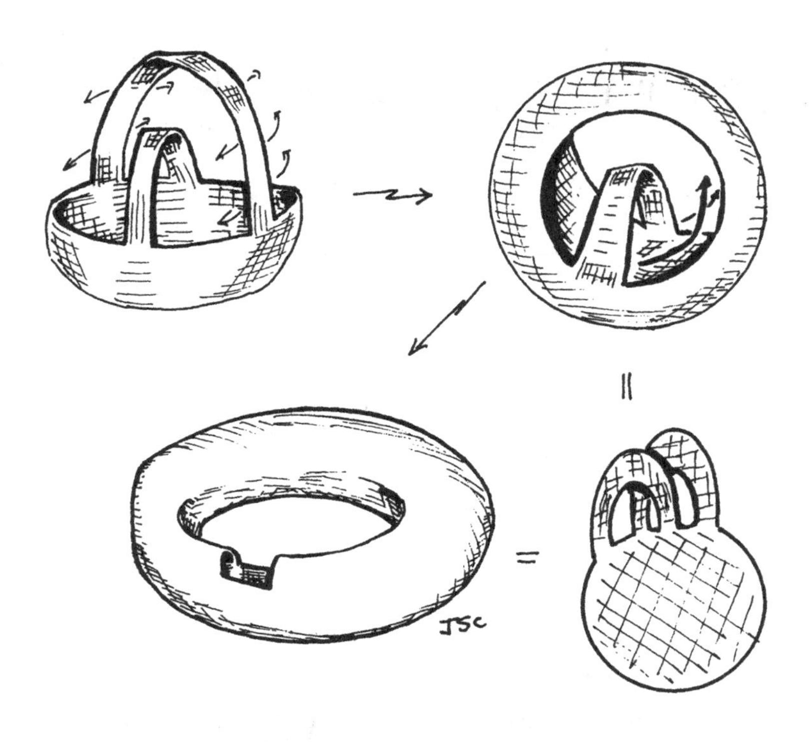

Figure 1.12: A torus with a hole in it is a basket shaped thingy

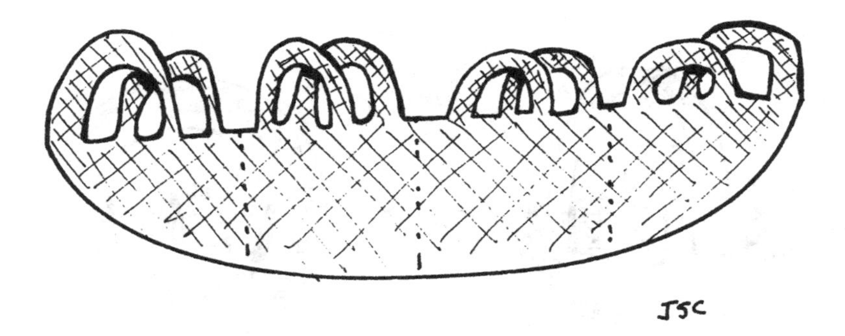

Figure 1.13: Gluing punctured tori along segments of their boundary

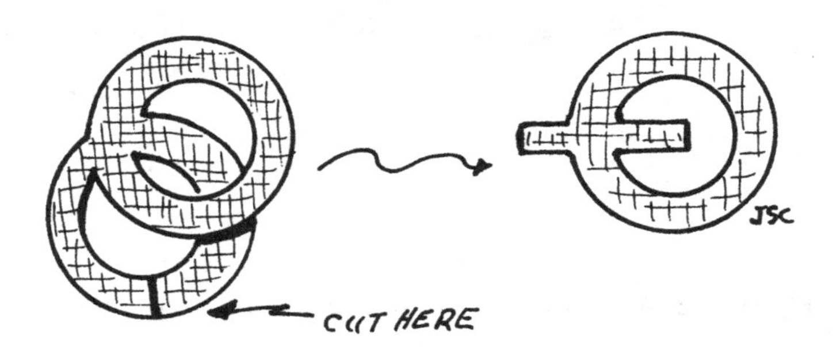

Figure 1.14: A substantial arc on a punctured torus cuts it into an annulus

of the other. There will be more discussion on turning our point of view upside-down in the Chapters 3, 4, and 7.

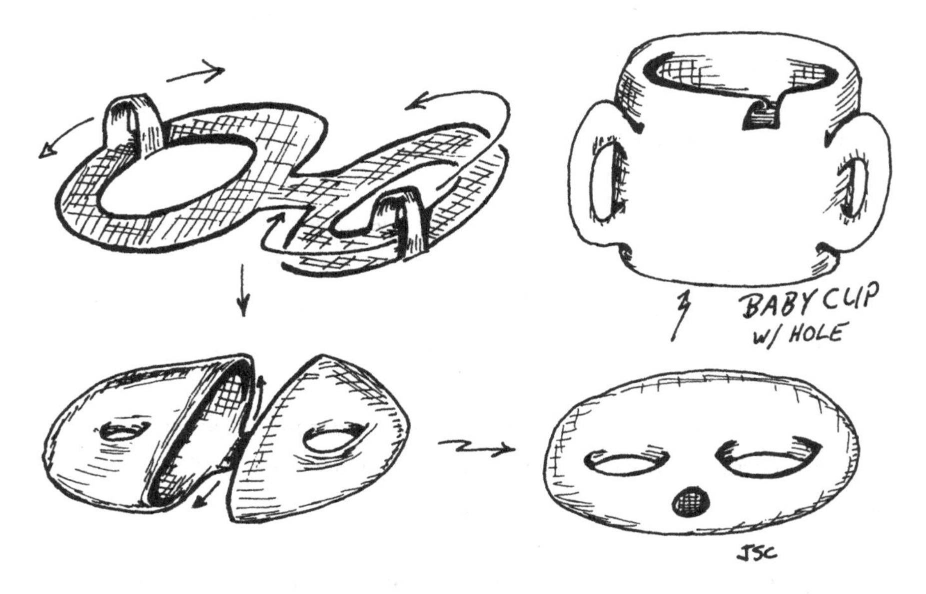

Figure 1.15: Two punctured tori, when glued, form a punctured baby cup

Now what happens in the case when we have a lot of punctured tori glued together along segments of their boundaries? If there are two glued together, then the rank is 4 as Figure 1.15 indicates. In general if there are k tori glued together, then the rank is $2k$. Let me explain. The handles near each basket come in pairs. Say there is a left handle and a right handle. Then if we cut the left handle of each pair, we get a planar surface with $(k+1)$ boundary components. (Again this is mathematical induction, or 99 bottles of beer on the wall.) The planar surface that results has rank k. So we made k cuts to get a planar surface of rank k, and k more cuts to get to the disk. The total number of cuts made is $2k$, so the rank of the original is $2k$.

It is time for one more remark before we proceed. Namely, in the process of cutting we had a strange kind of surface as an intermediate stage. Let's consider the rank 4 surface that is the result of gluing two punctured tori together along a pair of segments, one on each boundary circle. After the first cut, we had a toroidal part and two boundary components. This intermediate surface has not appeared in the list of surfaces we have so far. The list we have constructed goes:

First Family: Planar surfaces, classified by rank. Disk, annulus, pants, shirt, *etc.*

Second Family: Connections among tori. Once punctured torus, a baby cup, *etc.*

We can combine the surfaces in the two families as follows. Cut a torus with a cookie cutter; as many holes as you like can be removed. Each such hole increases the rank by one. On the other hand, we can glue together a planar surface and a punctured torus along a segment of the boundary circles of each to get the same result.

Earlier, I said that the rank and two other data determined the topological type of a surface. One of these data is the number of boundary circles. The other will be discussed subsequently. Here is the classification so far:

Theorem 4 Classification of Orientable Surfaces with Boundary. *The orientable surfaces with boundary are classified by rank and the number of boundary circles. If the rank is one less than the number of boundary circles, then the surface fits into the plane.*

Let us examine the status of this statement. First, I have asserted that the number of boundary circles and the rank of a surface are topological invariants. That is, if two surfaces have either a different number of boundary components or they have differing ranks, then the surfaces are not topologically equivalent. That assertion follows from the intuitive property of continuous functions. The next assertion is that if these numbers are the same, then the surfaces are topologically the same. Indeed, if the rank is one less than the number of boundary components, then the surface is planar. The goal of the current section, then, is to show that if two orientable surfaces have the same rank and they have the same number of boundary circles, then they are topologically the same.

Before we continue, I had better say something about the adjective "orientable." So far it is an undefined term. It is the third datum that we need to classify surfaces with boundary. The term will remain undefined for now, but let me assert that a surface is either orientable or it isn't. I won't consider non-orientable surfaces in depth until the next chapter, so the term doesn't much matter at this point; I just use it to avoid making a mistake in the statement and proof of Theorem 4. I will emphasize where the assumption of orientability comes into play at the moment that it is used.

Suppose that a surface has two boundary circles. Then there is a substantial arc that has its ends on the different boundary circles. (This is different from what we did before. Before, we started with a planar surface and showed a substantial arc connected different boundary circles. Now we want to see that an arc that connects differing boundary circles *is* substantial.) If a cut-line is going to separate anything, it will separate a pair of points immediately to its left and to its right. (See Figure 1.16.) I will show that if the foot and the head of the cut-line are on different boundary components, then these points are not separated.

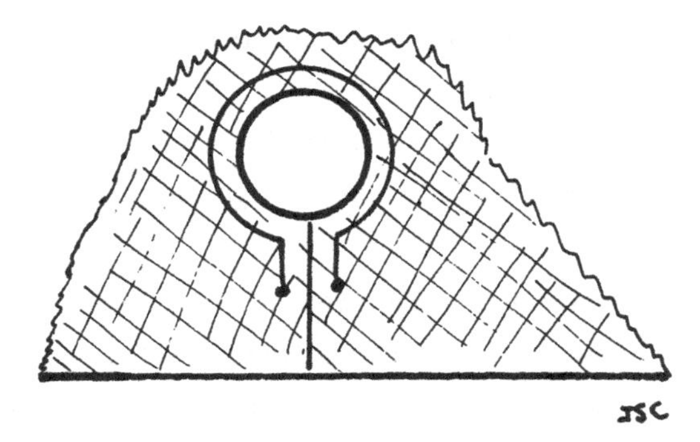

Figure 1.16: An arc between boundary circles is not separating

Starting from the point to the left of the cut-line, trace an arc parallel to the cut-line towards the foot until you get to the boundary. Now follow the boundary circle until just before you reach an end point of the cut-line. We have put the foot and the head of the cut-line on different boundary circles. So when the path that follows the circle reaches the cut-line, it reaches the cut line at its foot. Moreover, it reaches the cut-line to the right of the foot. Now trace an arc parallel to the cut-line that travels up to the point on the right hand side.

In the preceding paragraph, we constructed an arc that connects points on either side of the cut-line. The path didn't intersect the cut-line, and it was contained entirely within the surface albeit near the edge of the surface. We constructed this path under the assumption that there were two boundary circles. Thus we have shown that if there are two boundary circles, there is a substantial arc that joins them. There could be more boundary circles as well. But whenever there is a pair of them, there is a substantial arc between them. We want to examine an arbitrary surface that has a given rank and a given number of boundary circles.

Now let us exploit a property of golden fleece. Namely, when we cut the fabric, it will remember which segments were attached. Cut the surface along a collection of substantial arcs until you get a disk. Along the boundary of the disk there are segments that can be fused together in pairs to re-form the substantial arcs. Two arcs that can be so fused are called **mates**.

Suppose each cut line is **oriented** in the sense that one end of the cut-line is called the foot and the other end is called the head. We think of the cut-line as an arrow pointing from foot to head. When the surface is cut along the cut-line, each of the mates is similarly oriented. The arrow on the segment points the same way as the cut-line.

After the surface is first cut and before we have hammered the result into the shape of a disk, there may be bands that emanate outward like Medusa's hair. But when the reshaping has taken place, the arrows on the mates have a peculiar property: If one points clockwise, the other points counter-clockwise. This is where the assumption that the surface is **orientable** is made. The surface is orientable if and only if the mates of any substantial arc are oriented clockwise and counter-clockwise along the boundary of the disk that results from the cut.

The mates can be labeled with letters so that those with the same letter come from the same substantial arc. We will recognize the surface by looking at the sequence of letters along the boundary of the disk and by finding a way to rearrange them into a certain standard order.

To describe the standard order, let's first consider the case of a planar surface. In Figure 1.17, I depicted a shirt, two systems of cut-lines, and the result of cutting along these segments. On the disk that is the result of cutting, there is a pair of mates that are adjacent on the boundary of the disk.

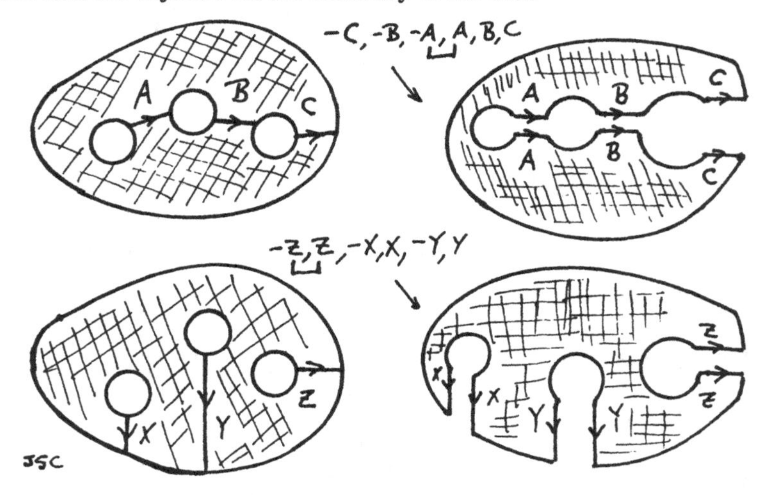

Figure 1.17: Substantial arcs on shirts

In Figure 1.18, the standard double torus is cut along its substantial arcs, and in this case the mates on the disk have the property that a pair of mates are separated by another pair of mates. That is, the labels along the boundary appear in the order $a, b, -a, -b$. (The minus sign $(-)$ indicates that the arrow on this segment is pointing clockwise. In the right-handed universe, the natural direction for circles is counter-clockwise.)

If, upon cutting along the substantial arcs, the two members of each pair of mates are either adjacent or are separated by at most one segment, then we will recognize the surface. Namely, each pair of adjacent mates corresponds to an arm-hole, and each quadruple of mates in the order $a, b, -a, -b$ corresponds to a punctured toroidal piece (a basket shaped thingy.) We can even recognize a few more cases because if

Figure 1.18: Cutting the standard double torus

there are arcs in the order $a, b, -b, -a$, then these contribute to two planar holes — say, an arm-hole and a neck-hole.

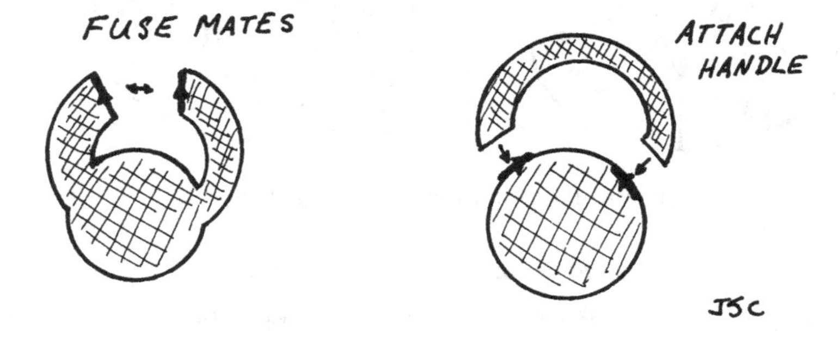

Figure 1.19: Attaching a handle between mates is the same as fusing the mates

So what should we do when, between any two mates, there is a large collection of other arcs? We will find means of sliding these arcs over each other so that the topology of the surface doesn't change. To deform the surface in this way, we need to shift our point of view.

Suppose mates with the same label fuse to form a substantial arc. One way of achieving the same topological surface is to attach a thin strip between the mates as in Figure 1.19. This operation is called **attaching a handle.** If two handles are attached to a disk (or another surface), then one can be slid over the other as in Figure 1.20. The handle slides can be performed in either order, and the figure indicates why this is a topological equivalence. There are essentially three types of handle slides. The first type affects sequences of mates as follows:

$$a, -b, b, -a \rightleftharpoons b, -b, a, -a.$$

The second type affects sequences of mates as follows:

$$a, b, -a, -b \rightleftharpoons b, -a, -b, a.$$

The third type affects a sequence of three pairs of mates as follows:

$$a, b, c, -a, -b, -c \rightleftharpoons b, c, -b, -a, -c, a.$$

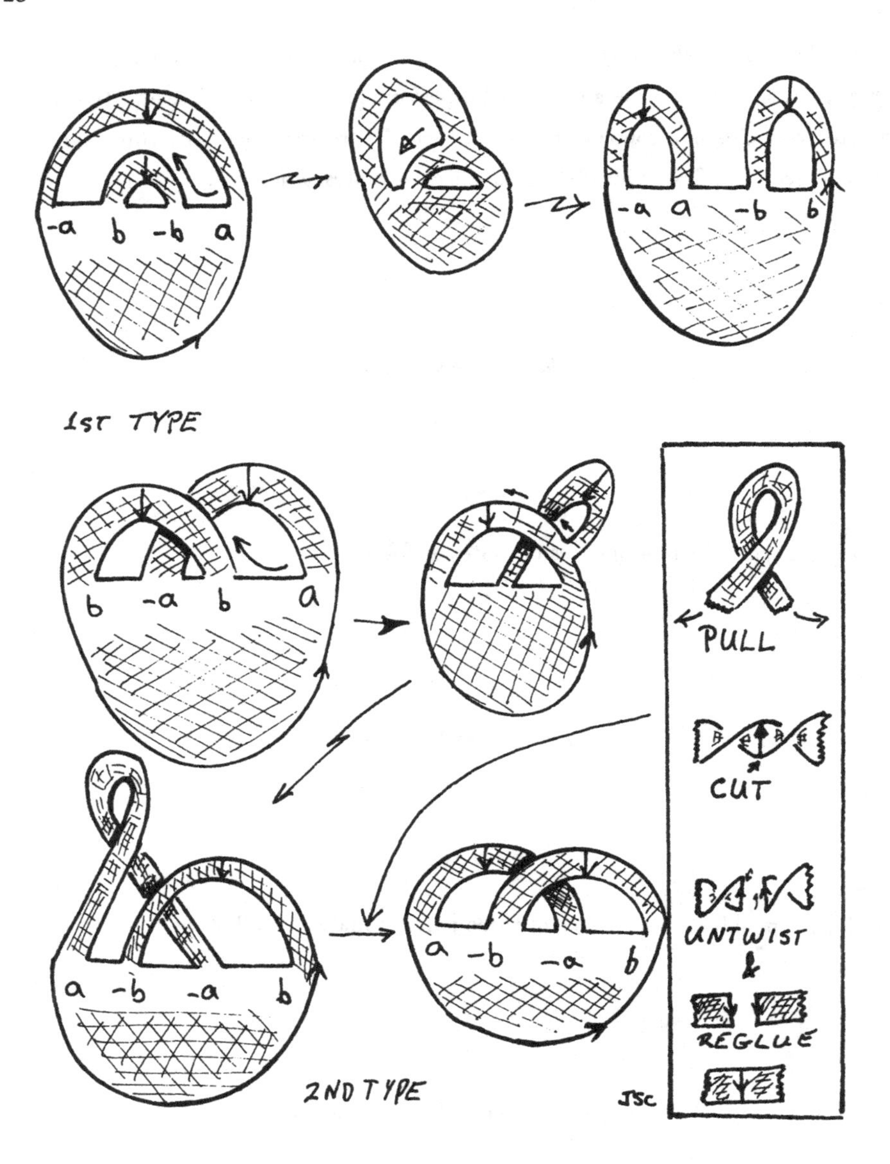

Figure 1.20: The handle slides

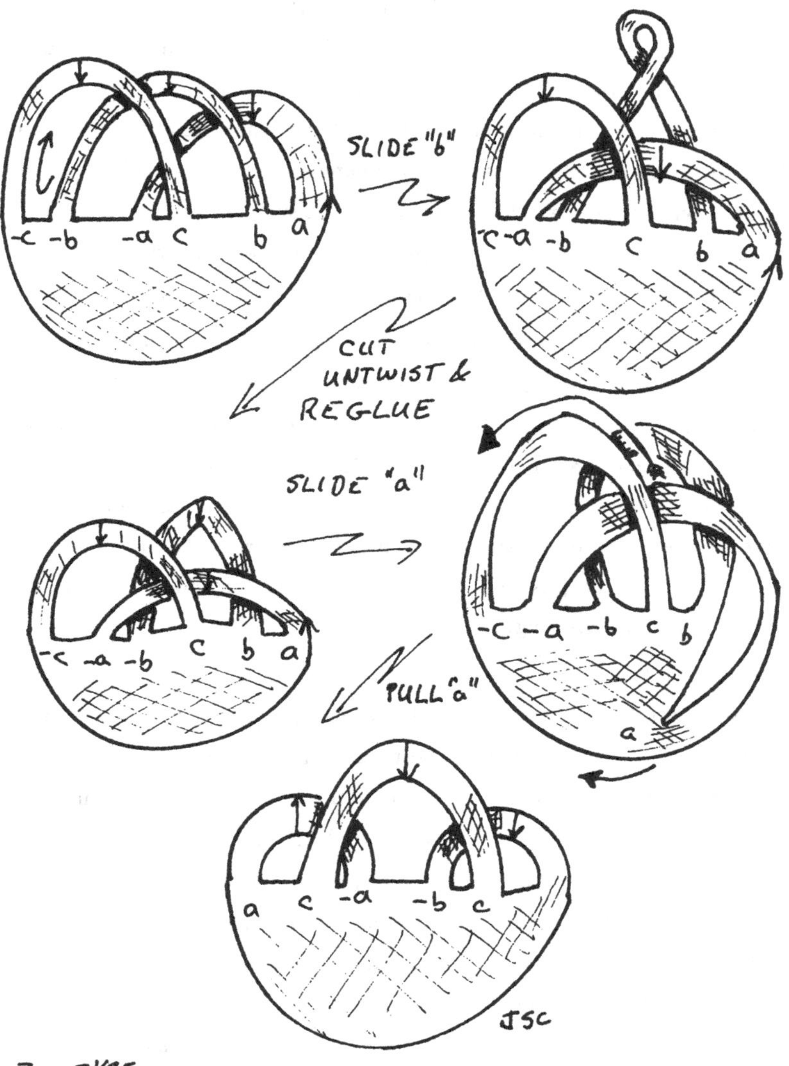
-c -b -a c b a
SLIDE "b"
-c -a -b c b a
CUT
UNTWIST &
REGLUE
SLIDE "a"
-c -a -b c b a
-c -a -b c b
a
PULL "a"
a c -a -b c
JSC

3RD TYPE

If the second or third type of handle slide is performed in space, then the handle that is being slid gains a full twist. We cut the surface, temporarily, and reglue after removing the twist. This operation of cutting and regluing is like picking up the phone, and letting the headset dangle until the kinks are out of the cord. The operation, although discontinuous in the ambient space (because it involves cutting), is continuous on the surface. That is, points on the surface that were close before the cut remain close after the fusion is complete. Thus by the properties of golden fleece, the net result of the operation is continuous.

Recall, if each pair of mates were adjacent on the boundary of the disk or were separated by only one other arc to be fused, then we could recognize the surface: We count the adjacent pairs — the number of these is the number of cookie cutter holes, and we count the number of quadruples of the form $a, b, -a, -b$ — the number of these is the number of toroidal pieces. Next we assume that there are no adjacent pairs, nor are there any alternating quadruples of the form $a, b, -a, -b$. We will show in the next three paragraphs that we can perform handle slides to get to either an adjacent pair or an alternating quadruple.

We will show this by induction, but the induction is kind of tricky. Each pair of mates subtends two arcs on the circle. Consider, for a given pair of mates, the subtended arc that contains the smallest number of other arcs to be fused. Call this number the **complexity** of the pair of mates. For example, in the case of a punctured torus the mates $a, -a$ have complexity 1 because the mates in order around the boundary are $a, b, -a, -b$. Every pair of mates has a complexity, and we want to consider a pair of mates with the smallest complexity that is bigger than 1. The complexities of two different pairs of mates may be the same; for example the complexity of $b, -b$ is also 1 on the punctured torus. If there are two different pairs of mates that both have smallest complexity, just consider one such pair.

We would like to show that either the complexity is 1, or the complexity can be made smaller by handle slides. If we could show that, we would be finished: If we can make the complexity of a given pair of mates smaller, then we can eventually reduce it to 0 or 1. If the complexity is 0, then the substantial arc corresponds to a cookie cutter hole. If the complexity is 1, then the substantial arc made from fusing these mates corresponds to one of two handles in a basket handle pair.

To see that we can always perform handle slides to reduce the complexity, we suppose that handle a has the smallest complexity that is bigger than 1 in each of the Figures 1.21 through 1.23. In case 1, sliding handle a over handle b will cause its feet to be closer than they were before. The complexity of handle b is smaller than the complexity of a so it must have complexity 1. In case 2, sliding a over b shortens a. In case 3, left foot of handle b slides over handle c, and this move shortens a. If the right foot of c is on the arc subtended by a, then c has length 1. Then after b slides over c, case 1 applies. Therefore the handles play leap-frog until they are in standard position.

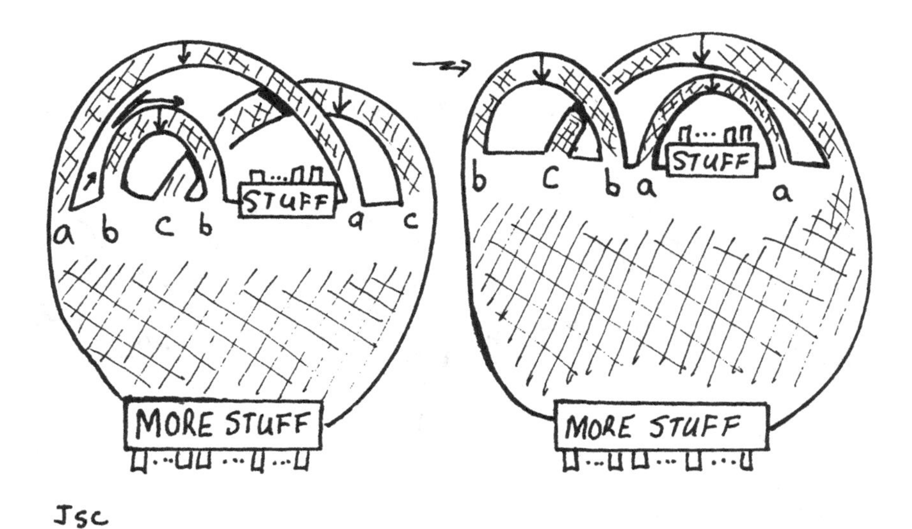

Figure 1.21: Case 1

The complexity of a given arc is like the number of bottles of beer on the wall. Taking one down and passing it around is achieved by sliding handles. The song is finished once each pair of mates has complexity 1 or complexity 0.

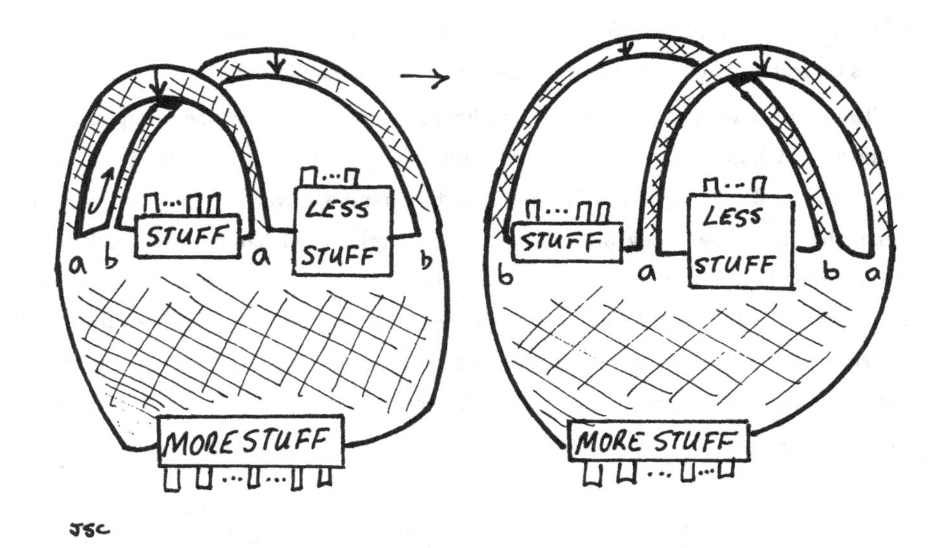

Figure 1.22: Case 2

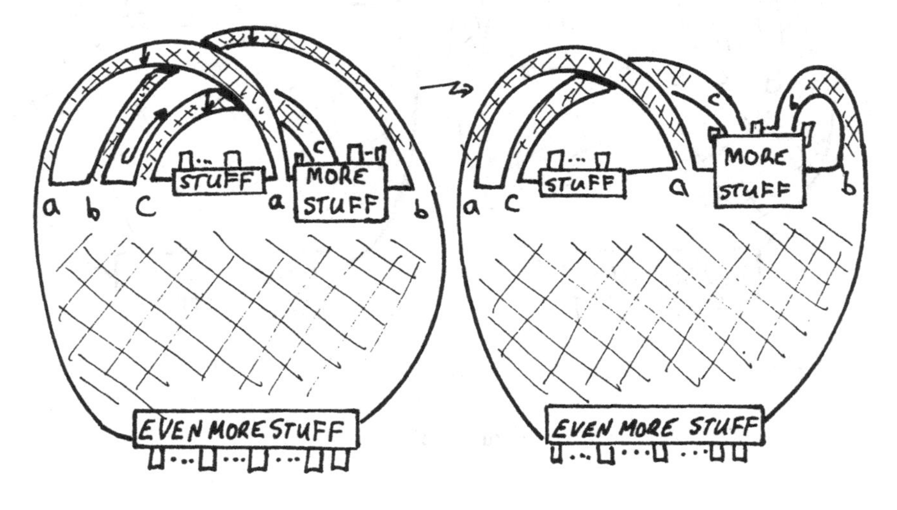

Figure 1.23: Case 3

Let's complete the classification of orientable surfaces with boundary. Given an orientable surface with boundary, compute its rank and the number of boundary components. Find a system of substantial arcs. If there are k boundary components, there are at least $k-1$ substantial arcs that connect the boundary components. There are more substantial arcs if the rank is larger. Cut along the substantial arcs to get a disk on which pairs of mates arcs are identified to re-achieve the surface. Then slide handles to put the surface into standard position. Therefore, given the rank and the number of boundary curves, we can identify the surface as being a standard surface with these invariants.

There are two points to make. First, you might wonder about sliding handles. Is there no way to perform this step intrinsically? Second, is it possible or even necessary to determine whether a surface is orientable to carry out this process?

In reply to the second question, I will treat the non-orientable surfaces carefully in the next chapter, but we really don't need to assume that the mates point in opposite directions. If we have a single substantial arc and the result of cutting along that arc is a pair of mates a, a read with that orientation, then the surface with which we started was a Möbius band. Because handle sliding is trickier in the non-orientable case than it is in the orientable case, the general discussion of non- orientable surfaces is postponed until Chapter 2.

In reply to the first question, Figure 1.24 indicates that handle slides can be understood as intrinsic topological deformations. The act of sliding handles can be understood in a rigorous algebraic setting as follows. First, the term "rank" refers to the rank of a certain vector space that is associated to the surface. Second, handle sliding corresponds to choosing a new basis in that vector space. Third, the rank of a vector space is independent of a choice of basis; you need not be concerned that the entire classification was based on choosing a collection of substantial arcs. A different choice would have resulted in the same rank. For those who don't know what a vector space is, ignore this paragraph. Better yet, after you finish this book, go and find out.

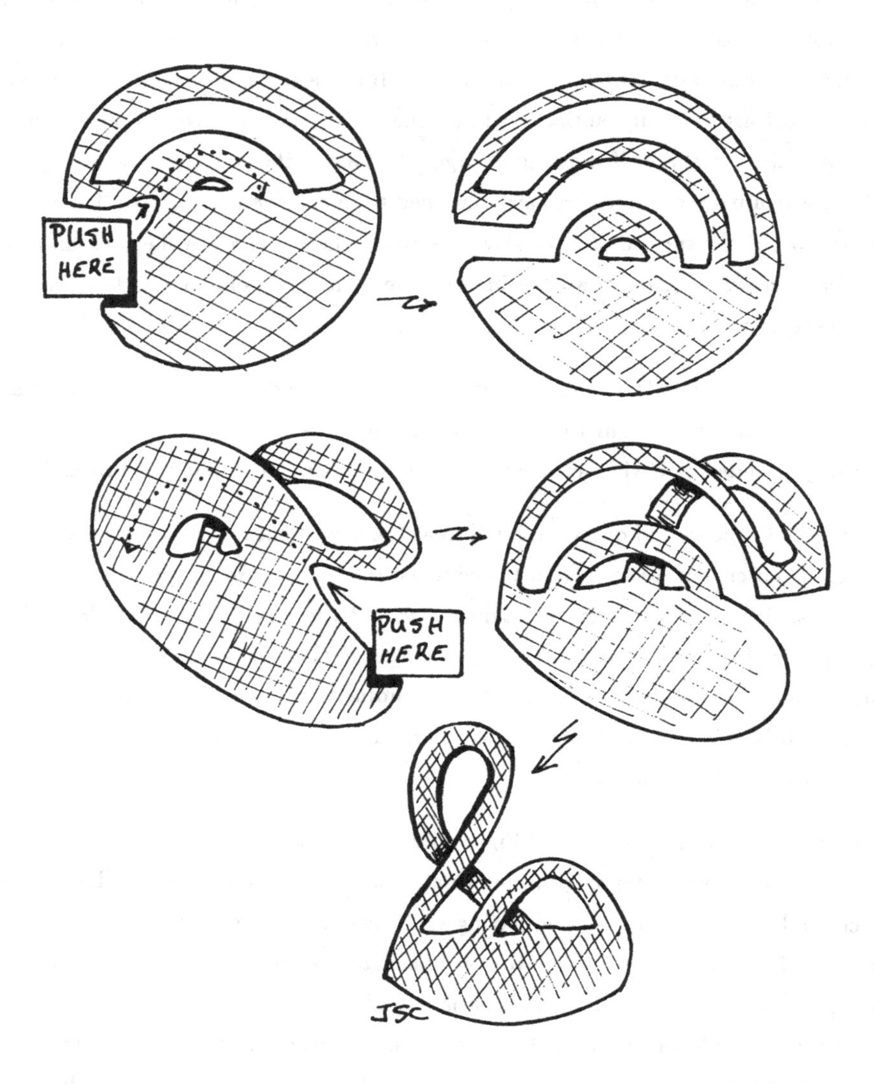

Figure 1.24: Sliding handles intrinsically

The Closed Orientable Surfaces

Consider, once again, a disk. Observe that its boundary is a circle, and observe that if two disks were glued together along the entire length of their boundaries, the result would be a sphere. This sphere is like an orange peel, and the disks that were fused are the two halves of the peel that you see after you cut and eat the meat of the orange. (You may get up from the book and eat an orange now, I'll wait.)

We now turn to filling all of the boundary circles of a surface with disks. In doing so we will create all the topological types of surfaces which surround you: the surfaces of bagels, chairs, the interior walls of a house, a colander, a screen door, *etc.* Having examined the case of a disk, let's consider other planar surfaces.

Consider a planar surface and an embedding of it into the plane. The surface exists in the abstract, and we choose a realization of the surface in the plane. In this way we could, if we want, put any of the surface's boundary circles on the outside: Planar surfaces made of golden fleece are fine articles of clothing for children who occasionally put their heads through the arm-holes. All of the boundary circles, except one, are found on the inside of the exceptional one, and none of the others are nested. Call the exceptional circle **the outside circle**, and call the other boundary circles **the inside circles** .

In the case of the planar surfaces, it is easy to glue disks to all of the boundary circles. The inside circles all bound disks in the plane, so gluing a disk to one of these is tantamount to erasing the circle. This leaves only the outside circle. After the inside circles are glued to disks, the resulting surface is a disk — the disk bounded by the outside circle. We have already observed that gluing two disks together along their entire boundary gives a sphere.

Theorem 5 *The result of filling all of the boundary circles of a planar surface with disks is a sphere. The sphere is the only surface such that cutting along any simple closed curve yields a pair of disks.*

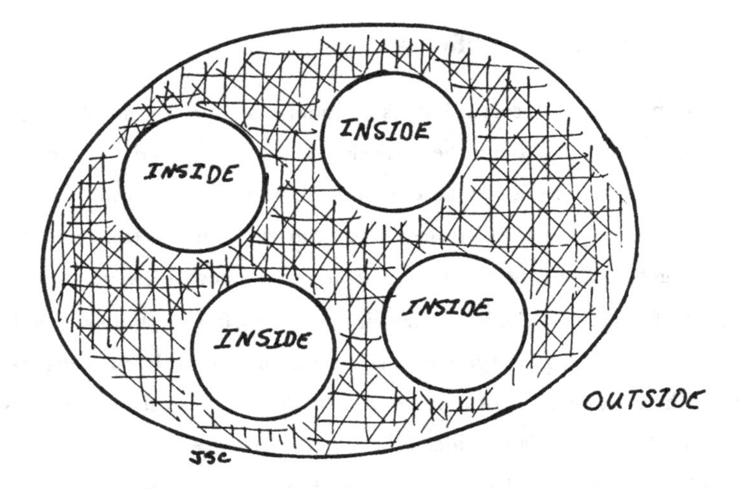

Figure 1.25: Inside and outside circles

The Jordan Curve Theorem tells us that a simple closed curve in the plane separates the plane in two pieces, one of which is a disk. But there is also a Jordan Curve Theorem in the sphere, and the second sentence of Theorem 5 is a statement of it.

The plane and the sphere are similar in many respects. From Santa Claus's point of view, they are the same because he doesn't have to deliver toys to the south pole. For Santa, the south pole might as well be an infinite distance away from the north pole. That is, the sphere can be obtained from the plane by adding a single point "at infinity," or from Santa's perspective by putting a person on the south pole at Christmas eve.

However, the sphere and the plane are different surfaces! Here is a simple reason. An embedded line separates the plane into two. For example, the x-axis in the coordinate plane separates the plane into up and down. On the other hand, no line in the sphere can separate it. In order to separate the sphere we need a closed curve, and the line is not closed. (A line can be separated by removing one point, but two points must be removed to separate a closed curve.) Thus the sphere and the plane are different.

The sphere, then, is the surface that is obtained by sewing disks to each boundary circle of any planar surface.

Again I need to distinguish some terms. The **sphere** is the surface of a ball. The sphere is intrinsically 2-dimensional because we can determine any point on the sphere by specifying longitude and latitude. The **ball** is the solid earth. The pair of terms (ball, sphere) are analogous to the terms (disk, circle). In either pair, the latter is the boundary of the former; points that move in the former have one more degree of freedom than do points that move in the latter.

⋆⋆⋆

Now consider the surface of a tire inner-tube. The surface forms a torus. If several, say four, cookie cutter holes were cut out, we would have a surface of rank five with four boundary circles. There is only one such orientable surface by Theorem 4. On the other hand, if four disks were sewn around the boundary of the rank five surface that has four boundary circles, then the result would be topologically the same as the tire.

The surface of a baby cup as depicted in Figure 1.26 is called **a genus two surface**. If a single hole were cut from it, the result would be a rank four surface with one boundary circle. So if a disk were sewn to the rank four surface with one boundary component, the result would be a baby cup, topologically.

Now we consider the general case of a surface that has one boundary circle. The rank is an even number. Let's suppose the rank is bigger than zero (so we don't start with a disk). When the boundary circle is sewn to a disk, we get the surface of an ordinary object such as a tire, a baby cup, a chair, or a colander (Figure 1.26). If the rank of the surface that we started with was $2n$, then the result of sewing on the disk has n holes.

Sewing a disk to each boundary circle is called **closing the surface**, a surface that has no boundary circles is a **closed surface,** and, as we have assumed throughout this section, the surface is orientable.

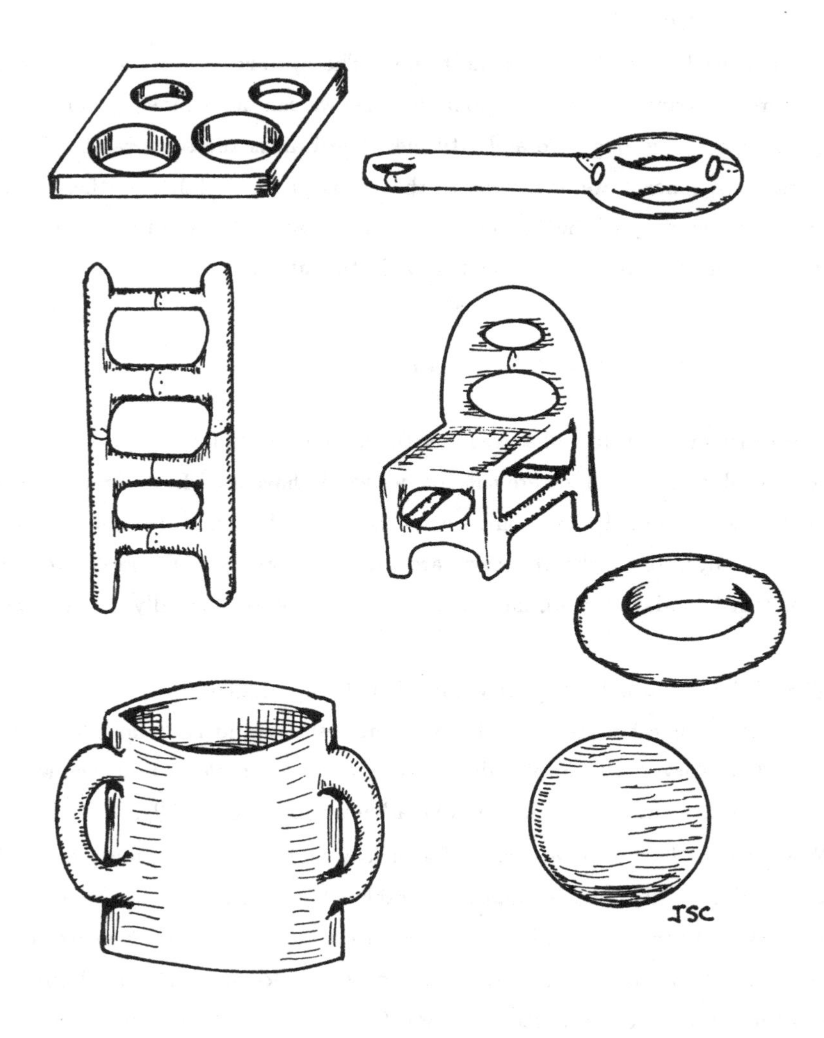

Figure 1.26: Some closed orientable surfaces

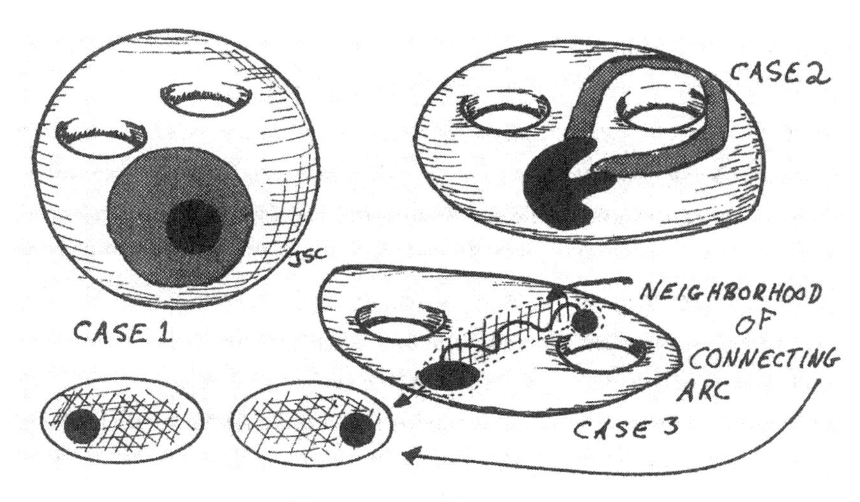

Figure 1.27: Disks on a surface

If we start with an orientable surface of high rank and a large number of boundary circles, we can sew a disk into all but one boundary circle. The result is a surface whose rank is an even number. When the surface is closed, the result is, again, the surface of an ordinary object. What remains to be seen is that the topological type of the closed surface depends only on the difference between the rank and the number of boundary holes, and not on the location of the holes. From another point of view, we can start with two closed surfaces, and cut a cookie cutter hole in each. If the ranks of the bounded surfaces are the same, then the closed surfaces are the same. This last statement requires proof.

Define the **rank of a closed orientable surface** to be the rank of the surface that results when a disk is removed from it.

Theorem 6 *Two closed orientable surfaces are topologically equivalent if and only if they have the same rank.*

Suppose we are given a closed orientable surface. Does the rank of the surface depend on which disk is removed? No. There are three cases to consider. (1) One disk completely overlaps another. (2) The disks that are removed intersect, but one is not contained in the other. (3) The disks do not intersect.

In the first case, the surface that is missing the big disk is stretched over the surface that is missing the little disk. The stretch is achieved by the malleability of the surface. In the second case, the disks overlap, but one does not engulf the other. So one of the disks can be shrunk, using the first case, to a disk that does not intersect the other. Now the second case has been reduced to the third case which is also easy.

In the third case, there are two disks that are disjoint on the surface. We choose an arc that connects these disk, and we look in the surface at a neighborhood of the arc and the terminal disks. The neighborhood is also a disk. And if either little disk is removed from the neighborhood, the result will be an annulus. Either annulus is glued to the bigger surface along the boundary of the bigger neighborhood, and the annuli are homeomorphic. So the surfaces that result from removing either disk are homeomorphic.

Therefore, if two different disks are removed from a given surface, then surfaces of the same rank result.

Now suppose that two closed surfaces are given that have the same rank. We want to see that these are topologically the same. To measure the rank, we remove a disk from either and compute the rank of the bounded surfaces. The bounded surfaces have the same rank and both have one boundary circle. So the bounded surfaces are topologically the same; *i.e.* there is a homeomorphism between the bounded surfaces. We want to construct a homeomorphism between the original closed surfaces. The bounded surfaces are contained within the closed surfaces, and so there is a map between rather large subsets of the closed surfaces — the correspondence is given everywhere but on the small disks that were removed. But the disks are homeomorphic, so the given homeomorphism can be extended over the entire surface. Therefore the closed surfaces are topologically the same.

We have completed the classification of closed orientable surfaces.

Since the rank of a surface with one boundary circle is always an even number, we define **the genus** of a closed orientable surface to be 1/2 of the rank. The genus can also be used to classify closed surfaces. The genus measures the number of holes in a solid object that is bounded by the surface.

You might wonder if the rank of a closed surface can be defined without cutting a disk from the surface. Yes. Closed surfaces do not have substantial arcs because they have no boundary, but they can have substantial circles. A **substantial circle** is a simple closed curve in the surface that does not separate. Spheres do not have substantial circles, but tori and higher genus surfaces do.

Figure 1.28 illustrates some substantial circles. There is a theorem called Poincaré duality that implies that a substantial circle in a bounded surface will intersect a substantial arc in a point and that substantial circles can be paired by their intersections. Poincaré duality applies to much more general situations (as we sketch in Chapter 7), but the essence of the theorem is seen directly on the orientable surfaces.

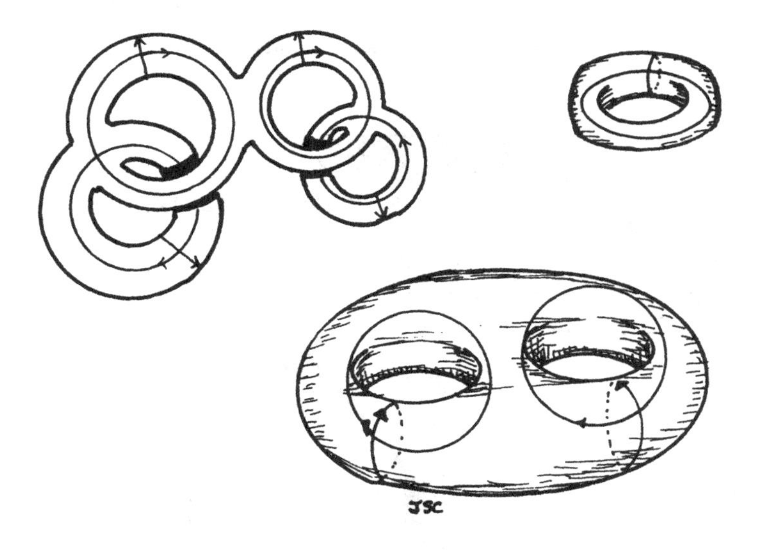

Figure 1.28: Substantial circles in surfaces

1.3 The Intersections of Surfaces in Space

In the next chapters, we construct certain functions from abstract surfaces into space. We need these functions to give realistic descriptions of surfaces that are not orientable. Without some restrictions the intersections of surfaces in space could be very complicated. But there is a hypothesis called "general position" that forces the surfaces to look, at least on a small scale, like one of the four pictures in Figure 1.29.

Topologists consider general position to be a general situation as follows. If surfaces are not intersecting in general position, we can shake the surfaces, and they will be in general position. There is an extraordinary theory that allows us to say, "Just shake the surfaces." In this theory, people consider a space that consists of functions between other spaces, and find within the function space a certain set of functions (the general position ones) that is so big that every function is arbitrarily close to one in general position. That is every smooth function can be smoothly approximated by a general position function. Having said that, let us understand the pictures.

In Figure 1.30, several different views of branch points are illustrated. Different versions of these will be used as we go along.

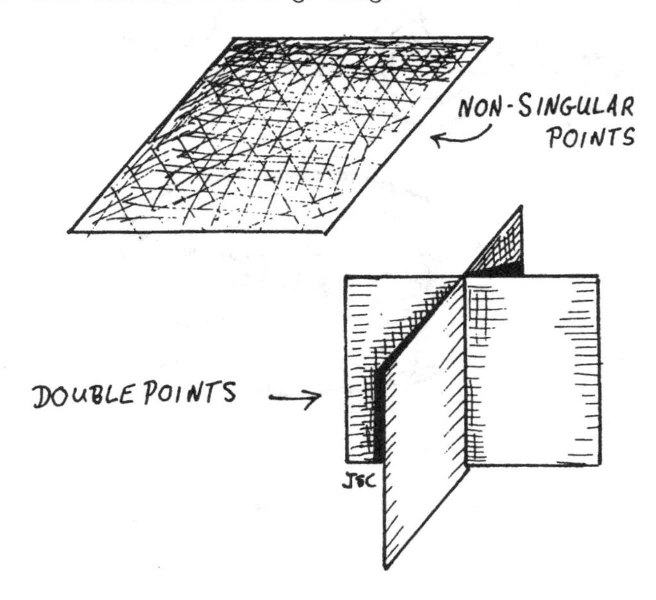

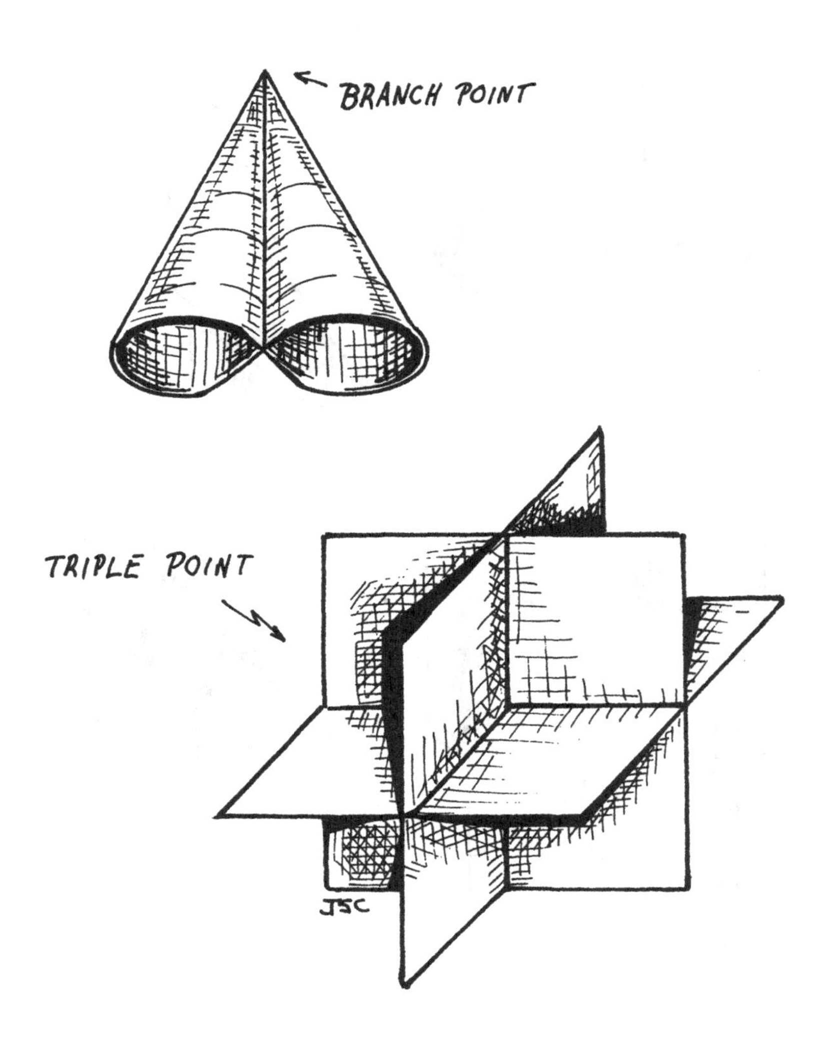

Figure 1.29: General position intersections of surfaces

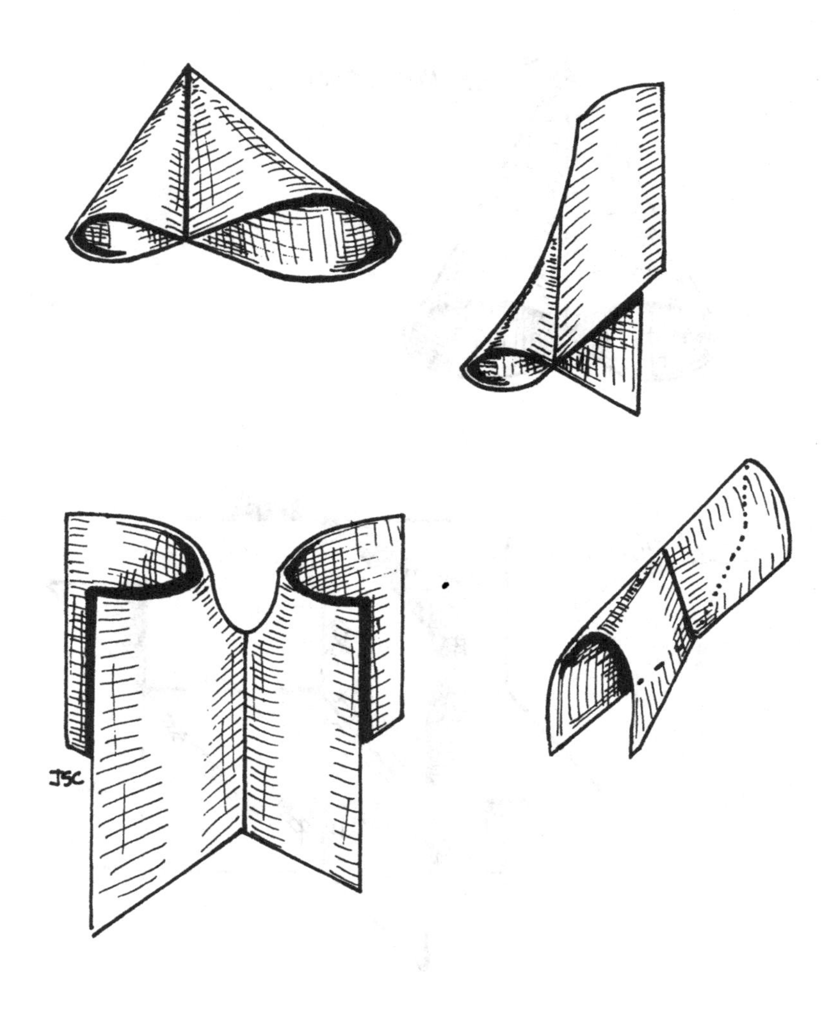

Figure 1.30: Various views of branch points

1.4 Notes

If you want to learn more about continuous functions, I suggest M. A. Armstrong's book [4].

The material on the classification of surfaces was synthesized from [4], [57], and [51]. The book [43] gives a simliar proof of the classification theorem in a more rigorous framework and is very accessible. Meanwhile [31] discusses the notions of separation at an introductory level. Weeks's [74] book discusses geometry and topology at an elementary level. Francis's book [34] has obviously influenced the techniques of drawings that I am using. Rudy Rucker's book [69] has a lot to say about space and the 4th dimension. And Tom Banchoff's book [6] discusses the notion of dimension in a large variety of contexts. At a graduate level [40], discusses general position in a very clear exposition.

Chapter 2

Non-orientable Surfaces

2.1 Surfaces with Boundary

We begin with the most famous example in topology.

The Möbius Band

Take a piece of ordinary notebook paper, and cut from it the vertical strip that is determined by the right margin marker. This is the red line that runs vertically up the paper. Put a half twist in the strip, bring together the ends as in Figure 2.1, and tape them. The result is a paper model of **the Möbius band.** Now conceptualize this surface as being made of golden fleece: infinitesimally thin and infinitely malleable.

The Möbius band is not orientable. The ends that were taped form the mates of a substantial arc, and when the strip is identified with the disk, both mates point in a counter-clockwise direction.

Fact. *In a Möbius band the notions of left and right do not make sense.*

Drape your piece of paper over the corner of a table, and put a pen or pencil on it so that the stylus points down. Now gently roll the paper so that an arc will be drawn on the paper. For convenience, start drawing at the substantial arc that was taped. Keep rolling the surface until the taped edge appears again. Observe that your pen is on the "other side" of the surface. Next periodically draw arrows perpendicular to the center circle that point to the left on the surface. When you return to the tape, the left arrow points to the right from the point of view of the first arrow that you indicated. Hold the paper up to the light to see this, or compare your calculation with Figure 2.1.

A non-orientable surface is a surface in which left and right don't make sense globally. *Each non-orientable surface has a Möbius band as a sub-surface:* For if one can find a loop in the surface upon which left and right don't make sense, then a neighborhood of this loop in the surface is a Möbius band. Furthermore, when a non-orientable surface is cut along a substantial arc that intersects the orientation reversing loop once (and the other substantial arcs have been cut), the resulting disk has a pair of mates that both point in the same direction along its boundary. Therefore, a surface on which left and right don't make sense is non-orientable according to the meaning given in Chapter 1.

⋆⋆⋆

I had trouble with the notion that left and right were switched on a Möbius band for quite some time because the first thing that my brothers taught me about the Möbius band was that it was one sided. The way they illustrated one-sidedness was to continue drawing the center circle until the pen returned on the paper to the beginning point. When I included the left-pointing arrows perpendicular to the center circle, I would draw them twice around the surface. But a Möbius band is made of golden fleece, and paper has thickness. I told you to stop drawing when you got back to the tape because this circle goes only once around the surface made of golden fleece.

⋆

A *non-orientable space* is one in which left and right can only make sense locally. A global notion of left and right is inconsistent in a Möbius band. If our world were non-orientable, then you might wake up some morning and find yourself inside the mirror. Or I might. The Möbius band is the prototypical non-orientable space because Möbius bands always appear in non-orientable spaces. Möbius bands and

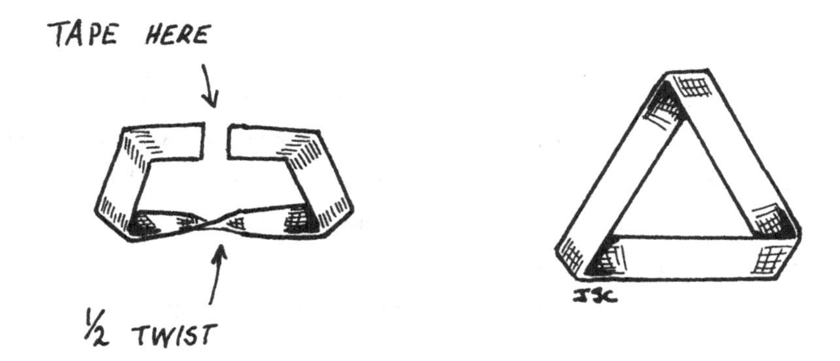

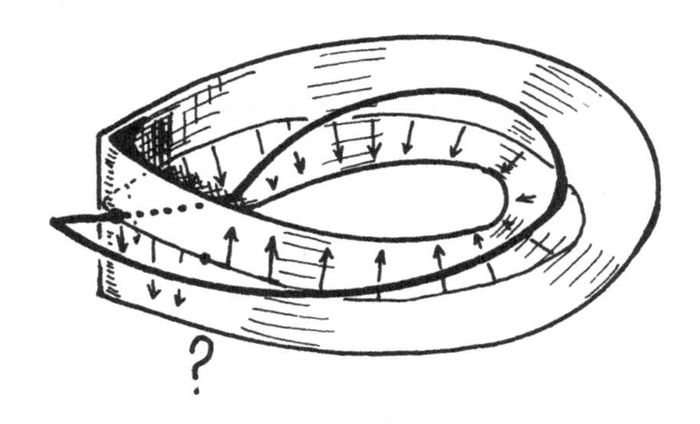

Figure 2.1: A Möbius band is one sided and non-orientable

non-orientability pervade science fiction. I have often felt that the southern tip of Manhattan Island is non-orientable. This feeling is due to the fact that on several occasions when I was sure that I was headed north, I found myself entering the Holland Tunnel — a most disquieting experience.

⋆

The symbol for recycling is modeled after a Möbius band. There is a nice idea being put forth in that symbolism. The process — product to waste to product — seems non-orientable to those of us who grew up in the last half of the twentieth century. In fact, this process is the way that nature functions. We will see in Chapter 4 that non-orientable surfaces can exist in orientable spaces, and they are substantial there. So the non-orientable process of recycling is a substantial part nature's way.

⋆

An exhibit of a Möbius band travels with the IBM museum show "Mathematica." It illustrates one-sidedness by having an arrow travel around the surface on a monorail. I first saw a similar exhibit at the Museum of Science and Industry in Chicago when I was no more than six years old. The last time that I was there, the exhibit wasn't. It is a shame. The scientific wonders to which a child is exposed can influence his or her adult life.

⋆⋆⋆

Now take your paper model and cut it along the center circle that you drew. The result is connected and orientable. Did you see that the result is an annulus that has a full twist in space? If not, check the classification of orientable surfaces. Many people are amazed that the Möbius band, when cut, is not separated. I trained you in Chapter 1 not to be amazed by this, but I hope you are anyway. In Chapter 1, we saw that arcs need not separate surfaces — those that don't are substantial. And I mentioned that substantial circles intersect substantial arcs (in the non-planar case) in a point. We constructed the Möbius band as a surface with one substantial arc. It

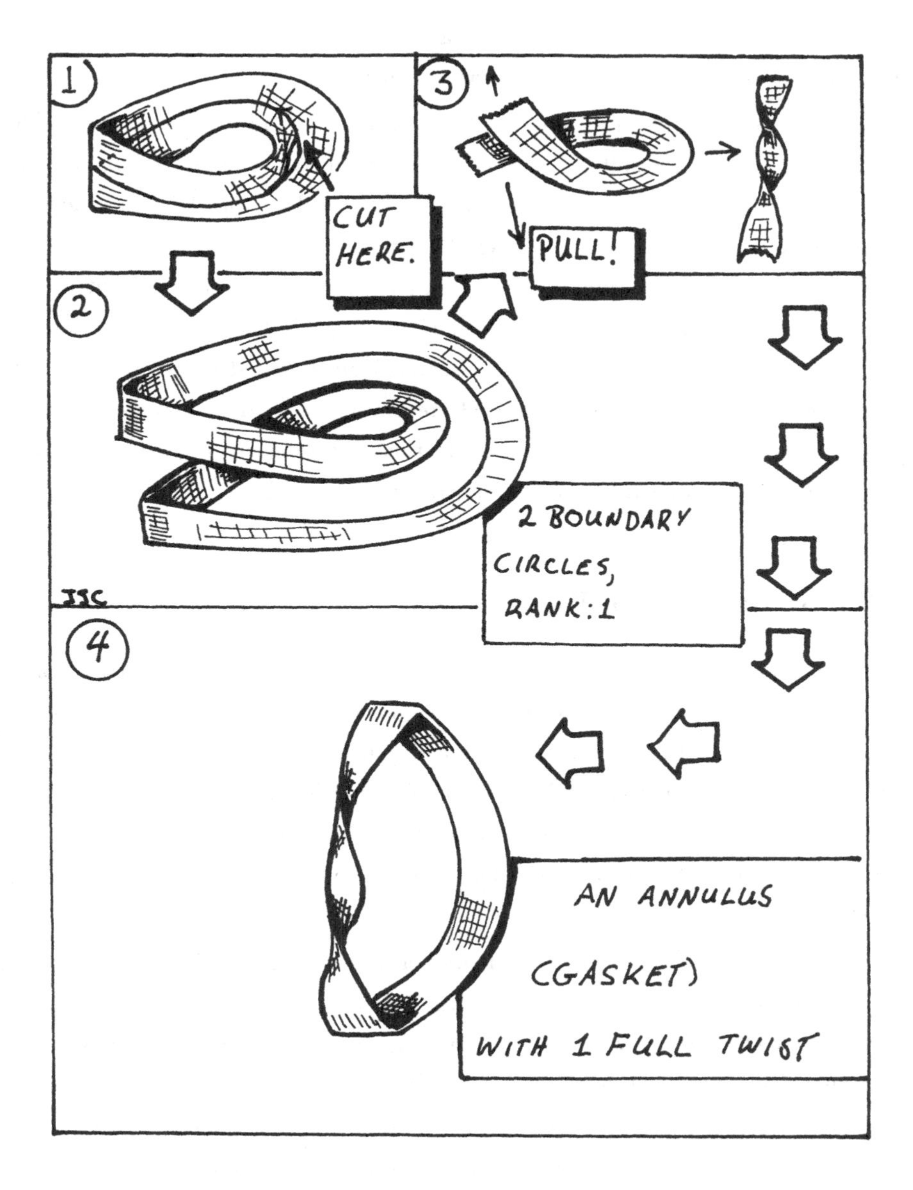

Figure 2.2: An annulus with a full twist and the Möbius band

is a non-planar surface, so it should have a substantial circle. I think the interesting thing is that when it is cut it becomes an annulus.

Here is perhaps the most amazing thing of all. Cut the twisted annulus along its center circle. The result is two annuli that are linked. The twisted annulus is a model — albeit an overly simplistic model — of closed circular DNA. Inside all of your cells you have closed circular DNA. The model for DNA replication is that the molecule splits down the middle just as you cut your twisted annulus. But there is a problem because the DNA has to become disentangled. The actual closed loops of DNA have many twists. The need for disentanglement caused biologists to discover certain enzymes called topoisomerase. By regulating the action of topoisomerase in cancer cells, replication can be slowed. In this way, topology is used to help cure cancer. (I am indebted to De Witt Sumners for this story.)

Möbius Bands with Holes

Suppose that for your paper model you used the left, rather than the right, margin of your notebook paper. In my notebook paper there are three holes. The surface that you started with, then, would have had four boundary circles. The boundary of a Möbius band is a circle, and the notebook paper holes are like the cookie cutter holes we had before. We can compute rank and the number of holes on non-orientable surfaces as well. The three indices: rank, number of boundary holes, and orientability classify all surfaces.

There are two differences between non-orientable and orientable surfaces from the point of view of the classification theorem. The first difference is that the mates of a non-orientable surface will point in the same direction when the substantial arcs are cut. The substantial arcs can be labeled with arrows even if the surface is not orientable. Indeed, to reattach the mates, we need to know their orientations. The second difference is the form of the standard surface. In the orientable case, every pair of mates could be made to either have one or no other mates between them. When either case happened, the presentation of the surface couldn't be simplified

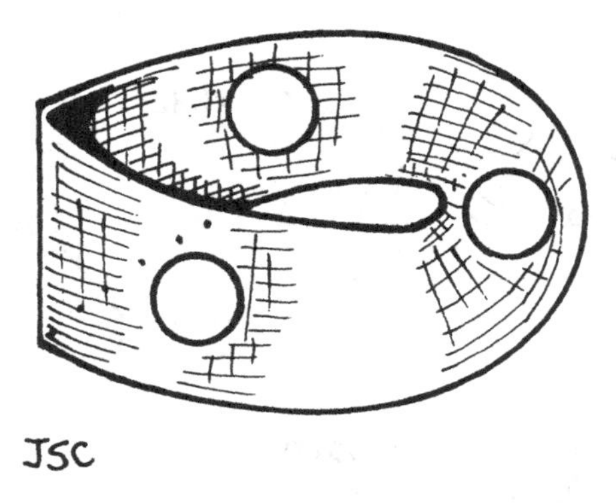

Figure 2.3: A Möbius band with holes

further. In non-orientable surfaces there is a way to further simplify a pair of mates that have complexity one. The next section illustrates this.

Handle Slides

Non-orientable surfaces are peculiar with respect to handle sliding. Let's examine that peculiarity. I have indicated in Figures 2.4 and 2.5 that a pair of mates with complexity one on a non-orientable surface can be transformed to a pair of mates with complexity zero.

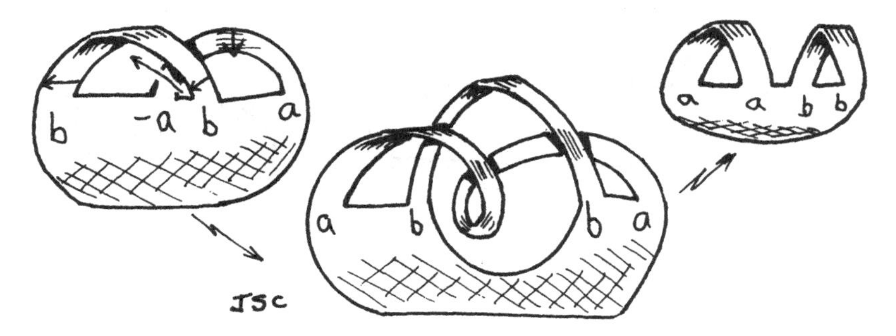

Figure 2.4: Sliding handles on a non-orientable surface

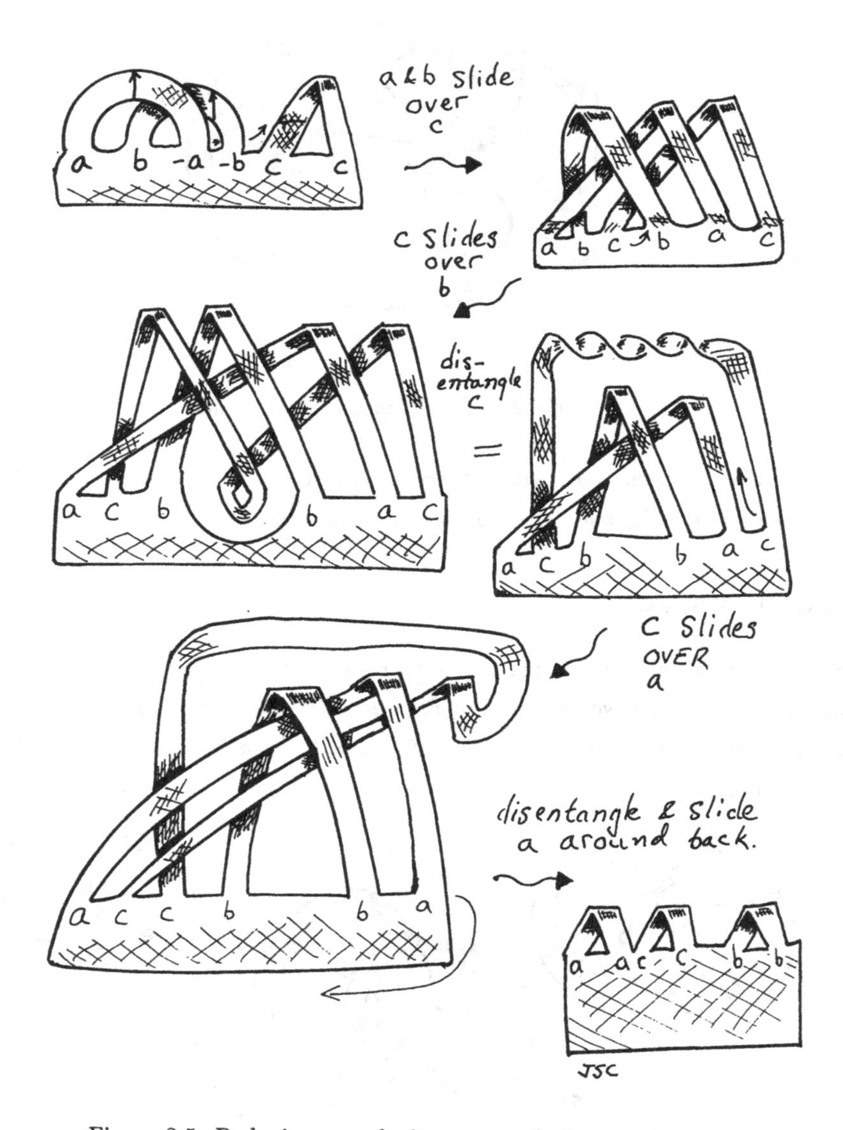

Figure 2.5: Reducing complexity on a rank three surface

Theorem 7 *A non-orientable surface with boundary that has rank larger than one has a standard form in which every pair of mates has complexity 0.*

We have in place all the tools to complete the classification of surfaces with boundary — orientable or not.

Theorem 8 *Two surfaces that have boundary are topologically equivalent if and only if they have the same rank, the same number of boundary circles, and are either both orientable or both non-orientable.*

The proof of Theorem 8 follows along the same lines as the orientable case. Given a surface with boundary, we find a system of non-intersecting substantial arcs and cut the surface along these to get a disk. Along the boundary of the disk, we read the sequence of mates in counter-clockwise order. We slide handles until every pair of mates has complexity at most one. If, for each substantial arc, the mates that comprise it point in different directions, then the surface is orientable. In that case Theorem 4 applies. If there is a pair of mates that both point in the same direction, then further handle slides can be performed until each pair of mates has complexity zero. The details of that last step will be delegated to the energetic reader.

Suppose two surfaces are given that are either both orientable or not. If they have the same rank and the same number of boundary components, then their standard forms are the same. In this way, we have completed the classification of surfaces with boundary. After we look at some examples in detail, we will classify closed non-orientable surfaces as well.

2.2 The Projective Plane

The Möbius band has one boundary circle and rank one. Those two facts characterize the Möbius band among all surfaces because orientable surfaces with one boundary circle have ranks that are even numbers. The disk has one boundary circle. What happens when we glue a disk and a Möbius band together along their boundaries? The result is a space that is called the **projective plane**. An explanation of this term will be given shortly, but first I want to show that the projective plane cannot fit into space without intersecting itself. So those of you who were considering constructing a paper model should wait for a while.

Any closed curve separates the plane. That fact is a consequence of the Jordan Curve Theorem. A similar statement holds in ordinary 3-dimensional space: *An embedded closed surface separates 3-space into two regions.* An **embedded surface** is one that does not intersect itself in space. Embedded surfaces have no double lines; they might be called simple in analogy with the term simple closed curve, but they can be too complicated to warrant that tag. The complications occur because the topological type of the regions into which space is separated is difficult to discern. A great deal of effort is currently (1992) being directed at understanding the topological type of space that an embedded torus determines. Knot theory is the study of these spaces; this subject is touched upon in Chapter 3.

An embedded surface separates space into two pieces. So a simple closed curve that intersects the surface at least once must intersect it an even number of times. That fact is obvious: If you leave the room in which you are seated, in order to re-enter you must either go through the door, window, or a trap door in the floor or ceiling. If the walls of your room were a projective plane, then it would be hypothetically possible for you to leave the room and return to it without crossing the threshold. To prove that last statement, I will construct a loop in space that only intersects the projective plane once.

Consider the arc of ink that you drew around the center of your paper Möbius band. You started at the line of tape and ended at your next encounter with the tape. Place a thread along the ink arc. The ends of the thread are on "either side" of the Möbius band, and the thread does not form a closed curve. Poke a small hole in the paper Möbius band, and let one end of the thread pass through the hole so that the ends of the thread can be tied together. The thread passes once through the surface of the Möbius band, and the Möbius band is contained inside the projective plane. If the projective plane were embedded in space, the loop that intersects the

Möbius band once would also intersect the projective plane once. To return to the room without crossing the threshold, just follow the string around the Möbius band.

No further intersections occur between the thread and the rest of the projective plane because the thread lies close to the center circle of the Möbius band and has no opportunity to pass over the boundary.

The projective plane doesn't embed into space. How are we going to understand the projective plane if we can't see it sitting whole and in front of us? In the rest of this section, we will learn how to see the projective plane.

The projective plane is a popular example in math, and consequently there are several apparently different descriptions of it. Let's turn to one of these.

⋆⋆⋆

John Hollingsworth was my first topology instructor. He told our class about the **quotient topology** in which a new space is made from an old one by gluing points in the old space together. The space that resulted is necessarily the continuous image of the old space. In this book, we used the quotient topology when we glued disks and surfaces together along their boundaries, or when we glued segments of the boundary of a disk to get a new surface.

Hollingsworth's homework question for us was to recognize the result of gluing diametrically opposite points on the 2-sphere together. That is, think of the sphere standardly embedded in space. A diameter of the sphere punctures the sphere in two points; glue these points together for every pair of points on the sphere. So the north and south poles are identified, and Perth, Australia is identified with a point not far from Bermuda. Santiago, Chile is identified with a point in central China. The region diametrically opposite to Hawaii is in Botswana.

Hollingsworth's exercise required thought rather than technique. I started working on the problem on a weekend in which two friends invited me to go skiing. (I won't go into the sordid details of that road trip.) During my hours awake in the back of my friend's baby blue Volkswagon beetle, I pondered the exercise deeply. I didn't solve the problem completely on the trip, but I did develop some strong visual imagery about it. Some of that imagery appears in the figures of this section. Other images are too abstract to be helpful.

This quotient map can be thought of in two pieces. Using loose geographical terms, the arctic and the antarctic regions are glued to each other, while the tropical regions are identified among themselves. In the tropical regions, we can cut along the prime meridian and the international dateline and glue the western tropics to the eastern tropics upside down and with a flip. Points in the northeast of the tropical strip are glued to points in the southwest of the tropical strip. So points along the northern section of the prime meridian are glued to points on the southern section of the international dateline. Therefore, the tropical annulus maps to a Möbius band.

Let's go through the tropical step one more time. The tropics form an annulus. Embed the annulus into space as an annulus with one full twist. The twisted annulus maps onto the Möbius band in a rather natural way now. Figure 2.7 indicates the details. In fact in this figure, I distributed the full twist in the annulus to a pair of half twists.

The quotient of the sphere is the quotient of the annulus glued to the quotient of the polar regions. The latter quotient is a disk; the former is a Möbius band. The total space that results is a projective plane.

⋆

It is interesting to note that the annulus with a full twist is the result of cutting a Möbius band along the latter's substantial center circle. There is magic trick in this process somewhere. Let's see ... You cut the Möbius band down the middle without separating it; tape the result back together and get a surface with half of the area of the original. Hmm.

⋆

John Cage said that there is music all around us if only we had the ears to hear it. A similar statement about math is also true. Alas, people are more prone to close their eyes to math than they are to close their ears to music. I visualized the quotient map with my eyes closed in the back of Charlie's bug. I kept my mind's eye open.

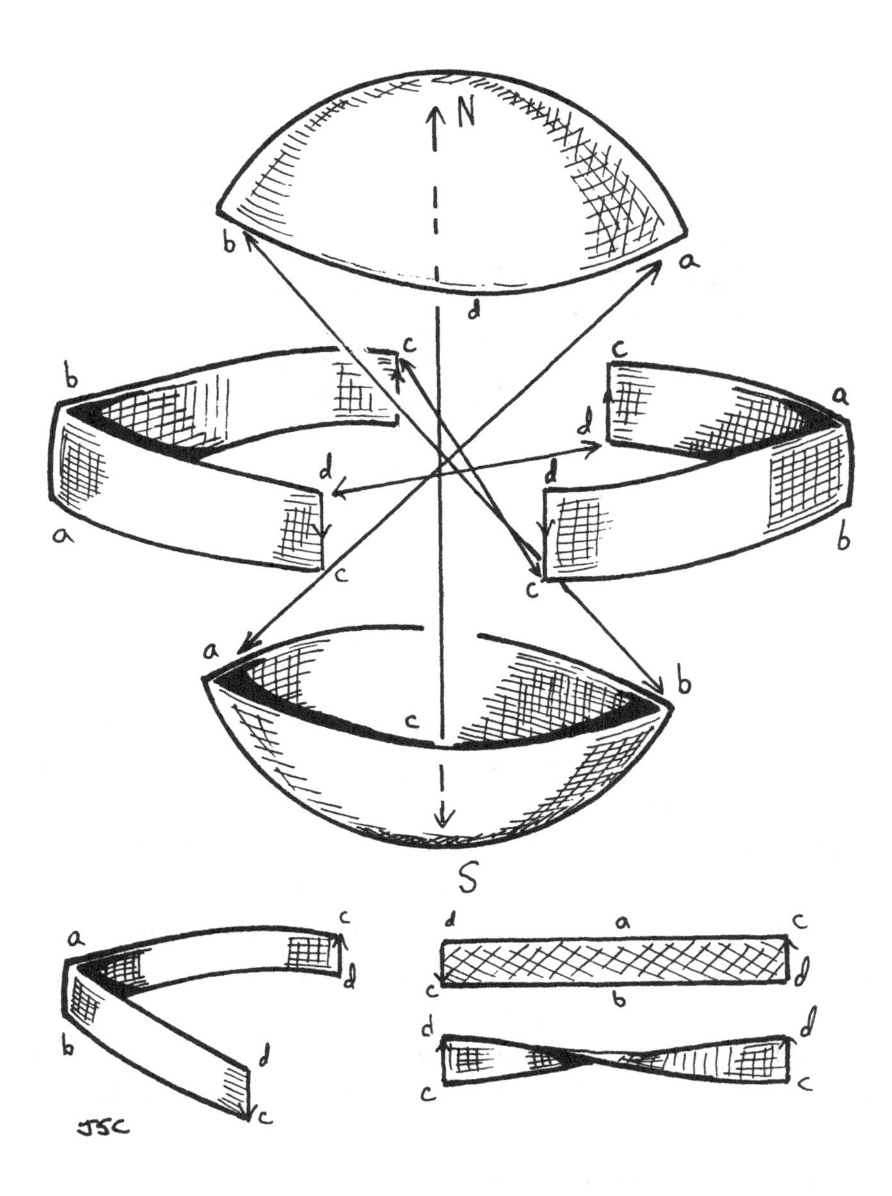

Figure 2.6: Identifying opposite points on the 2-sphere

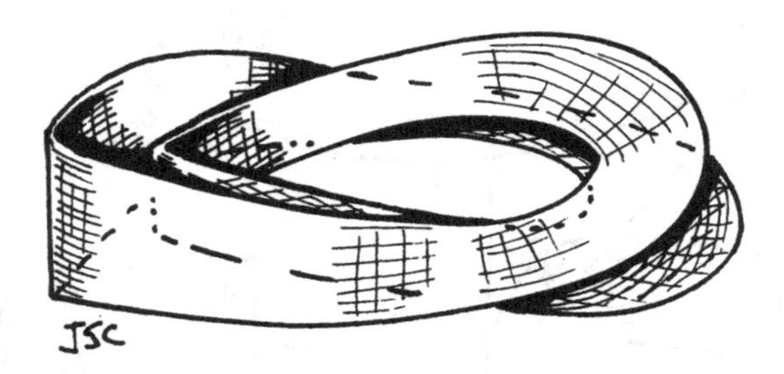

Figure 2.7: The annulus maps 2 points to 1 on the Möbius band

Boy's Surface

So far I have given two descriptions of the projective plane; both are too abstract. In fact, I have proven for you that the projective plane cannot be embedded in space, so that anything but an abstract description seems at present impossible. We are going to see in this section, and the next two, that concrete descriptions do exist. We can map the projective plane into space as long as we are willing to let the image intersect itself. At the end of Chapter 1, I illustrated the local pictures of general position intersections. In this section, we will glue these local pictures together in some fashion to see the projective plane.

The model of the projective plane that we will construct now is called **Boy's surface.** It was discovered by Werner Boy — the only student of Hilbert who studied geometry. Boy's paper [8] appeared in 1903. In relatively modern history, Francois Apery [2] found a collection of very simple equations that describe Boy's surface. In the notes to this chapter, I have included a program written for *Mathematica* that uses Apery's parametrization to graph Boy's surface on the computer screen. John

Hughes [47] has made a short computer animated film of Boy's surface that illustrates some of the remarkable symmetry and beauty of the projective plane. (This is not the same John Hughes who films stories about young people living in the northern suburbs of Chicago.)

We start at a triple point. Three coordinate disks intersect in space. Each disk is embedded; their union intersects. The boundary of the intersection consists of three great circles on the sphere: the prime meridian, the equator, and the meridian that passes through Baton Rouge, Louisiana. In Figure 2.8, the disks are depicted with square boundaries, but squares are the same as circles in topology.

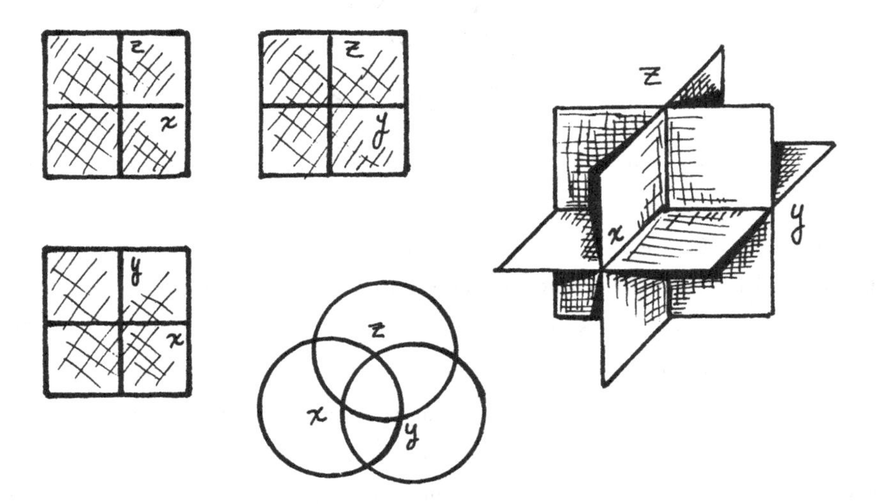

Figure 2.8: Boy's surface has a triple point

In the next step as depicted in Figure 2.9, we stretch the horizontal disk and glue together two edges of the vertical strips. The resulting surface is a pair of disks that are intersecting in space. One disk intersects itself and intersects the other. The horizontal disk remains embedded.

In the third step (Figure 2.10), the horizontal disk is stretched and glued to the self intersecting vertical disk. The union of two disks along segments of their boundaries

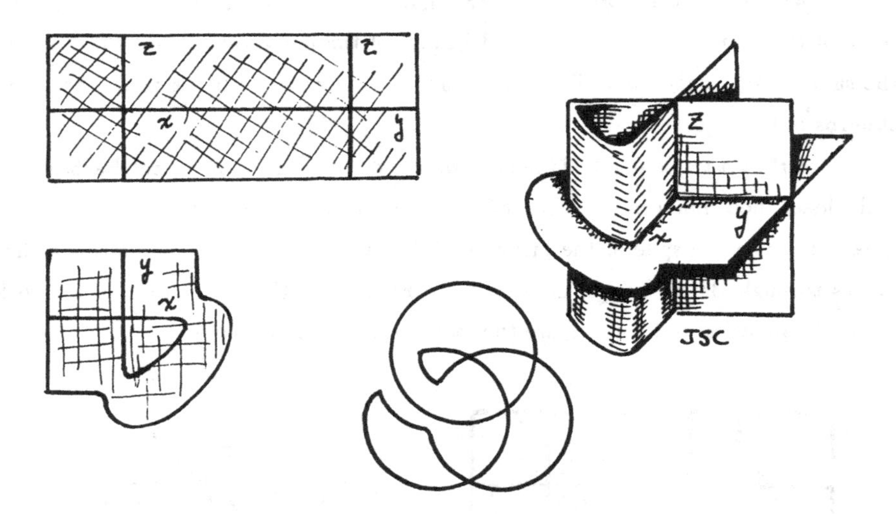

Figure 2.9: Glue two disks along an arc on the boundary

is still a disk. Figure 2.10 contains an illustration. The disk appears inside a 3-dimensional ball, and the curve that appears on the boundary of the ball has two double points. That curve is called the $\alpha\gamma$-curve (alpha gamma) because it is similar to those Greek letters juxtaposed.

The Möbius band is the result of gluing segments along the boundary of a disk together. The disk that we used in our model was a thin strip of paper; nevertheless it represented a disk. In Figure 2.10 a disk intersects itself. In Figure 2.11 two segments of its boundary have been stretched and glued together. The surface that results is non-orientable, and no effort was expended in making it so. So Figure 2.11 illustrates a Möbius band that is intersecting itself in space. The boundary circle of the Möbius band is now a simple closed curve in the sphere.

The circle on the sphere that is the boundary of a Möbius band is a lot like the seam on a baseball or a tennis ball. On the surface of a baseball, two disks each with two lobes are stitched together. On the surface of the sphere illustrated in Figure 2.11, the disks determined by the boundary of the Möbius band have three

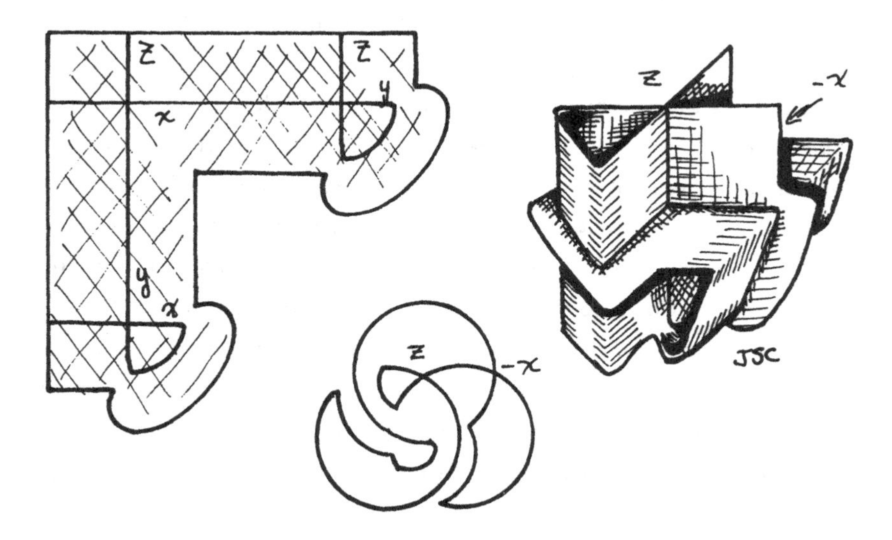

Figure 2.10: A disk in the ball with a single triple point

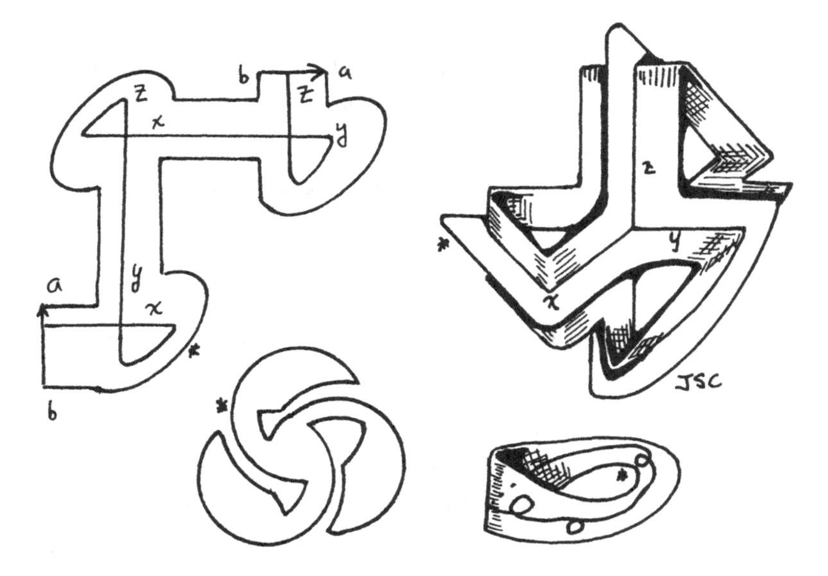

Figure 2.11: A Möbius band that intersects itself in the ball

lobes. In Figure 2.12 one of these trilobite disks is glued to the boundary of the Möbius band.

Boy's surface is the continuous image of the projective plane mapped into space. It is not embedded, and it does separate space. (Observe that the numeral 8 also separates the plane even if it is not an embedded circle.) In fact, Boy's surface determines a definite inside and outside. Surfaces must separate ordinary space whether or not they be embedded. Later on we'll find spaces that embedded surfaces do not necessarily separate. Compare that to surfaces: Substantial circles do not separate surfaces that have non-zero rank, and there are no substantial circles on the sphere. Similarly, there are no substantial surfaces in ordinary space. It seems reasonable, then, to think that there might be spaces with substantial surfaces and to think that substantial surfaces can be used to help distinguish spaces. They can, but they do not classify spaces.

The triple point of Boy's surface is worthy of comment. A theorem proven by Banchoff [5] (but apparently part of the folklore before the proof appeared in print) implies that if the projective plane is mapped into ordinary space it either has triple points or branch points. The image can have both, as in the next section, but if the projective plane is mapped into space without branch points, then it must have an odd number of triple points. In particular, **there must be at least one triple point in the image of the projective plane if there are no branch points.**

The Roman Surface

In the next model of the projective plane, both branch points and triple points appear. The model is simple to describe, but perhaps it is not so simple to see that the result is the projective plane.

To the three intersecting coordinate disks, we add four triangles. The easiest way to describe how these are added is to use the Cartesian coordinate system. Each coordinate plane can be given by an equation: the xy-plane has equation $z = 0$; the xz-plane has equation $y = 0$, and the yz-plane has equation $x = 0$. ("Everything is what it isn't, isn't it?" Harry Nilsson) These three planes divide space into eight

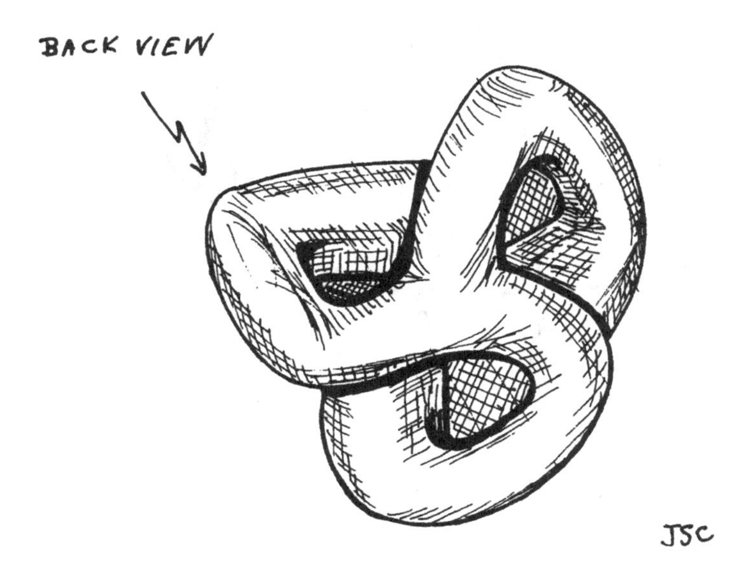

Figure 2.12: Boy's Surface: A map of the projective plane

regions that are determined by the possible signs of x, y, and z. For example, the points above the plane $z = 0$ are the points in which $z > 0$. A point in space might be above the xy-plane, to the right of the xz-plane, and in front of the yz-plane. In this case all three of its coordinates are positive. To locate the regions of space in which a point lies, we specify a triple of plus or minus signs. For the record, here are the possibilities: $(+,+,+)$, $(+,+,-)$, $(+,-,+)$, $(+,-,-)$, $(-,+,+)$, $(-,+,-)$, $(-,-,+)$, and $(-,-,-)$.

The round coordinate disks intersect the sphere in three great circles, and these intersect each other at 90° angles. (In spherical geometry the sum of the angles is always greater than 180°.) We add to the intersecting disks the four triangles on the sphere that have coordinates with an even number of minus signs in them. The result is Steiner's Roman surface.

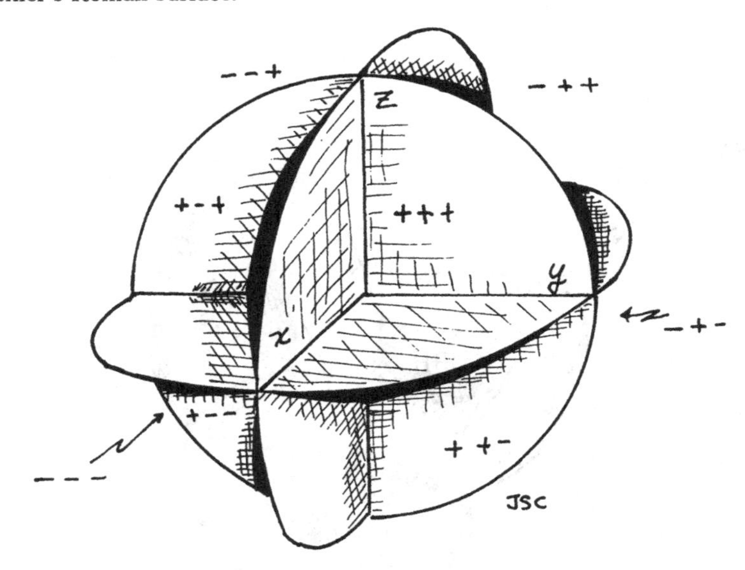

Figure 2.13: Intersecting disks and signs of regions

Let's see that this surface is the image of a projective plane. We'll start with the branch points on the surface. If you still have the piece of notebook paper from which you made the Möbius band, get it out; we need to construct a new model. Cut a

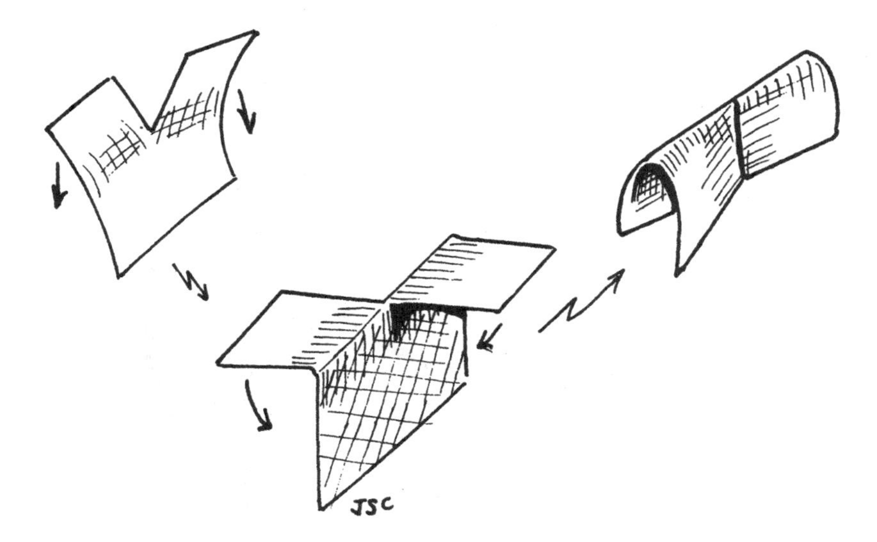

Figure 2.14: Folding paper to illustrate a branch point

sheet of paper halfway up the middle of the paper. Say that the cut runs vertically. Then put two horizontal folds in the paper at the end point of the cut. The fold lines make the flaps point in different directions. Figure 2.14 indicates how to do this. The arc along which you cut represents a double arc when the surface of the paper is brought together along this cut.

The folded paper represents a disk that is mapped into space with an arc of double points. One end of the double point arc is a branch point; the other end is on the boundary of the disk. This surface, called a **Whitney's umbrella**, is named for Hassler Whitney [75] who began the study of maps of surfaces. It is the prototype for all types of branching, just as the Möbius band is the prototype for all orientation reversing paths.

The Whitney umbrella is a disk with a branch point. In Figure 2.15, two more spherical triangles are glued to this disk. This figure indicates that the surface which is the union of the two vertical disks and the four triangles is a Möbius band. Consequently, the result of gluing in the horizontal disk is a projective plane.

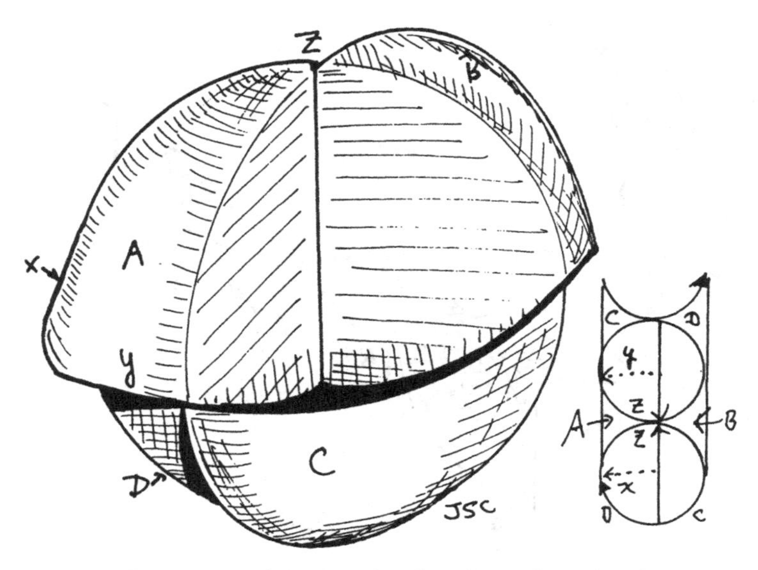

Figure 2.15: A Möbius band with two branch points

The Cross Cap

The cross cap map of the projective plane is very similar to the Roman surface, but it doesn't have a triple point. We begin the construction at Figure 2.15 which illustrates a Möbius band with a pair of branch points. The boundary of the Möbius band is the equator of the sphere. In the Roman surface, this equator bounds a disk on the inside of the ball, but in the cross cap, the equator bounds a disk on the outside. Figure 2.17 contains illustrations.

Figure 2.18 is an illustration based on one given in [34]. This is the first illustration of the cross cap in which *I* could see that the projective plane was being depicted.

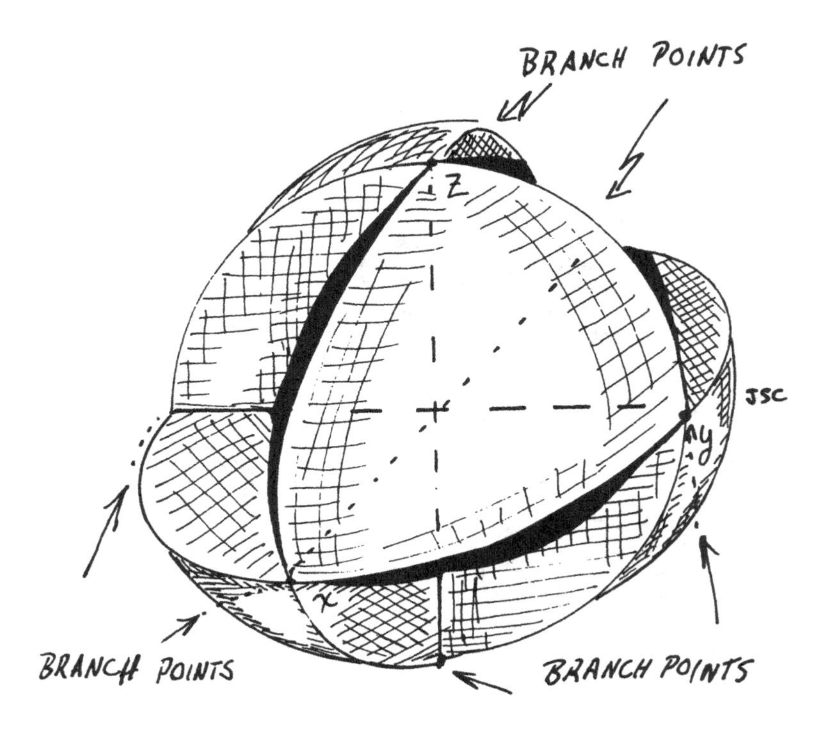

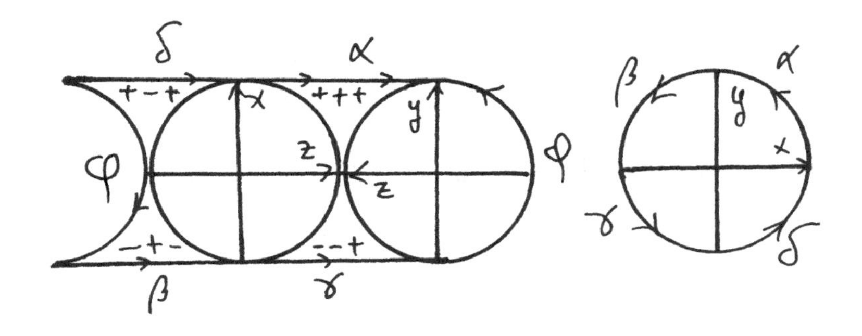

Figure 2.16: Steiner's Roman surface

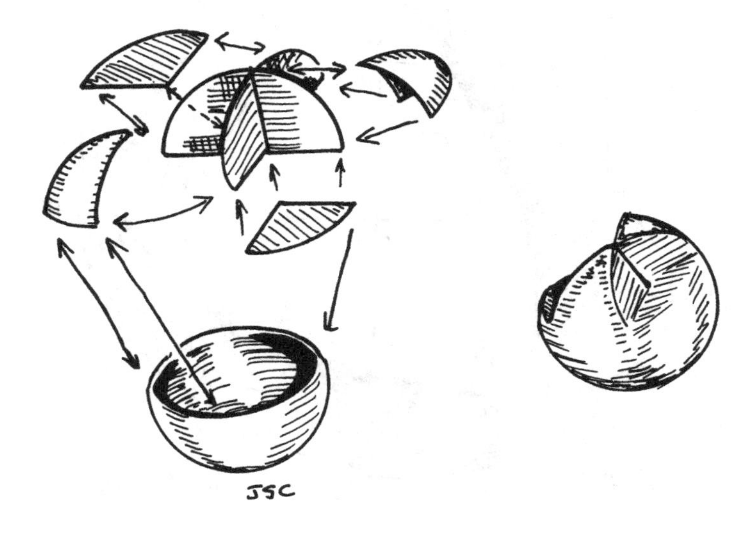

Figure 2.17: The cross cap view of the projective plane

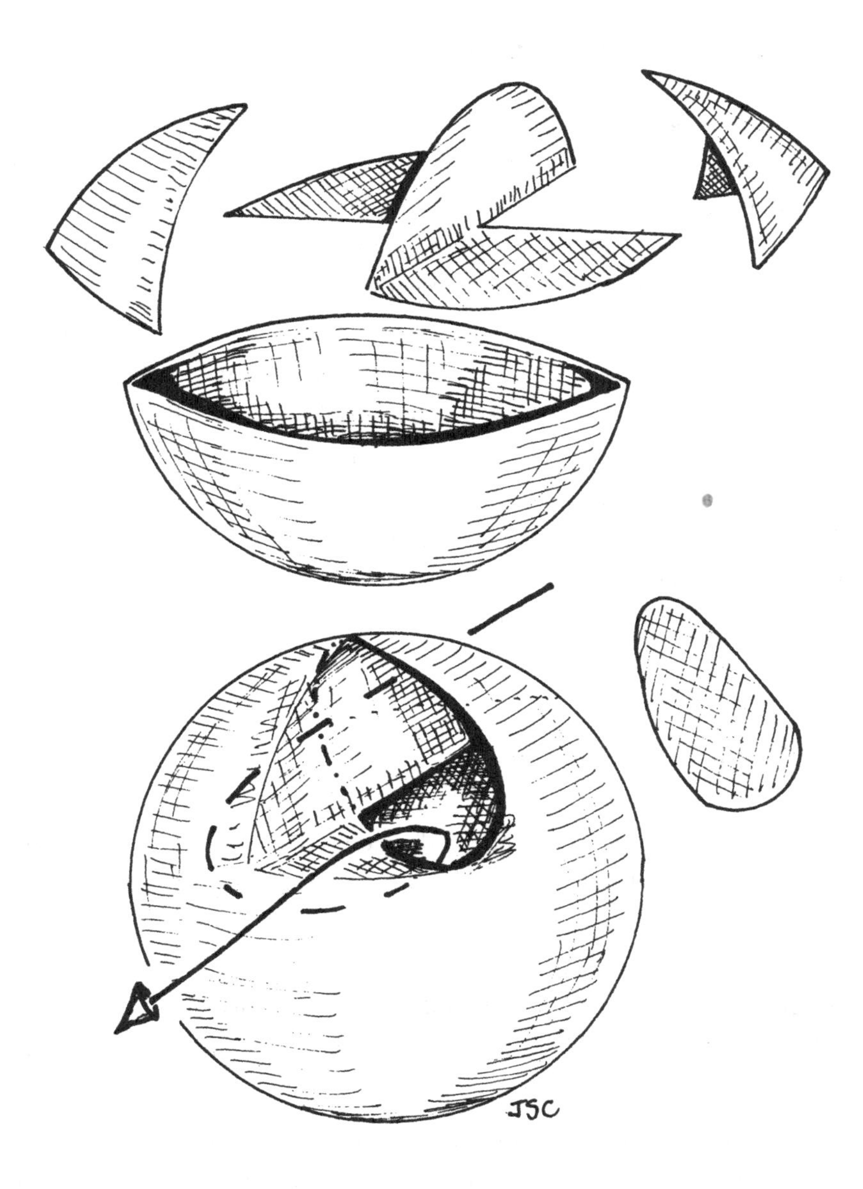

Figure 2.18: Francis's method of seeing that the cross cap is the projective plane

The Projective Plane in Visual Geometry

A standard trick in perspective drawing is to draw a horizon as a horizontal line. Parallel lines are depicted as lines that emanate from a point on the horizon. So in perspective drawing, parallel lines intersect! Each point on the horizon corresponds to some set of parallel lines. Look down a long corridor, down a long straight highway, or down a pair of railroad tracks. The lines seem to merge at infinity.

The projective plane can be thought of as adding a "circle at infinity" to the ordinary plane. Railroad tracks also merge to a point on the horizon behind you. So each set of parallel lines determines a pair of points on the circle at infinity, and those points are diametrically opposite on that circle. When we add the circle, you might think the result is a closed disk, but in fact it is the disk with diametrically opposite points being identified. Therefore, the plane together with the circle at infinity that is determined by parallel lines is the projective plane. This paragraph, then, justifies the term, for the projective plane is the plane that is projected onto our eyes.

2.3 The Klein Bottle

After the Möbius band, the Klein bottle is the most well recognized non-orientable surface. That it is more familiar than the projective plane strikes me as a cultural oddity. The projective plane is simpler than the Klein bottle because it has smaller rank. Nevertheless, some of the drawings of the Klein bottle that are illustrated here may already be familiar to you. I hope to show you some unfamiliar versions, and I hope that you will see that the pictures actually depict two Möbius bands glued along their boundaries.

The Klein bottle cannot be embedded into space for the same reason that the projective plane cannot be. The thread that intersects any Möbius band in a single point intersects a closed non-orientable surface in a single point. By definition, non-orientable surfaces always contain a Möbius band as a sub-surface.

The **Klein bottle** is the result of gluing two Möbius bands together along their boundary circles. To map the Klein bottle into space, we will find a way of making the Möbius bands intersect each other or themselves.

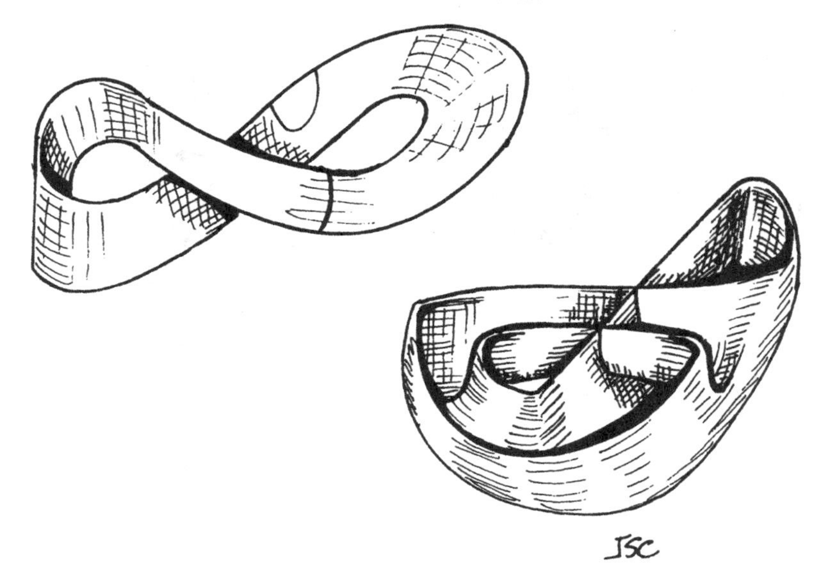

Figure 2.19: A Möbius band that intersects itself along an arc

Figure 2.19 indicates a map of a Möbius band into space that has a simple arc of self intersections. The half-twist that appears in the Möbius band can be either right-handed or left-handed with respect to the orientation of space. Figure 2.20 illustrates the standard model of the Klein bottle formed by gluing a left-handed and a right-handed self intersecting Möbius band together.

I have made a paper model of the standard Klein bottle using self intersecting Möbius bands, but the paper has really been too stiff to make the model be very pleasing to the eye and the hands. An easier model is obtained by using a slinky toy. The slinky represents an annulus or a cylinder. The boundaries of the cylinder are glued to each other by means of a reflection, so that the cylinder passes through itself as in Figure 2.21.

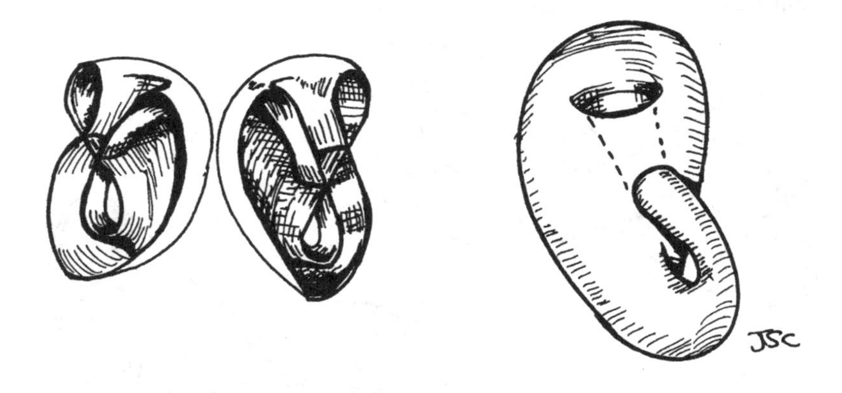

Figure 2.20: The standard map of the Klein bottle

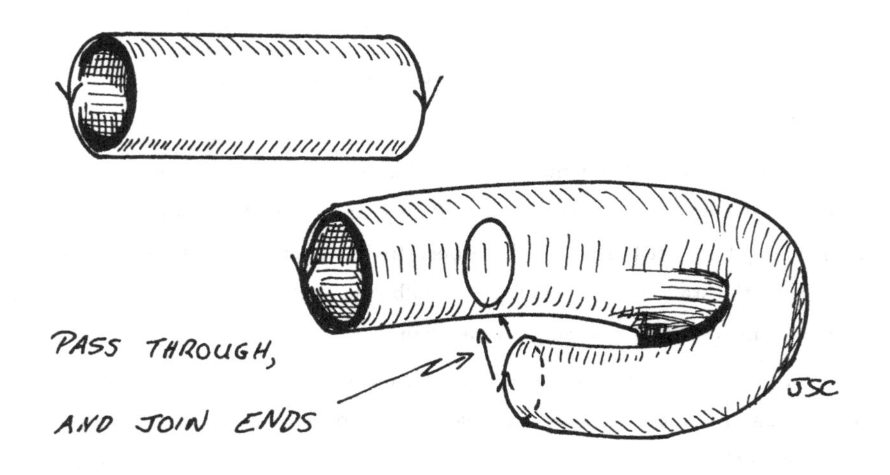

Figure 2.21: The slinky toy model of the Klein bottle

While in graduate school at Yale, Gail Ratcliffe knitted a model of the Klein bottle out of green yarn. The model was placed in the math library in Leet Oliver Hall. I don't know if it remains among the memorabilia stored there.

The Twisted Figure 8

I first learned about the following model of the Klein bottle in Banchoff's paper [5]. Consider a standard Möbius band embedded in space. For each point on the center circle draw a line segment perpendicular to the Möbius band. The union of these segments is another Möbius band that intersects the first. The intersection is along their center circles. Now bend the boundaries together. The result is the twisted figure 8 model of the Klein bottle. Figure 2.22 contains details of the construction. I have also seen this figure as a *Mathematica* computer graphic example. It shouldn't be too hard to generate as a parametric 3-dimensional plot.

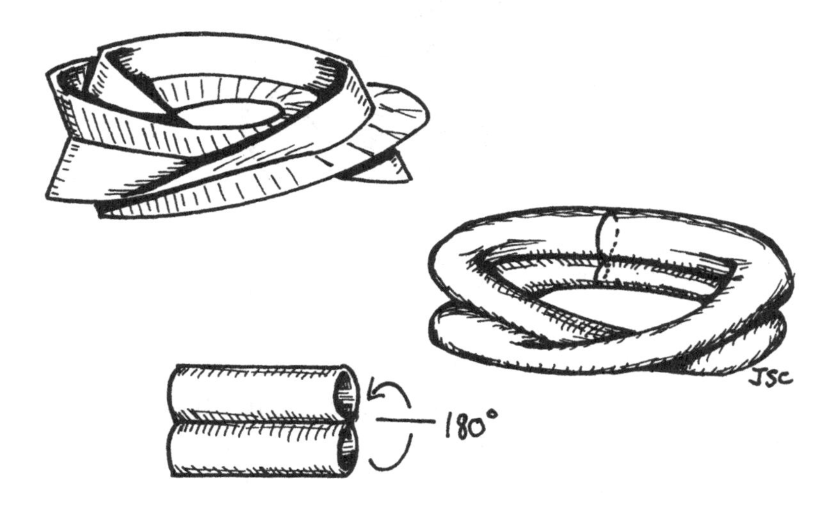

Figure 2.22: Banchoff's Klein bottle

The Pinched Torus

In Figure 2.23, yet another view of the Klein bottle is depicted. This figure starts from two Möbius bands intersecting along their substantial arcs. They are glued together so that the arc of intersection has two branch points. The figure also indicates that a neighborhood of the arc of intersection is the image of a map of an annulus. Recall that the Klein bottle can be obtained from the annulus by gluing its boundary circles together in an orientation reversing fashion as in the slinky example.

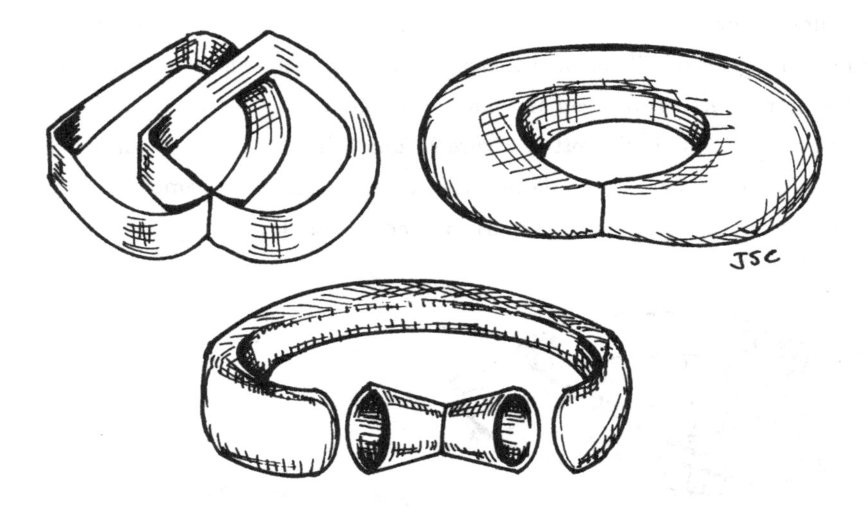

Figure 2.23: The pinched torus map of the Klein bottle

In Chapter 3, I will lay the ground work for imagining surfaces in 4-dimensional space. In Chapter 6, we will see that all of the models of the Klein bottle that have been constructed here are different projections of the same embedding of the Klein bottle in 4-space. On the other hand, we will show that Boy's surface does not embed into 4-dimensions, even though the projective plane does have an embedding. The models of the Klein bottle and of the projective plane that I have depicted have rather simple self intersections. Therefore, they are fairly easy to visualize in 4-space.

2.4 The Closed Non-orientable Surfaces

The closed non-orientable surfaces can be constructed in two apparently different ways. First, we can take many copies of the projective plane, remove disks and glue the results together. The projective plane minus a disk is a Möbius band. So this amounts to cutting disks from Möbius bands and gluing the results together along their boundaries. Second, we can take a projective plane or a Klein bottle and add coffee cup handles to the result as in Figure 2.24. To see that these two constructions can be used to give the same surfaces, we will remove disks, compute rank, and slide handles as in the classification of orientable surfaces.

The **rank of a closed non-orientable surface** is the rank of the bounded surface that results from taking a disk out of the surface. Recall that the rank of the bounded surface is the number of substantial arcs in the surface, and that this definition makes sense even when the surface is non-orientable.

Theorem 9 Classification of Non-orientable Surfaces *Two closed non-orientable surfaces are topologically equivalent if and only if they have the same rank.*

The proof of Theorem 9 is nearly complete. We know that two non-orientable surfaces with one boundary circle are homeomorphic if and only if they have the same rank because the handles can be slid until each handle has a single half twist in it. The feet of the handle are the mates that form substantial arcs, and the arrows on the mates point in the same direction. The homeomorphism between the closed surfaces is constructed as in the orientable case: We have a homeomorphism between the surfaces with boundary, and we can extend that over the entire closed surface because the missing piece of either surface is a small disk.

We compile all of the characterization theorems of surfaces into one.

Theorem 10 Classification of surfaces *Two surfaces are topologically equivalent if and only if they have the same rank, the same number of boundary components, and are either both orientable or both non-orientable.*

The statements of Theorem 10 and Theorem 8 are the same. Theorem 10 includes the possibility of the boundary being empty or equivalently the surface being closed.

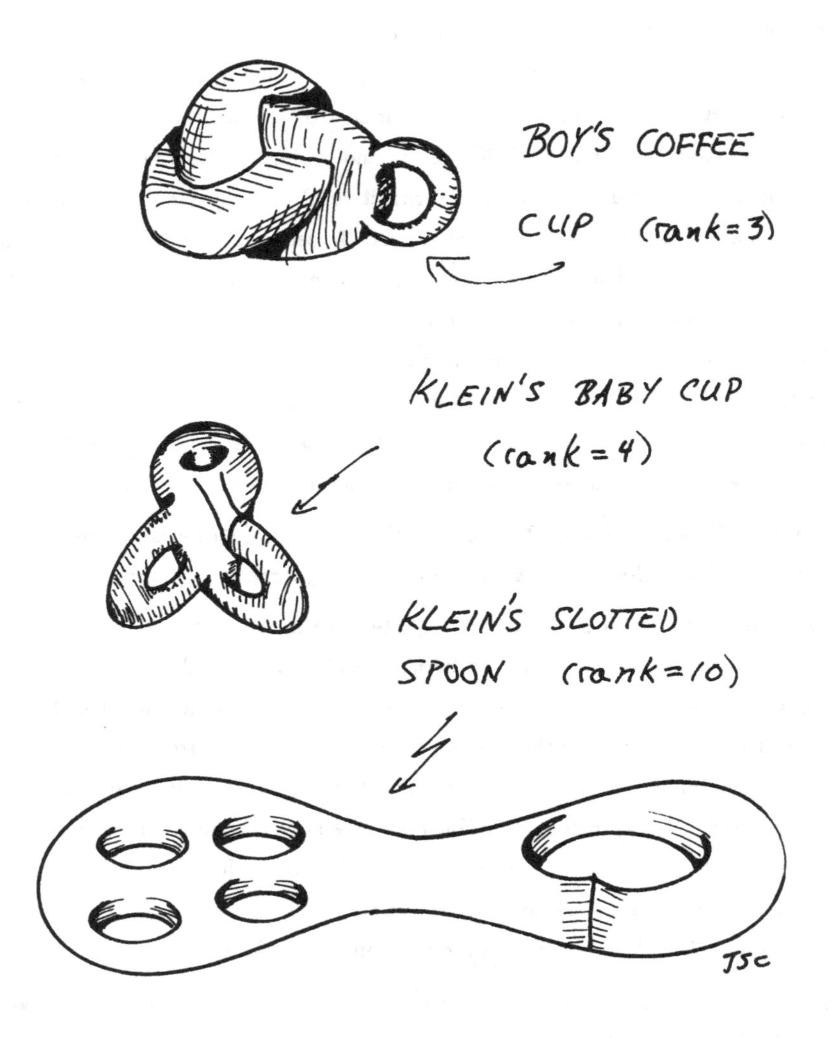

Figure 2.24: Boy's coffee cup, Klein's baby cup, and Klein's slotted spoon

2.5 Notes

The current chapter has been dedicated to understanding three important examples: The Möbius band, the projective plane, and the Klein bottle. These three non-orientable surfaces are related, and they will open the door for the higher dimensional explorations that will begin in Chapter 3 and continue in Chapter 6. Chapters 4 and 5 will provide a bridge between those two chapters and the current one by showing how to construct surfaces in new 3-dimensional spaces, and by examining the relationships among the triple points, branch points, and rank of the surface. We will see immediately and unequivocally that the spaces discussed in Chapter 4 are different because they will contain substantial surfaces — surfaces that do not separate.

⋆

Here are some references to non-orientability in fiction: "Through the Looking Glass," by Lewis Carroll explores non-orientability by putting Alice in a backward world. In [77] the lead character has to undergo a couple of inversions before he is set back with the correct orientation. In [76] a villain is turned inside out when he tries to escape via a dimension inverter.

⋆

For computer users I am including a program that will run using the software package *Mathematica* . This program depicts Apery's view of Boy's surface. The parameters phi and theta are spherical coordinates; they are the mathematical version of longitude and latitude. It is good to vary the range of the parameters to view different aspects of the surface. The symbol, Pi, is the area of the circle that has radius 1.

```
(* Apery's parametrization of Boy's surface.
The variables x,y,z, are given in spherical coordinates.
    f[1], f[2], and f[3] are essentially Apery's
      quartic functions.
   Use ParametricPlot3D[ {f[1],f[2],f[3]},
     {phi, phi_min,phi_max},
       {theta,theta_min, theta_max},
          BoxRatios->{1,1,1},
           ViewPoint->{X,Y,Z}],
      where you determine values for X,Y,and Z *)



   x =Cos[theta] Sin[phi];


   y = Sin[phi] Sin[theta];


   z= Cos[phi];


   f[1]= 1/2 ( (2x^2 - y^2 - z^2) + 2 y z (y^2 -z^2)
          + z x (x^2 -z^2) + x y (y^2 -x^2) );


   f[2] = Sqrt[3]/2 ( (y^2 -z^2) + z x (z^2 - x^2) + x y (y^2 -x^2) );


   f[3] = ( x + y +z) ( (x + y + z)^3  + 4( y - x) (z-y) (x-z));
   ParametricPlot3D[ {f[3],f[1],f[2]},
          {phi,0,Pi},
            {theta,0, Pi},
        BoxRatios->{1,1,1}, ViewPoint->{-2.2,1,1.9},
          PlotPoints->{20,30} ]
```

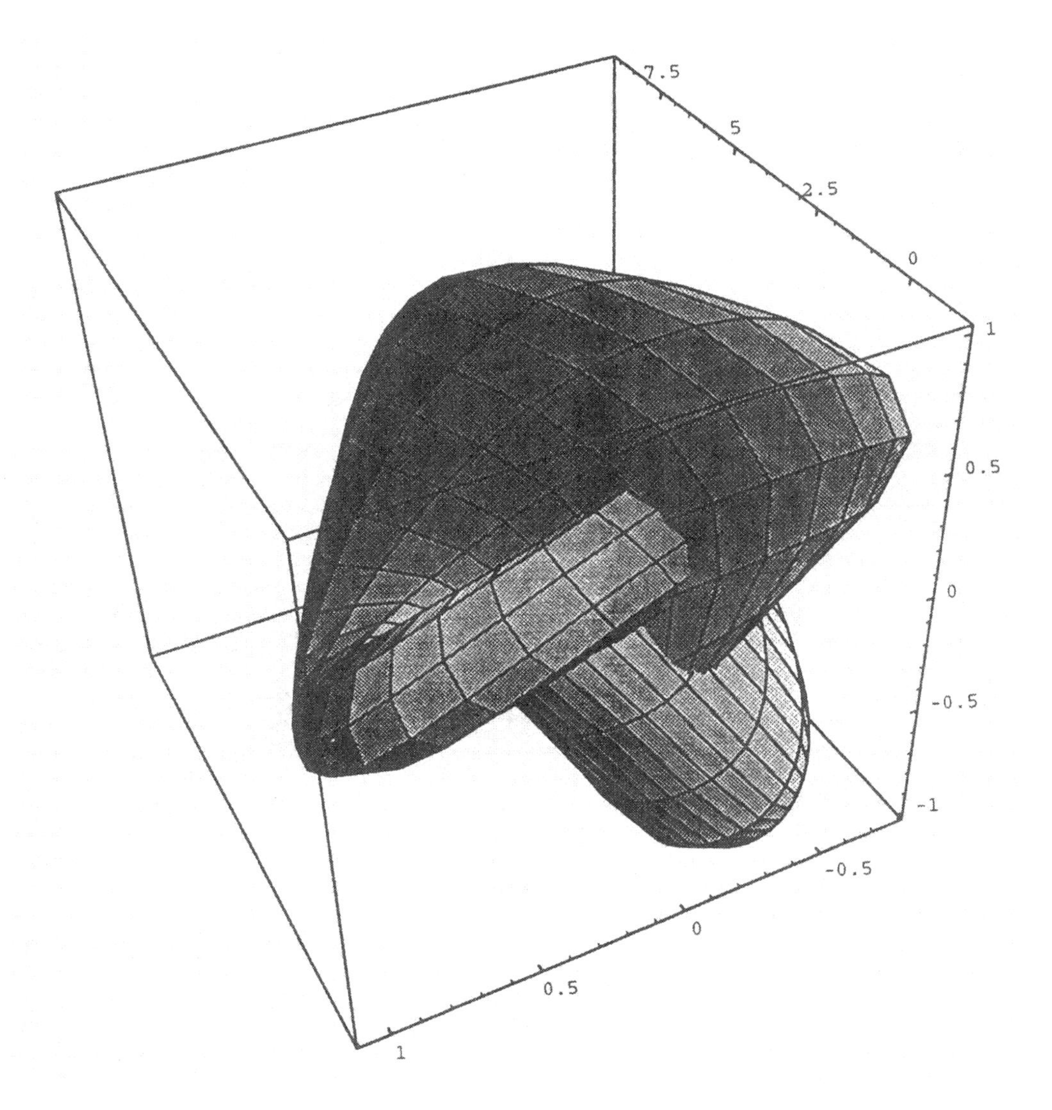

Figure 2.25: *Mathematica* illustration of Boy's surface

Chapter 3

Curves and Knots

In this chapter, we examine disks in space that are bounded by closed but not simple curves. We will see how to go from closed curves in the plane to knots in space, and how to use the disks bounded by the curves to gain information about the knot. We begin with closed curves.

3.1 Curves on the Sphere

The $\alpha\gamma$-curve has two double points. The curve separates the sphere into four regions. Two of these regions are triangles, and two are monogons or tear-drops. When the sphere is projected onto the plane of the page, the region on the outside may be either a triangle or a tear-drop. Figure 3.1 indicates the choices. Only two possible outside regions appear because the curve is symmetric.

(I won't make a distinction in this chapter between curves on a sphere and curves in the plane. This means that I will write the word "sphere" while drawing the pictures on the plane. I hope that Figure 3.1 indicates why the distinction need not be made. If not, go back and read about Santa's view of the world.)

A fundamental question about curves on the sphere is: How many topologically different closed curves with n-double points are there on the sphere? Each **closed curve** is the continuous image of a circle. And the curves are **topologically the same** if and only if there is a homeomorphism of the sphere that takes one curve

Figure 3.1: Views of the $\alpha\gamma$ curve

to the other. Assume that the double points are in **general position**; that is, the curves have no points of tangency, and at most two arcs intersect at a single point.

In this case, we think of the sphere as being made up of polygons in which each vertex has 90° angles, but these polygons are not considered to be rigid objects. Two curves are topologically equivalent if the polygonal pieces correspond in a natural way. So if one curve broke the sphere into triangles and if the other had no triangular pieces, then there wouldn't be a homeomorphism.

I don't have **any idea** how to count the number of distinct curves on the sphere. Every method attacking the question of which I know, or of which I can conceive, is enumerative: There are algorithms that will generate each closed curve in the sphere, but there is not a closed formula for the number of these.

Here are some data: There is one simple closed curve; there is one closed curve that has a single double point; there are two curves with two double points; there are six curves with three double points; there are 19 curves with four double points; and Grant Cairns and Daniel Elton [10] have found there are 76 curves in the sphere with five double points. Dan Frohardt [37] found a way to count the curves with n double points that can appear in arbitrary orientable surfaces. Frohardt's technique was based on the Gauss word that is defined below. Figure 3.2 illustrates the types of curves for up to four crossings; arrows indicate the starting points and the directions.

The general method that mathematicians use to count a set of objects is to find a correspondence between the set and another easily counted set. There is a set of graphs that corresponds to closed curves, but a method of counting these graphs is not obvious — at least to me it's not.

There is a simple algebraic way of describing a closed curve called the Gauss word that was described by Gauss [39] in his unpublished notebooks. Mathematicians use the term "word" to mean any finite sequence of letters or numbers. This practice can freakout (disturb) people who are not hip to the lingo (aware of the terminology). Now that you can dig it, swing to the groove.

The **Gauss word** is defined as follows. Each double point on a closed curve is labeled with a letter. A direction and a starting point (different from a double

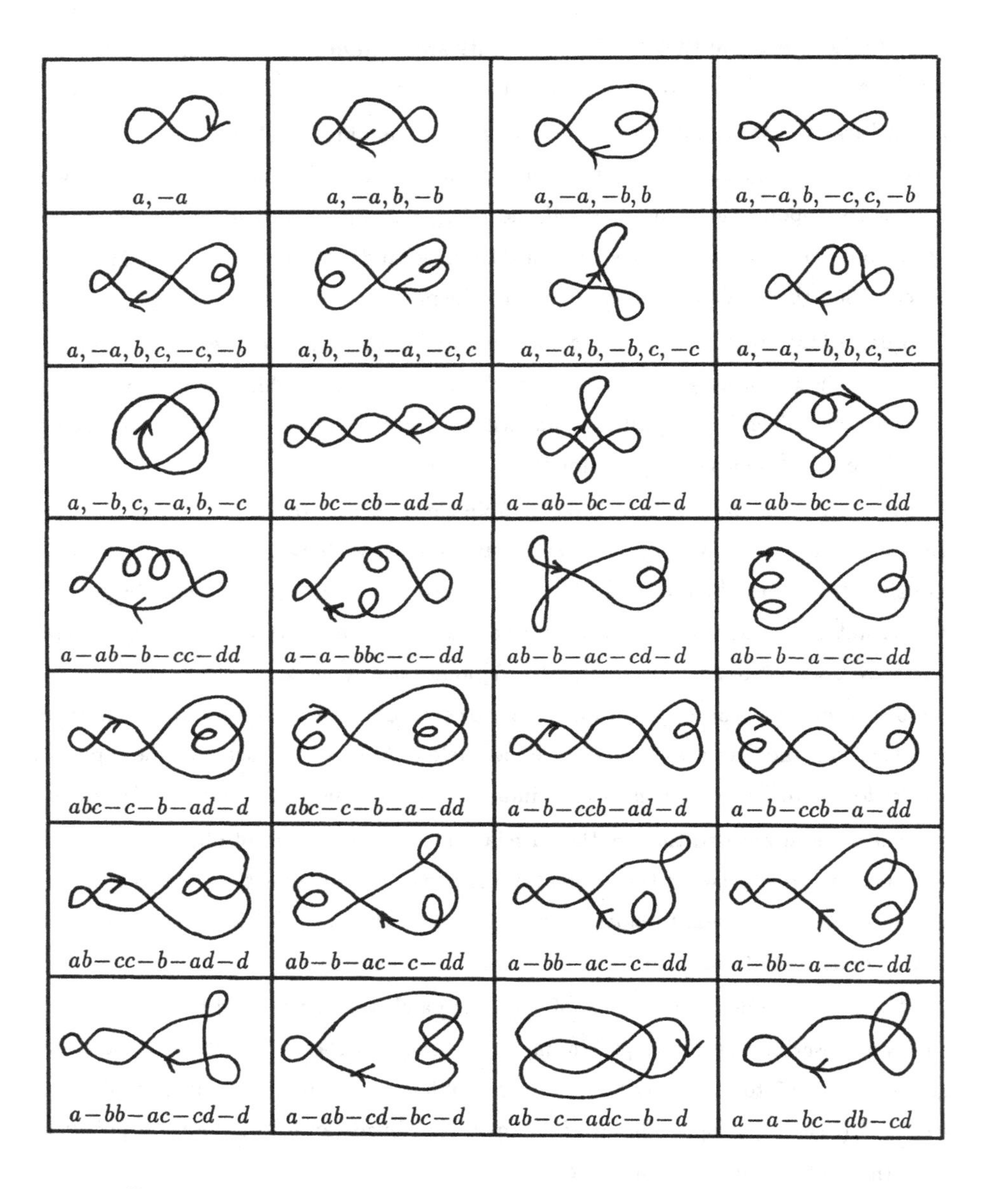

Figure 3.2: The closed curves in the sphere with up to four crossings

point) are chosen on the curve, so the letters will appear in alphabetical order as they are first encountered. We won't consider closed curves with more than twenty six crossings, so the Roman alphabet will be sufficient. (In the general case, numbers can be used.) We imagine an intelligent bug that starts at the starting point and travels along the curve in the direction of the arrow on the curve. Each time it gets to a crossing it records the associated label together with a $(\pm)$-sign. The sign is $(+)$ if the arc that runs across, does so, from left to right. If the crossing arc crosses from right to left, the sign is $(-)$. When the bug returns to the starting point, it has a record of all crossings in sequence. This record is the Gauss word.

Therefore, to every closed curve there is a sequence of letters associated such that each letter appears twice — once positively and once negatively. There are choices made in the process: choice of starting point, choice of direction, choice of labeling, and the choice of meaning in the $(\pm)$-signs. Different choices would result in different Gauss words, but that issue is irrelevant to the current discussion. The Gauss words are indicated on the curves with up to four crossings in Figure 3.2. The arrows on these curves indicate the direction and the starting point.

Gauss words are defined for curves on any orientable surfaces. We use orientability to distinguish left and right crossings, but we never need to assume that the curves are on the sphere.

Given a sequence of letters in which each letter appears once positively and once negatively, is there a curve for which this sequence is the Gauss word? Yes, provided that you allow curves on arbitrary orientable surfaces. The word $(a, b, -a, -b)$ is the word for a curve on a torus, and it cannot fit on the sphere. The word $(a, b, c, -b, -a, -c)$ fits on a surface of genus two, but not on a smaller genus surface.

⋆

(Once when we were talking, Dan Silver called the intelligent bugs **Massey bugs** because William Massey used them in arguments in his text [57]. I use the term Massey bug out of great affection and respect for Professor Massey who recently retired from the faculty at Yale University.)

There is a relation between the Gauss word and the geometry of the pieces into which a closed curve cuts a surface. In fact, there is an algorithm to compute the smallest genus surface in which a curve with a given a Gauss word can be mapped. In the algorithm, you construct from the Gauss word a surface in which the complementary regions are each disks. To use the algorithm, you determine at each vertex which of the subsequent vertices would follow if a Massey bug on the curve were to turn left. In Figure 3.3, the process is used to put the curve with word $(a, b, -a, -b)$ on the surface of a torus.

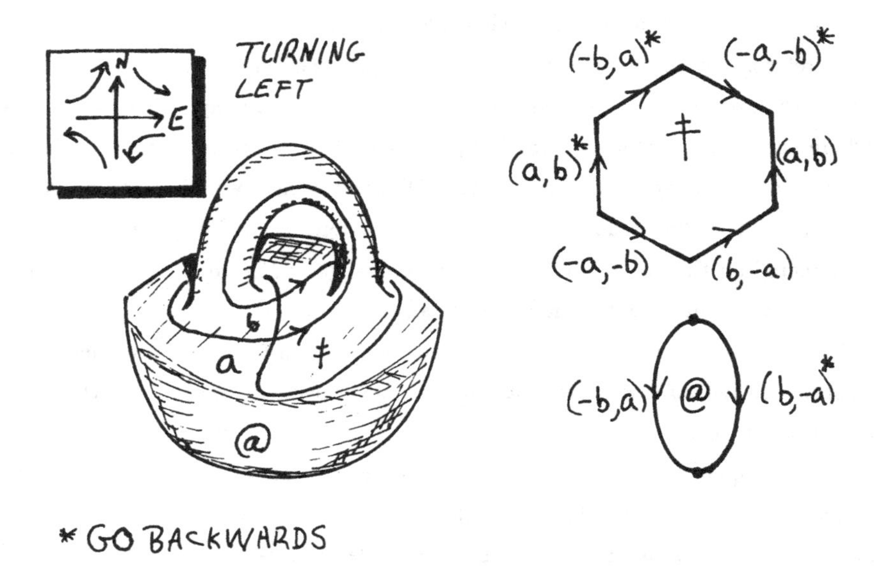

Figure 3.3: Constructing the torus in which the curve $(a, b, -a, -b)$ resides

The Gauss word resembles the sequence of mated arcs that we used in the classification of surfaces. Specifically, a word in which each letter is mentioned once positively and once negatively, represents a close curve and represents a surface in which the word depicts the sequence of mated arcs that appear in that order on the boundary of a disk. There is a relation between the curve and the surface that we explore in Section 3.3 and in Section 3.4.

3.2 Disks Bounded by Curves

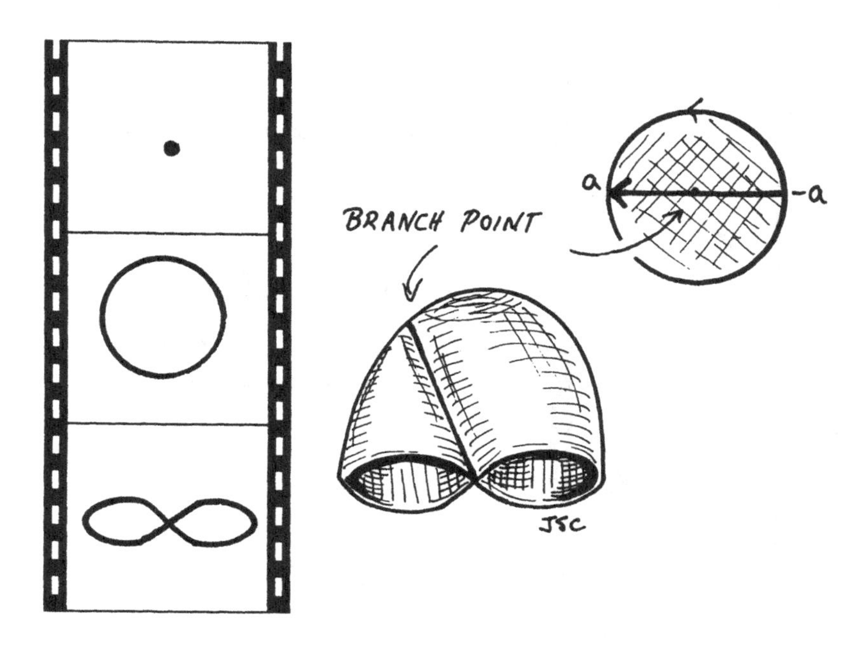

Figure 3.4: The cone on a Figure 8

Figures 3.4, 3.5, and 3.6 indicate disks bounded by the curves with Gauss words $(a, -a)$, $(a, -a, b, -b)$, and the $\alpha\gamma$-curve that has the Gauss word $(a, -a, -b, b)$. In the illustrations, two other pieces of information are indicated. First, a movie of the disk indicates slices of it by cross sectional planes. Second, the points on the flat disk that are mapped to the arcs of double points are indicated by non-simple arcs. Let's explore these alternate points of view.

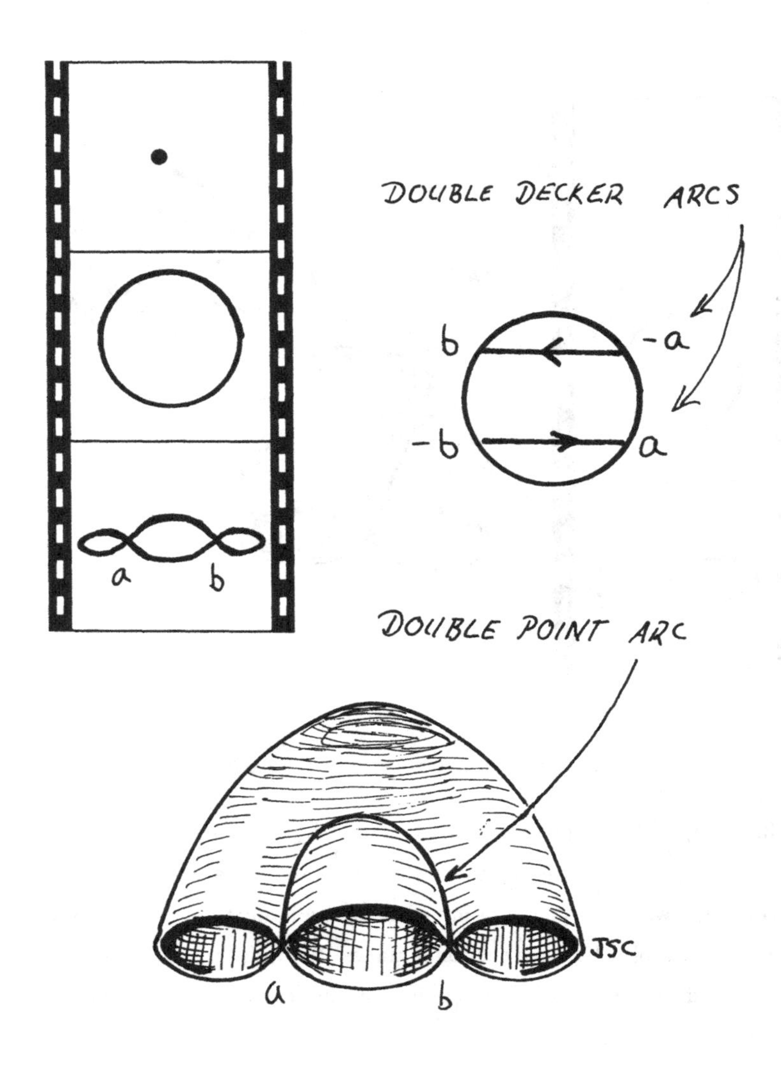

Figure 3.5: The disk bounded by $\alpha\alpha$

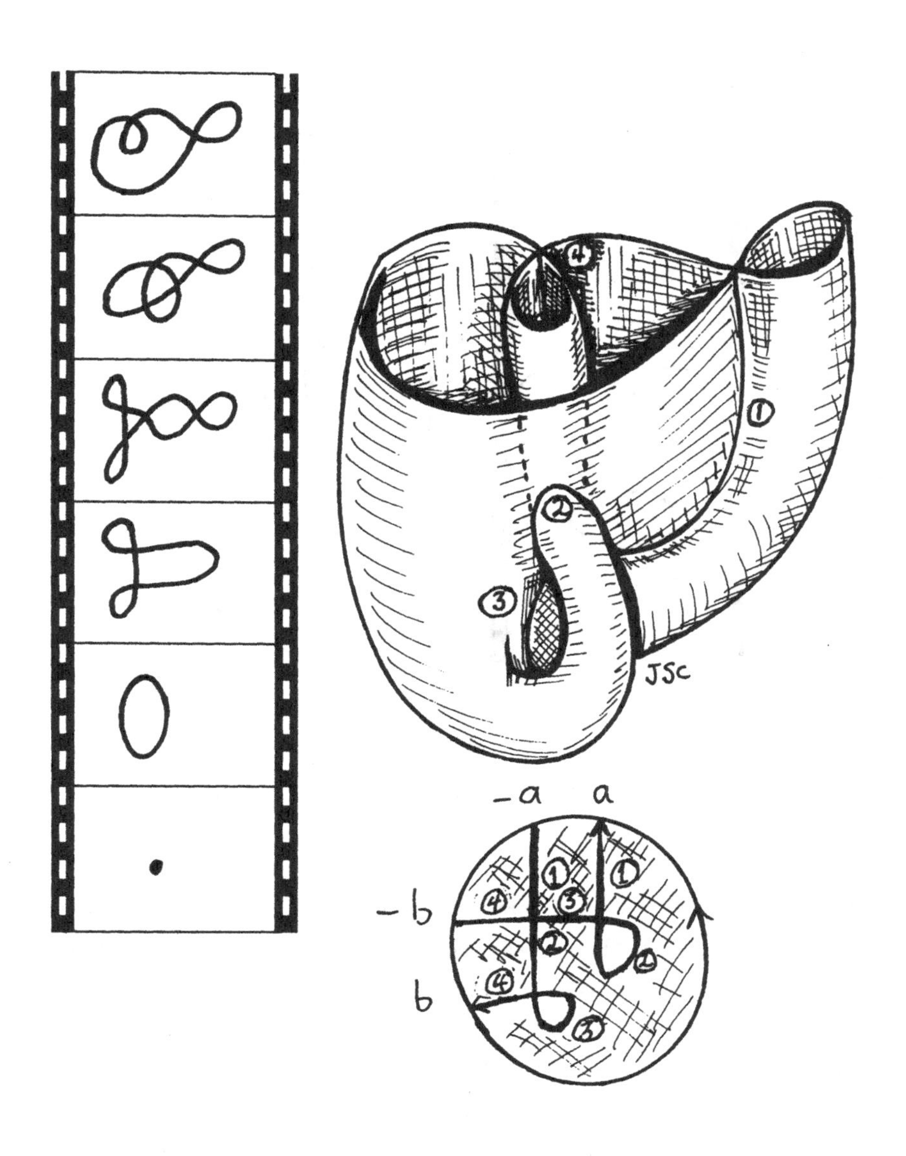

Figure 3.6: The disk bounded by $\alpha\gamma$

Movies

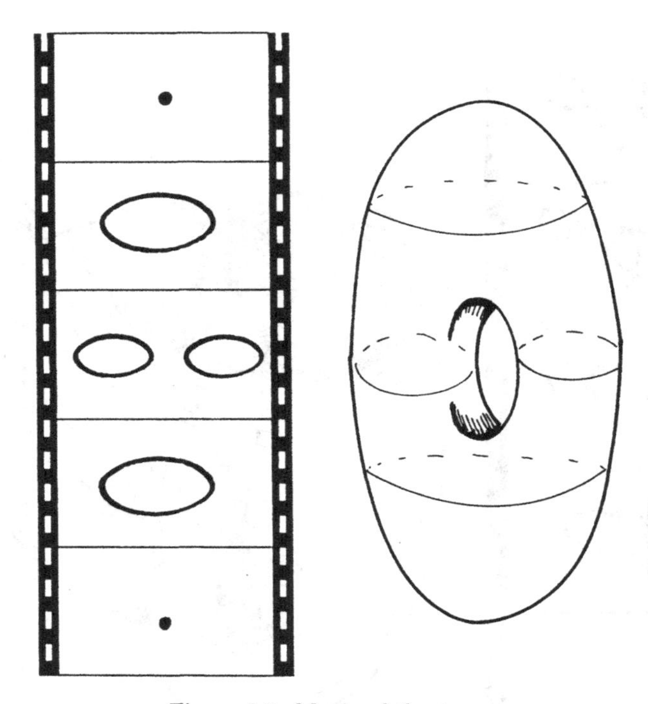

Figure 3.7: Movie of the torus

A surface in space may be cut by a plane. The intersection of a plane and a surface will usually be a closed curve. Occasionally the plane will intersect the surface at a point, occasionally the curve of intersection will have points of tangency, and occasionally three sheets of the surface will meet the plane at a triple point. But these occasions can be assumed to be isolated and rare. The operative assumption is "general position." If the plane intersects the surface at one of these points, we say the surface has a **critical point**. In the general position world, we can completely understand critical points by moving the cutting plane to levels just above and just below the critical points. Figure 3.8 indicates how to interpolate the surface near a critical point.

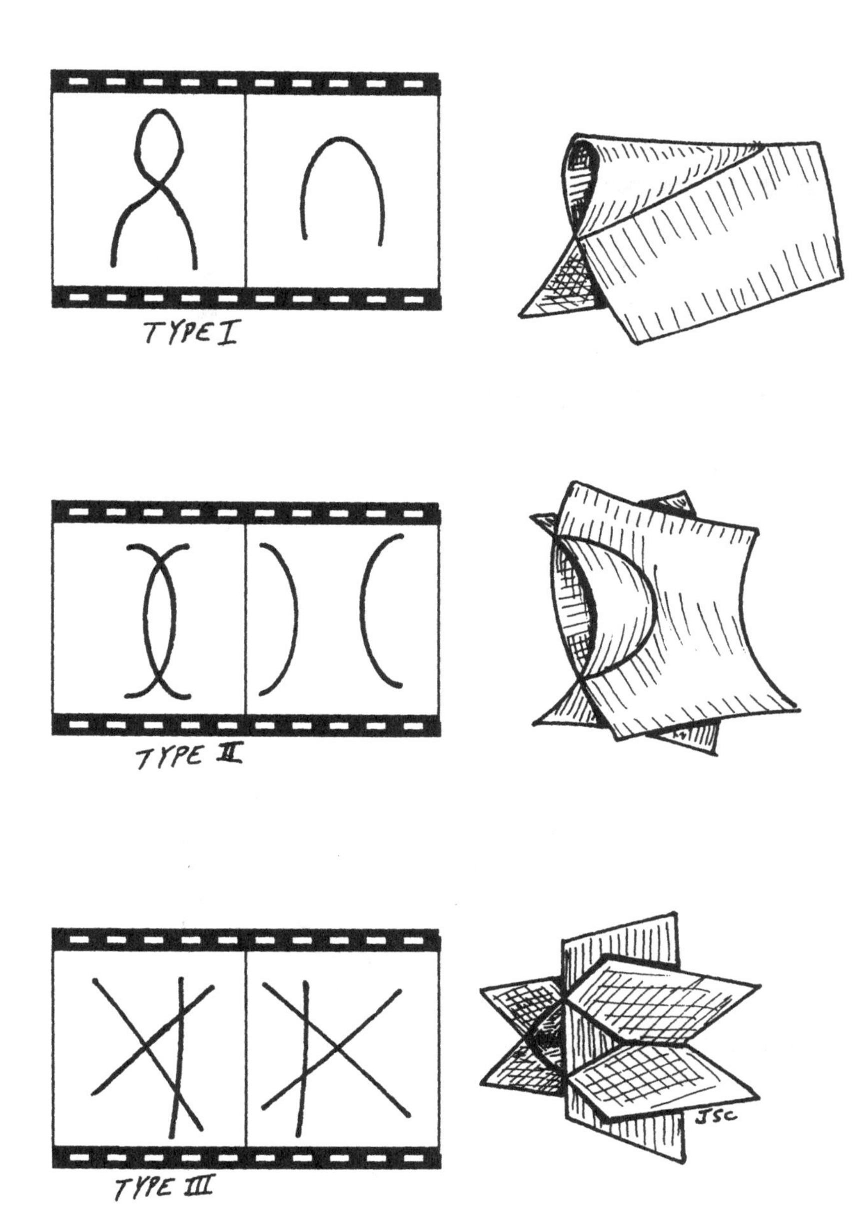
TYPE I
TYPE II
TYPE III
JSC

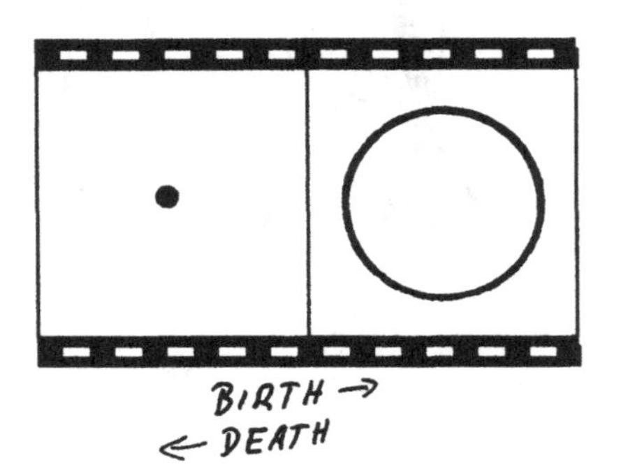

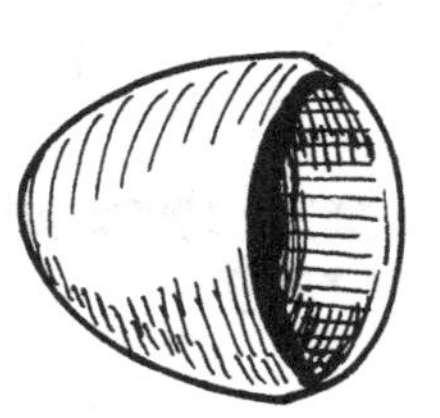

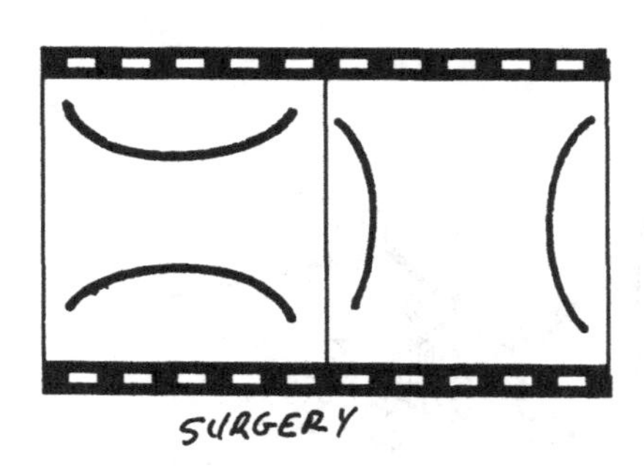

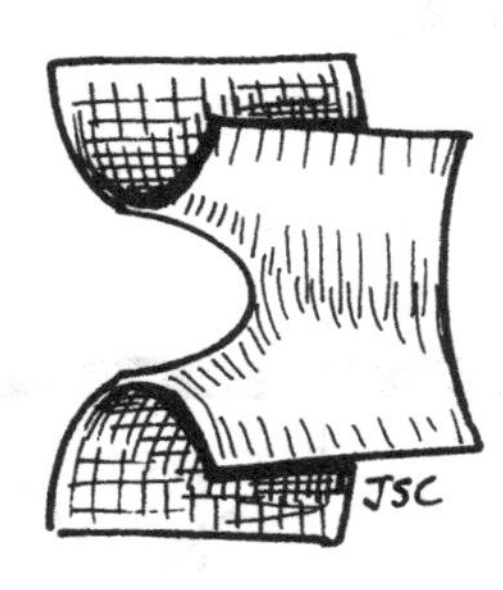

Figure 3.8: Critical points, cross sections, and interpolating

A **movie** of a surface in space is a sequence of closed curves such that successive terms in the sequence differ at most by a critical point. The closed curves are the intersection of the surface with non-critical planes. The **stills from the movie** are the pictures of the closed curves; in a still, more than one circle may appear. Figure 3.7 illustrates the movie of a torus balanced on a table. Figures 3.9 and 3.10 illustrate movies of the Klein Bottle and Boy's surface, respectively. Figure 3.11 indicates a movie of the Roman surface. Movies are useful in understanding surfaces in 4-dimensional space as we'll see in the subsequent section on knots and in Chapter 6.

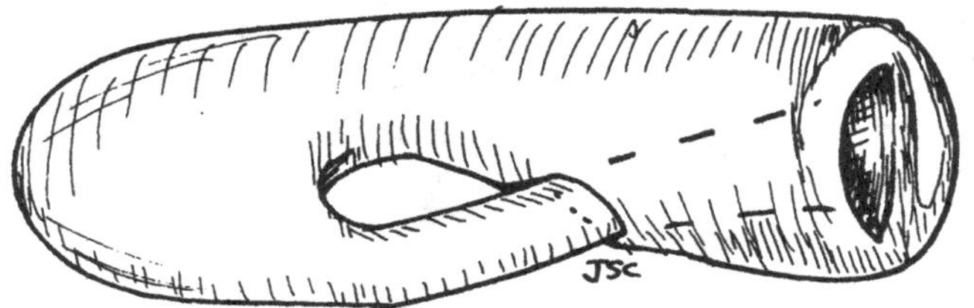

Figure 3.9: Movie of the Klein bottle

The side pictures along the illustration of Boy's surface in Figures 2.8 through 2.12 give another movie, but it is not a general position movie: The critical points of the double curves occur concurrently with the critical points of the surface. Such non-generic movies with be discussed in Chapter 6.

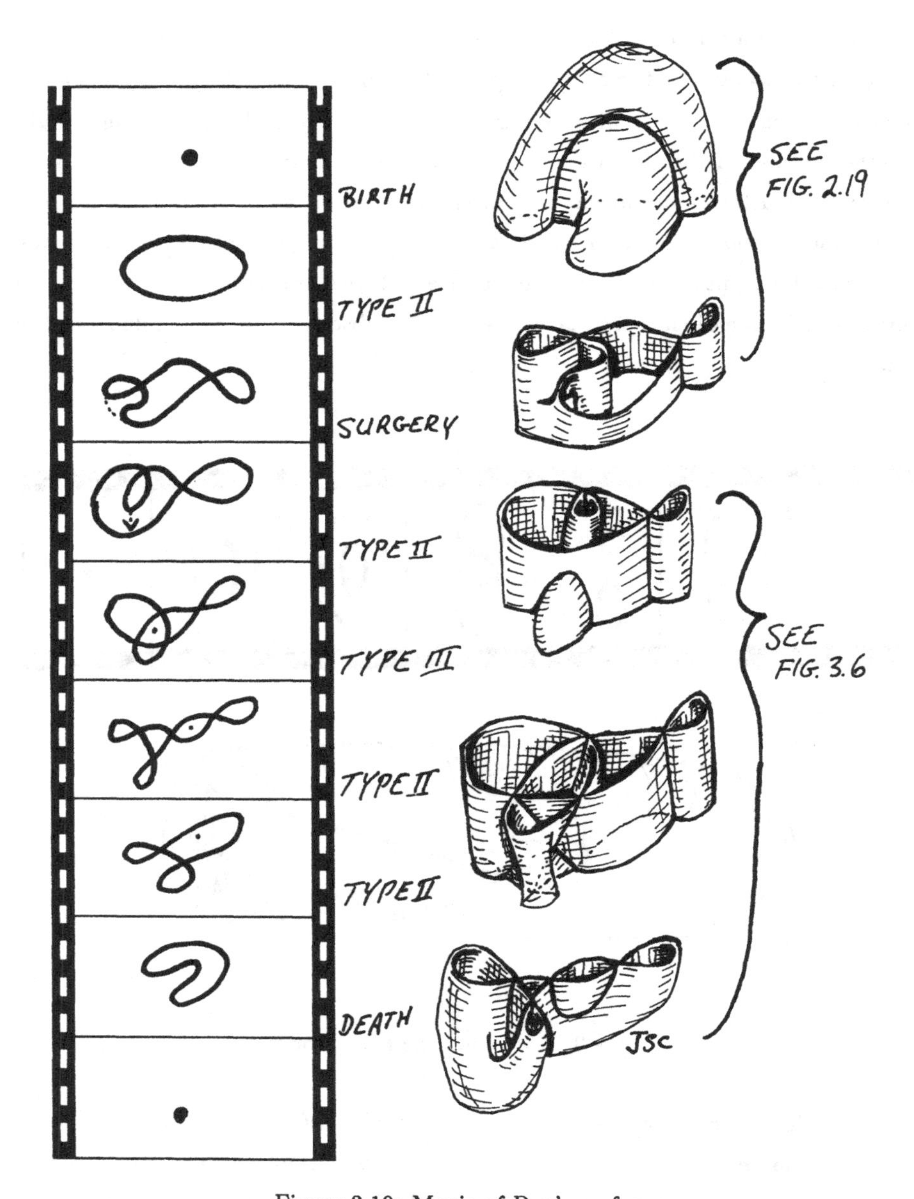

Figure 3.10: Movie of Boy's surface

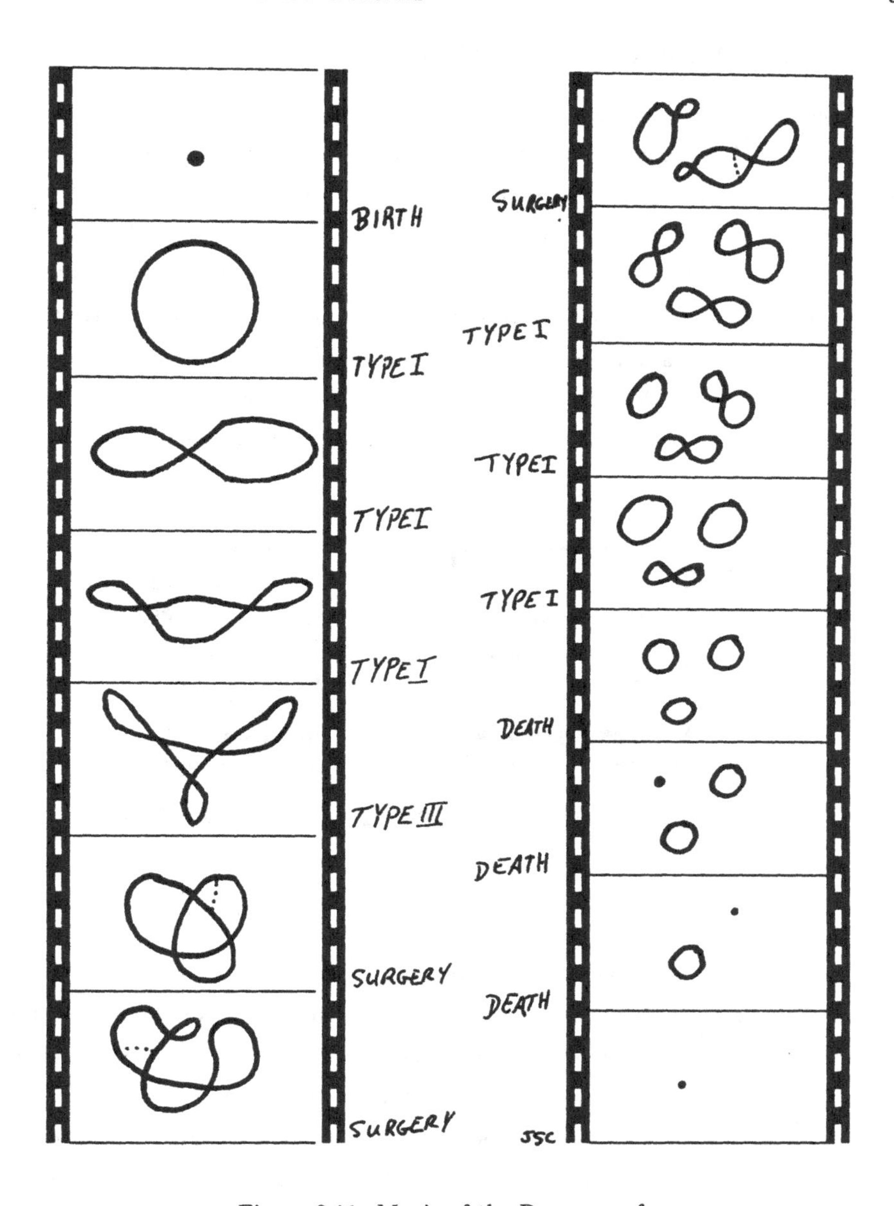

Figure 3.11: Movie of the Roman surface

The Double Points in the Preimage

In the figures of Boy's surface, the Roman surface, and the disks of Figures 3.4, 3.5, and 3.6, arcs and closed curves in the domains indicate where corresponding parts of the domains are to be mapped into space. In space, we have double arcs and double circles. When these arcs are pulled back to the surface, two arcs appear. The arcs in the surface are called **double decker arcs**; the arcs in space are called the **double point arcs.** The two double decker arcs are called **companions of each other.**

The double decker arcs are depicted with directions induced by the Gauss words. An arrow on a double decker arc points away from a negative crossing point and points toward a positive crossing point. Why is this convention natural?

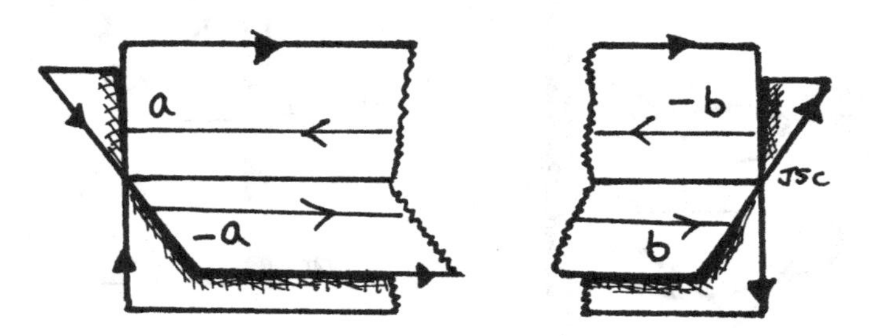

PUSH DECKER ARC OFF INTERSECTION TO SEE DIRECTION

Figure 3.12: Orientations of interior arcs

A neighborhood of a double arc in an orientable surface is also an orientable surface. The direction along the boundary of this neighborhood is induced from the arrow on the boundary of the surface. If a Massey bug were to travel along a double decker arc and if the boundary were oriented up at the beginning of the journey, then at the end the boundary would point down (See Figure 3.12). This up/down phenomenon occurs both along the deck on which the bug travels and along the companion deck. The sign on a double decker point on the boundary is controlled by the directions on both the given deck and the companion deck. After a little bit of thought or examining all cases, you'll see that negative points on the boundary are always connected to positive points on the boundary by the double decker arcs.

In Figure 3.13, a triple point is depicted as the intersection of oriented disks. Each companion in a pair of decker arcs intersects some other companion. We can compute the sign of this intersection using the same sign convention that we used in defining Gauss words. In the figure, the intersection number of each double decker arc is computed. It is interesting to see that the intersection numbers on companion pairs of arcs cancel.

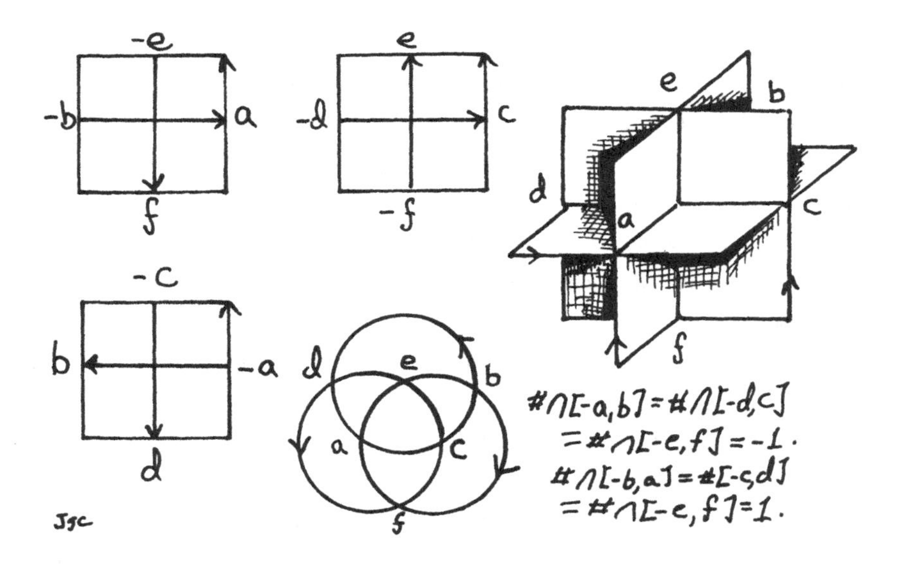

Figure 3.13: Companion arcs have canceling intersection numbers at a triple point

In a more complicated orientable surface bounded by a more complicated collection of closed curves the same cancellation will occur: The crossing points are the preimages of the triple points, and the arcs will be oriented at the triple points as in the case of the prototype. So each triple point introduces canceling pairs of crossing points on the double decker set. That fact warrants amplification.

Fact. *If an orientable surface is mapped into space with its boundary a closed curve on another surface, then the signs of the crossing points on each companion pair of double decker arcs cancel.*

It is an amazingly simple fact. It doesn't depend on the surface that is being mapped, it doesn't depend on the surface to which the boundary is mapped, and it doesn't depend at all on the space into which the surface is mapped. The fact allows us to come to an interesting conclusion about a certain closed curve in a surface of genus two.

A Curious Curve on a Surface of Genus Two

In Figure 3.14 a closed curve on a surface of genus two is depicted. The Gauss word for the curve is $(a, b, c, -b, -a, -c)$. This curve is **never** the boundary of a disk mapped into a space that has the genus two surface as its boundary [21].

Join points on the boundary of the disk by oriented arcs, and examine the intersections. The point labeled $-a$ is joined either to the point labeled a, b, or c. If $-a$ is joined to a, then either $-b$ is joined to b or to c. Thus there are four possible ways to connect these six points as if they were connected by double decker arcs. In none of these ways is the intersection number 0 for each pair of companion arcs. The Jordan Curve Theorem implies that there is no way to undo the bad intersections. So the given curve does not extend to a map of the disk.

Similar curves that don't extend to maps of disks can be constructed with an arbitrarily large number of crossings. This curve is the simplest one with that property. The proof that these curves don't extend is geometric, but purely algebraic consequences follow. An algebraic proof of the corresponding algebraic result using currently available techniques would be unduly complicated. The rule of thumb in the math business is that geometric facts are more easily solved when translated into algebraic terms. This example indicates that rules of thumb don't always hold true.

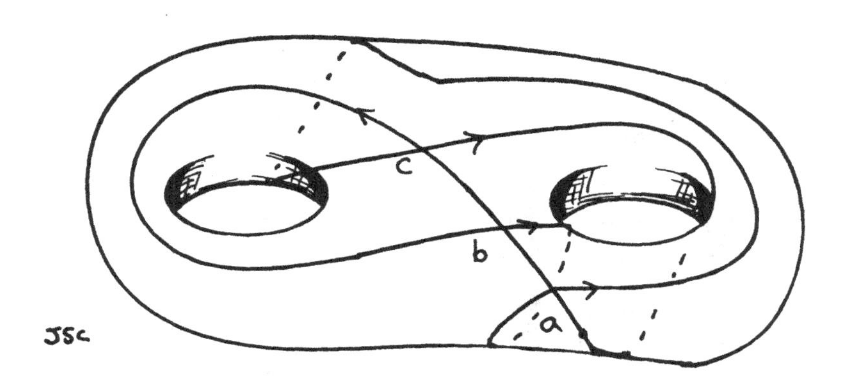

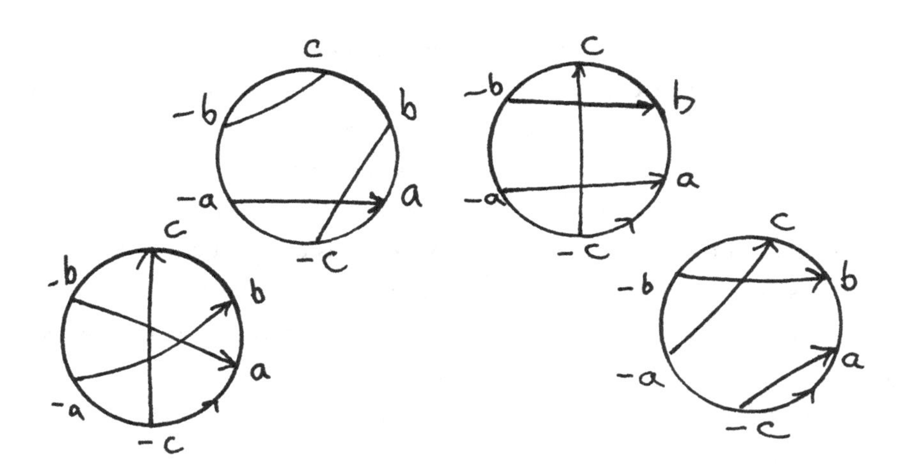

Figure 3.14: A closed curve that never extends to a map of a disk

3.3 Gauss Words and Surfaces

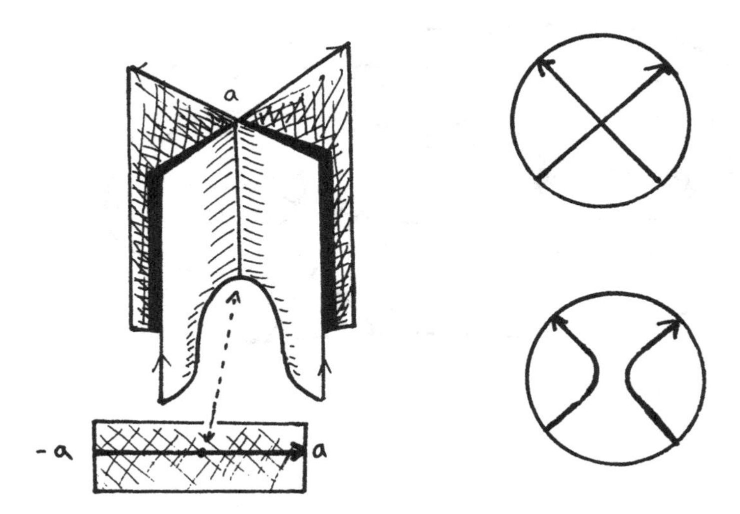

Figure 3.15: Smoothing crossing points

A sequence of letters in which each letter appears exactly once positively and once negatively represents a closed curve in some surface via the Gauss word. It also represents a surface via a sequence of mated arcs on the boundary of a disk. This latter surface can be associated in a natural way to the curve.

For example, consider the word $(a, -b, c, -a, b, c)$. The curve represented has three double points in the plane. The surface represented is an orientable surface of rank 3. If disks are sewn to the boundary, then a torus results. Next we will show that a once punctured torus can be mapped into a ball with the curve represented by $(a, -b, c, -a, b, c)$ as its boundary on the sphere.

Mapping this torus is relatively easy. We smooth each crossing point of the resulting curve in a way that orientations are maintained. The smoothing for one crossing is depicted in Figure 3.15. In this way a crossing point on the boundary is paired by a band to its companion, the band joining the double decker points is the connecting band for a pair of mates that is mapped into space with a branch point. After each crossing is smoothed, a collection of circles results. Each of these is embedded, and so each can bound a disk in the sphere.

If the circles are nested one inside another, then make a circle that is on the inside of all of the rest bound a disk first. Figure 3.16 illustrates.

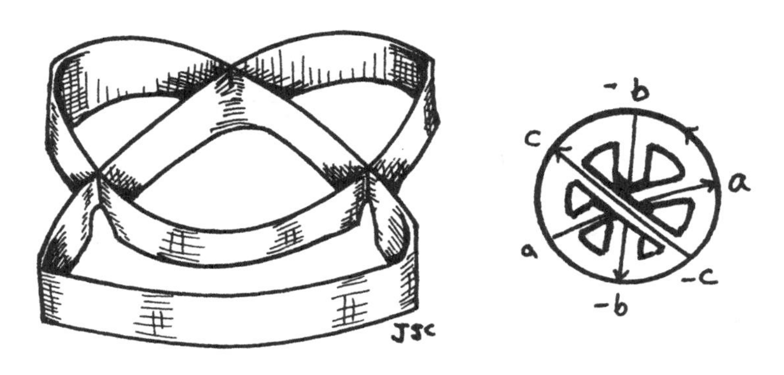

Figure 3.16: A punctured torus and the word $(a, -b, c, -a, b, -c)$

In general if a word is given, a closed curve represented by that word can be constructed in the same manner as above. Furthermore, a surface represented by the word is mapped into space and is bounded by the curve. The presentation that I have given takes place on the sphere, but a curve on any surface can similarly be bounded in some solid object. The surface constructed in this manner is called the **Gauss word surface.**

In this way, we have established a direct relationship between curves and surfaces that are represented by words.

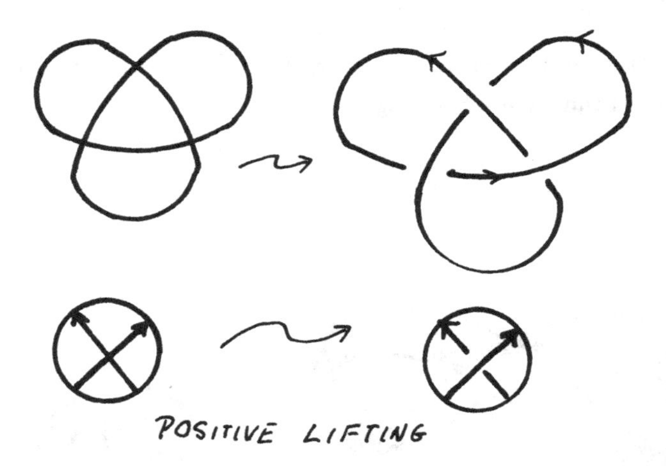

Figure 3.17: Constructing the knot diagram for the trefoil knot

3.4 Knots

In this section, we move from closed curves in the plane to simple closed curves in space by means of an easy artistic convention. The same convention is applied to surfaces bounded by the curve to schematize surfaces in 4-dimensions bounded by the knot. The surfaces need not be simple though, and the self-intersections of disks bounded by knots provide a rudimentary way of distinguishing knots.

In Figure 3.17 the knot diagram for the **trefoil** is shown. This knot is the simplest of all closed loops that are knotted. To go from a closed curve in the plane to a knot diagram, we delete a small length of one of the two intersecting arcs at each double point. The arc that is deleted is thought to travel under the other arc. The diagram is a realistic depiction of a wire in space because the over crossing wire eclipses the under crossing one.

Knot theory is the mathematical discipline in which the knotting of simple loops is measured. Two closed loops are thought to be the same if one can be deformed in space into the other without its arcs crossing. We try to determine if simple loops are the same by finding an algebraic or numeric quantity that can be assigned to each loop and that does not depend on the method that has been chosen to describe the loop. A simple loop is **not knotted** if it can be deformed into a simple closed curve in a plane.

A closed curve in the plane that has, say n, crossings can be lifted to 2^n potentially different loops in space — each crossing can be cut in 2 ways. Two of these embedded loops are known to be unknotted. So the closed curve that is the **shadow** of a knot cannot be used to detect knotting.

To form an unknotted curve with a given shadow, consider a Massey bug walking on the closed curve. When the bug first reaches a crossing, it lifts the arc along which it is traveling. The knot diagram that results is the diagram of an unknotted curve, for if you put a piece of string on the table with the same diagram, you could pick up the string without getting it entangled. To get the other unknot just push the diagram down at each crossing.

Figure 3.18: Making an unknot with given shadow

The Reidemeister Moves

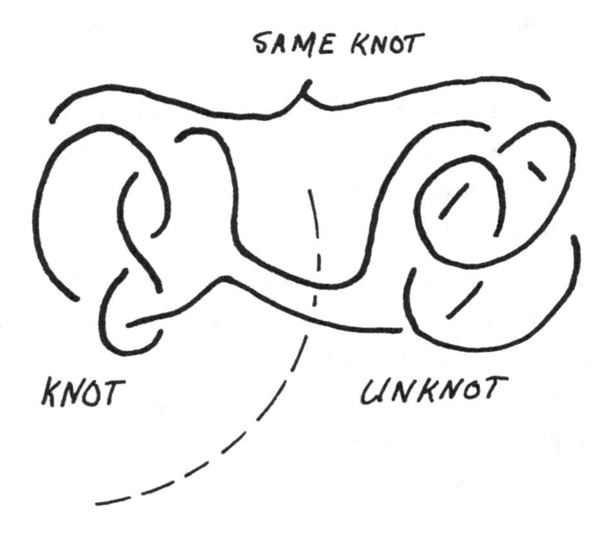

Figure 3.19: Fusing an unknot and a knot

The diagram of a knot can be used to detect knotting, but care must be taken because many different diagrams determine the same knot. For example, any closed curve in the plane could be used to give two diagrams of an unknotted loop. So an unknot has many different diagrams. Any other knot could be made to have a plethora of diagrams by fusing the diagram for the unknot and any given diagram of the knot as in Figure 3.19. There are, however, three moves such that two diagrams of the same knot can be transformed to one another by a sequence of the moves. These are called **the Reidemeister moves** and are depicted in Figures 3.20, 3.21, and 3.22.

The Reidemeister moves look like they take place on elastic strings, but they can also occur on rigid bodies. For example, if you and I sit on opposite sides of a table with a knotted wire between us, then your view and my view of the wire can be completely different. As we rotate the wire to see the other's point of view, we can do so slowly and watch the Reidemeister moves occur one by one. (Would that there

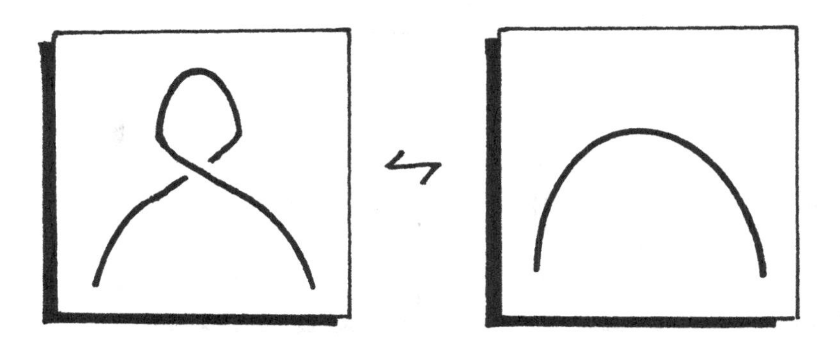

Figure 3.20: Type I Reidemeister move

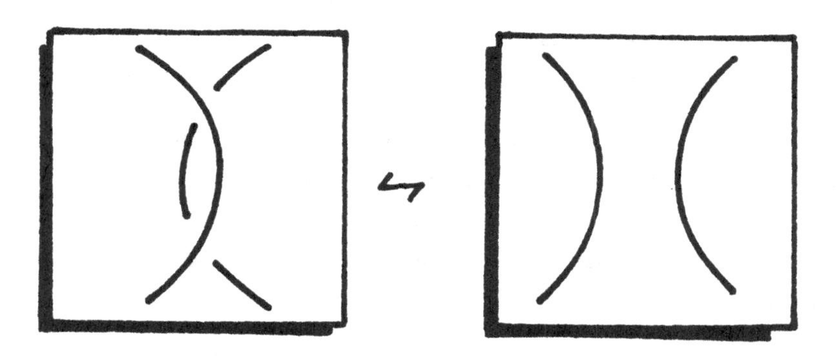

Figure 3.21: Type II Reidemeister move

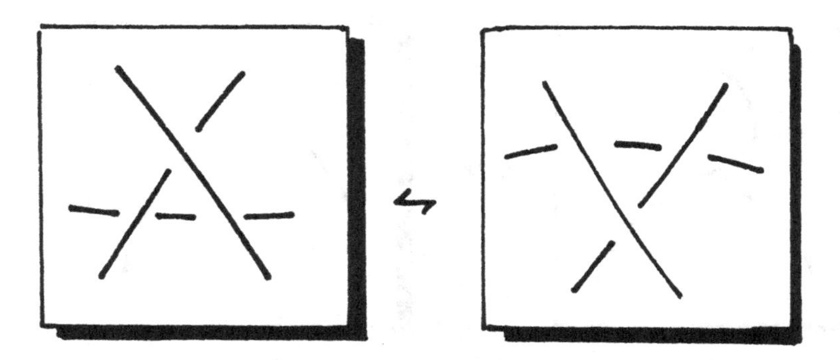

Figure 3.22: Type III Reidemeister move

were only three moves that were necessary for two people to see each other's point of view on all matters.)

The Reidemeister moves should be familiar. They are the same moves that we used to construct surfaces in space, but now we have included over/under crossing information. I will say more on this in the discussion about 4-dimensions.

If you want to define an invariant of a knot by means of a knot diagram, then you must show that the definition does not depend on the diagram that is chosen to depict the knot. To do this, you show that the definition is unaffected by each Reidemeister move. But once you do, you can use diagrams to study and compute your invariant.

I will outline the definition of one important invariant, called the Arf invariant, by means of knot diagrams. I will use the Arf invariant to prove that certain knots — the trefoil and the figure 8 knot (which has four crossings and shouldn't be confused with a figure 8 in the plane) — are indeed knotted. The invariant assigns either a 0 or a 1 to each knot so it doesn't tell you very much about a knot. But I have chosen to explain it because it is defined in terms of surfaces in 4-dimensions that are bounded by the knot diagram.

A First Look into 4-Dimensions

We went from a closed curve to a knot diagram by a simple schematic devise: We removed a small under crossing arc at the double points of the closed curve. In this section, a similar scheme will be used to depict a surface embedded in 4-space. The diagram of a knot is a 2-dimensional figure that depicts an object in 3-dimensional space. The diagram of a surface bounded by a knot is a 3-dimensional figure that depicts an object in 4-dimensional space. The fourth dimension in the current discussion will be thought of as a time parameter since Mr. Einstein's theories of space and time are relatively well known.

Each of the Reidemeister moves can be thought of as a motion of a knot in space. We interpolate the motion between two knot diagrams to produce a self intersecting annulus in the space between the two planes in which the diagrams are drawn. If a type I move occurs, the annulus has a branch point. If a type II move occurs, there is a minimum or a maximum on the double point curve. And if a type III move occurs there is a triple point in the surface. Then we break the surface along its double arcs in a way that is consistent with the bounding knot diagrams. Figure 3.23 indicates how this is achieved.

In Figure 3.23 the type I and type II traces have appeared in only one of two possible directions. The figures can be turned upside down to move knots in the other way.

In Figure 3.24 the remaining two types of critical points of surfaces are recalled. One is **a saddle point**, and the other is a **local maximum or minimum** depending on whether you look at it from top to bottom or *vice versa.*

The next order of business is to convince you that diagrams of broken surfaces really have something to do with surfaces in 4-dimensions. So let us consider what 4-dimensions might mean.

Ordinary 3-dimensional space can be thought of as a sequence of 2-dimensional planes stacked together like the pages of a book. An object in space can be cut into tomographic slices by the pages in the book. If that object is a surface that intersects itself, then these slices are the stills from movies that we have seen before.

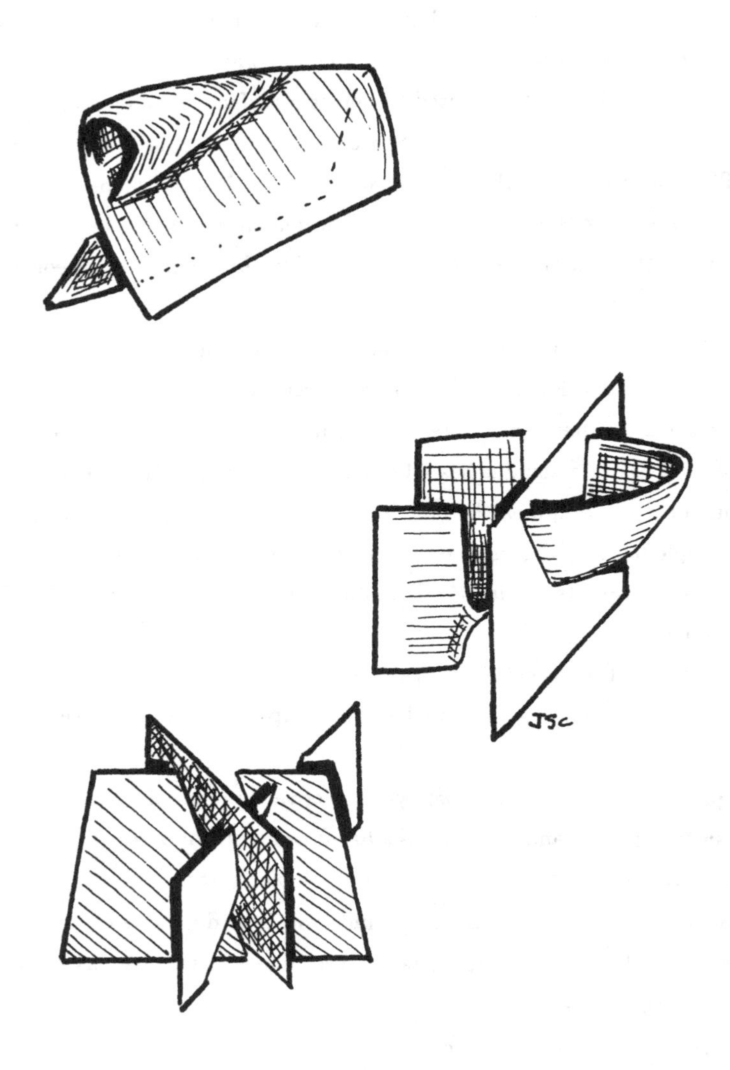

Figure 3.23: Interpolating the Reidemeister moves to get surfaces in 4- dimensions

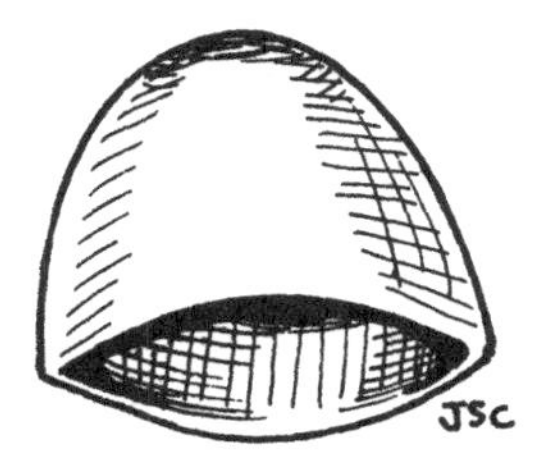

Figure 3.24: Saddles and local optima

Next we think of 4-dimensional space as a pile of 3-dimensional spaces, stacked in time. Billy Pilgrim, the hero of Vonnegut'sbook "Slaughterhouse Five or The Children's Crusade" [73], thought of life as a series of moments strung together. Because of his existence on Tralfamador, he was able to see the entire 4-dimensional view of his life as a single object. When he got unstuck in time, the sequence of his 3-dimensional moments was shuffled.

Similarly, we think of a surface in 4-dimensional space as a collection of curves in 3-dimensional spaces. That is, each 3-dimensional page in 4-space cuts a surface in 4-dimensions, and the intersection is a collection of closed loops. Because the surface is in general position with respect to these slices, we only need to examine slices that occur between successive critical points. The cross sectional loops shouldn't become unstuck in time; so the diagrams for successive cuts differ at most by a Reidemeister move, a saddle point, or a local maximum or minimum. The cross sectional pieces, then, are depicted as knot diagrams, and the movie of these diagrams defines the surface.

On the other hand, we have learned how to interpolate movies via the critical points of the surface and would-be self intersections. The interpolations give broken surface diagrams, so these depict surfaces in 4-dimensions.

Examples

In Figure 3.25, the disk bounded by the $\alpha\gamma$ curve is depicted as a broken surface diagram. This is the surface that results when twists in a string are canceled.

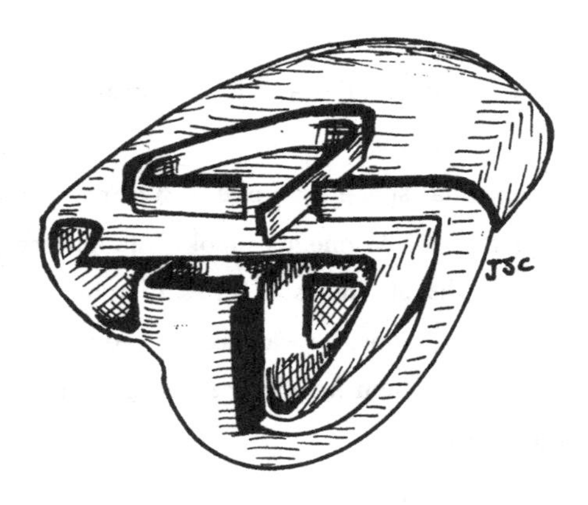

Figure 3.25: The embedded disk in the 4-ball bounded by the $\alpha\gamma$ curve

In figure 3.26 A disk in 4-space is depicted that is bounded by the **stevedore's knot**. The knot is known to be knotted yet it bounds a disk in 4-space.

(Stevedoring is the loading and unloading of ship cargo. I spoke to a man in Mobile, Alabama who has been a stevedore for over forty years. He told me that the bowline and the square knot were the most common knot used in stevedoring, and that the knot that mathematicians call the stevedore's knot is seldom, if ever, used. Furthermore, in modern stevedoring knots really aren't necessary. So much for choosing good terminology.)

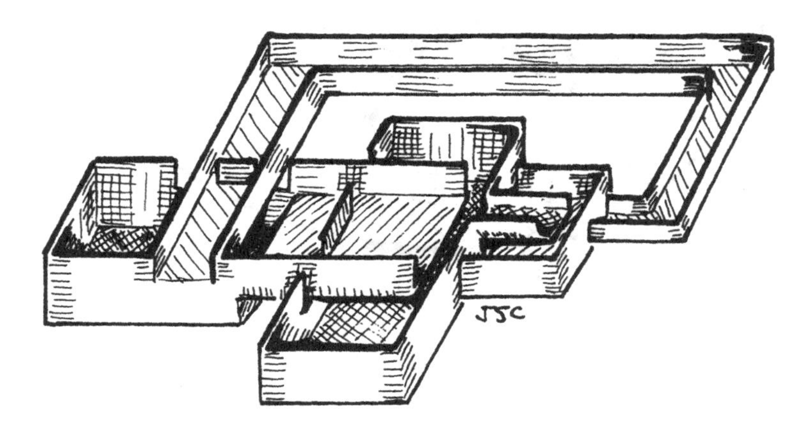

Figure 3.26: A slice disk for the stevedore's knot

Actually, many knots that are knotted bound disks in 4-space. Those that do are called **slice knots**. A loop is not knotted if it bounds a disk with no self-intersections in **3-space**. When a slice disk is pushed into 3-space, it will intersect itself unless the slice knot is unknotted. So even if a knot is slice, it can still be knotted.

(There is a technicality that requires mentioning. Every knot bounds a disk in 4-dimensional space, but these disks are not locally flat. Non-locally flat disks won't be considered here, but if I don't at least mention them, I will get sent to math jail for knowingly omitting a needed condition.)

Disks in 3-Space

Every knot, whether it be knotted or not, bounds a disk that might intersect itself in 3-dimensions. A disk bounded by the trefoil knot is depicted in figure 3.27.

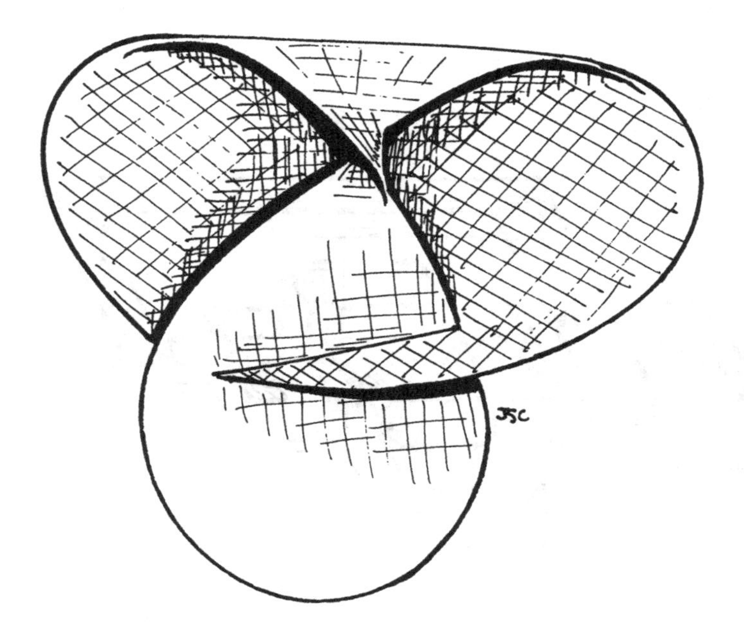

Figure 3.27: A disk bounded by the trefoil

A similar disk can be found from any knot diagram. There are two steps involved. First, we find a disk that is bounded by the closed curve that is the shadow of the diagram. This step is achieved algorithmically using the Gauss word for the shadow. I'll spare you the details, but mention that the algorithm involves measuring the complexity of pairs of oppositely signed letters in the Gauss word and choosing the pair of letters of smallest complexity. Complexity is defined just as it was for mates of substantial arcs. The Jordan Curve Theorem is used in proving that the algorithm works, so the closed curves to which it applies must fit into the plane. Second, we pull the lips of this disk up at each of the over crossings as in Figure 3.28.

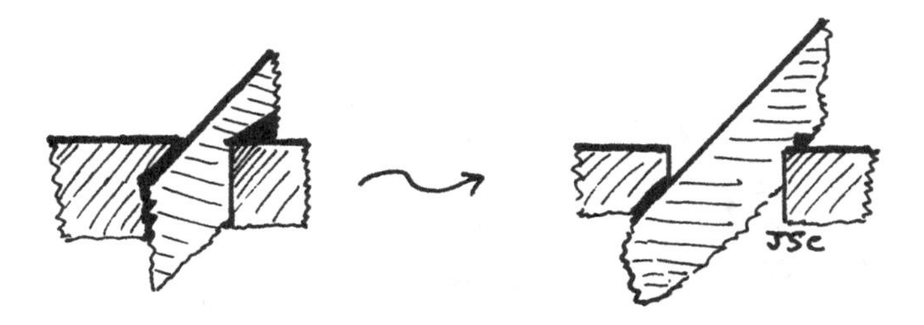

Figure 3.28: Lifting the disk at the over crossing

Pushing the Disks in 3-Space into 4-Dimensions

A disk in 4-space can be made from the disk in 3-space by means of a broken surface diagram that has a finite number places in which the breaks don't match up. These places represent double points of surfaces in 4-space. The prototypical example of surfaces not matching up is given in Figure 3.29.

The closed curves that are linking form the diagram for a pair of loops that topologists call the **Hopf link**. We call it that because (1) it is so common that it needs a name and (2) there is an example, that is well known to mathematicians, of a continuous map from a 3-dimensional sphere (see Chapter 4) to a 2-dimensional sphere that was constructed by Hopf. In this map, the preimage of any pair of points is a pair of linked circles. Banchoff *et al.* [7] have made a movie of the 3-sphere in which the Hopf link plays a leading role.

The intersection point of the disks in 4-space that are bounded by the Hopf link is represented as the point at which the broken surfaces change from horizontal to vertical. We know that these disks intersect in 4-space because we can write down equations that represent the disks. The simultaneous solution of these equations consists of a single point. (See **Chapter 4**.)

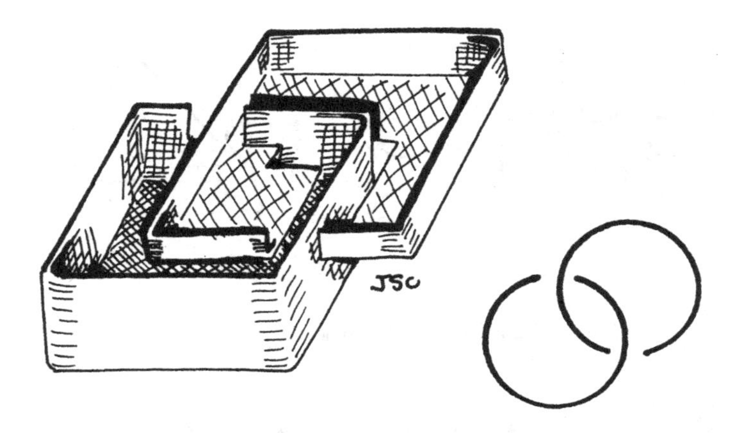

Figure 3.29: The Hopf link bounds two disks that intersect in 4-space

In Figure 3.30, some lower dimensional analogs are depicted in which the intersections of disks and arcs are seen as a single point on the inside of the ball. Often in topology, we use lower dimensional diagrams to depict phenomena in higher dimensional space. The rigors of the higher dimensional worlds are encapsulated within equations, but our intuition is encapsulated in the figures. Chapter 4 begins with a description of the 3-dimensional sphere. In that chapter we'll see why the pictures in Figure 3.30 effectively depict the intersection of the disks bounded by the Hopf link.

⋆⋆⋆

Given a knot diagram, we find a disk in 4-space bounded by the represented knot. The disk in 4-space intersects itself in a finite number of points. (Actually, we can make this number be no more than half the number of crossing points by using one or the other of the standard unknot diagrams.) The disk is schematically depicted as a broken surface with the intersections depicted as the places where the breaks don't match.

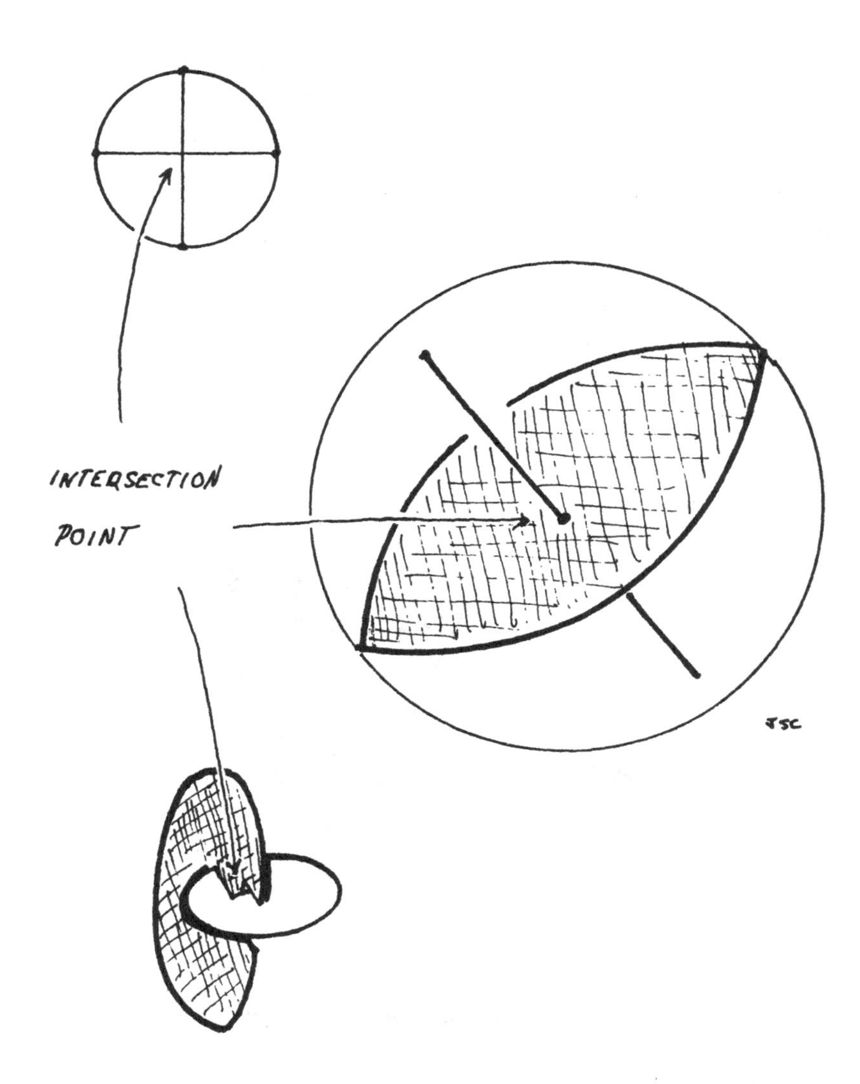

Figure 3.30: Polar and equatorial spheres in the visible dimensions

From the broken surface diagrams, we will find a way of assigning either a 0 or a 1 to each knot. If the number is 1, then we know that the knot does not bound a disk in 4-space. If the number is 0, we don't necessarily gain any information. However, if a knot is not slice, then it is not trivial. The trivial knot bounds a disk in 3-space, and a slice knot bounds a disk in 4-space. If a knot does not bound a disk in 4 dimensions, it can't bound in 3 dimensions either.

If, in the broken disk diagram for a knot, there are no points at which crossings change, then the knot does bound a disk in 4-space. If the disk depicted intersects itself, then we'll have to analyze the intersections to get further information. For example, in Figure 3.31, a self intersecting disk is illustrated.

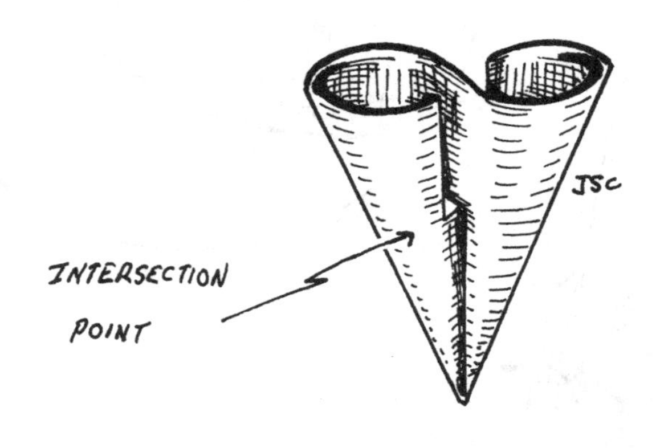

Figure 3.31: A self intersecting disk bounded by an unknot

The figure 8 in the plane when lifted to a loop in space certainly bounds an embedded disk in space because this curve is unknotted. But for some strange reason, I have shown a self intersecting disk in 4-space bounded by the figure 8 loop. The fact that the disk intersects itself means that I did something silly. So in the process of finding our condition, we'll have to quantify silliness.

There is another type of silliness that needs to be put to rest. The **mirror image** of a knot is obtained from a diagram by switching all of its crossings. A knot tied to its mirror image does bound a disk in the 4-ball. An illustration is depicted in Figure 3.32 for the trefoil. The fact that a knot and its mirror image form a slice knot rests on the fact that a reflection in one dimension can be achieved by a rotation in the next higher dimension. If we find a disk with a double point for a knot, then it is easy to find a disk with a double point for its mirror image. So an easy disk that is bounded by a knot and its mirror image would have twice the number of double points as the original.

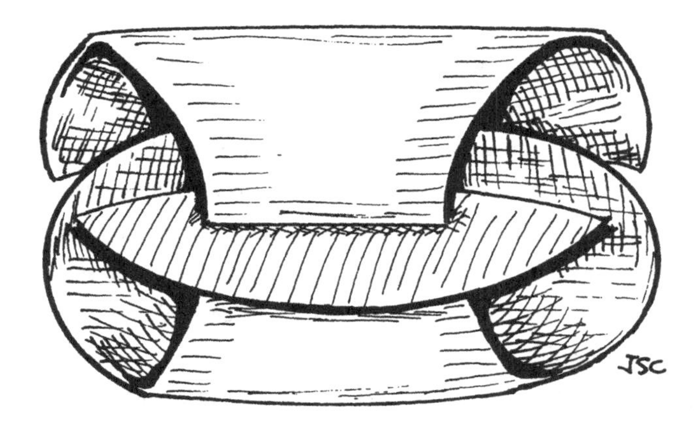

Figure 3.32: A knot and its mirror image form a slice knot

The second type of silliness is easy to handle. In short, we give up. We agree to count the number of essential (not silly) intersections only up to parity. The **parity** of a number is its eveness or oddness. If the parity is even, we assign a 0; if it is odd, we assign a 1.

We are trying to determine the number of non-silly self intersections on a disk in 4-space bounded by a given knot. No, change that: We are trying to determine the parity of this number. If a disk has an odd number of non-silly self intersections, then the knot is not slice. If a disk has an even number, the knot may or may not be slice. The invariant is weak in that sense, but it is enough to determine that some knots are knotted.

Eliminating Silly Intersections

The intersecting disks bounded by the Hopf link form the prototype for the double points of a disk bounded by any knot. The Hopf link also bounds a twisted annulus in 3-space. We want to know when double points on a disk are silly. The first step in quantifying silliness is to remove the intersecting disks at each double point and replace them with an annulus. In this way, we loose the disk but gain an embedded surface.

The silly self intersection was removed from the self intersecting cone in Figure 3.33. In this figure, the removal of an intersection point is also indicated in the abstract. For each such replacement, two disks are removed from a surface, and an annulus is sewn back. In this way, each double point is replaced by two branch points. The sewing can always be achieved in an orientation preserving manner. If there were one such double point on the disk, we would get our friend the basket shaped thingy — surface of rank 2 with one boundary circle. In general, we get such a toriodal piece at each double point.

Recall (from the end of Chapter 1) that the rank of a surface with boundary is the number of substantial arcs in the surface. For each substantial arc, there is a substantial circle. On a surface with one boundary circle, the rank is even. The substantial circles can be considered in pairs; a pair of substantial circles intersect in a single point.

We started from a self intersecting disk, and removed all of the double points to achieve an embedded surface that has as its rank twice the number of double points of the disk. For each pair of substantial circles, we can easily see one of them as the

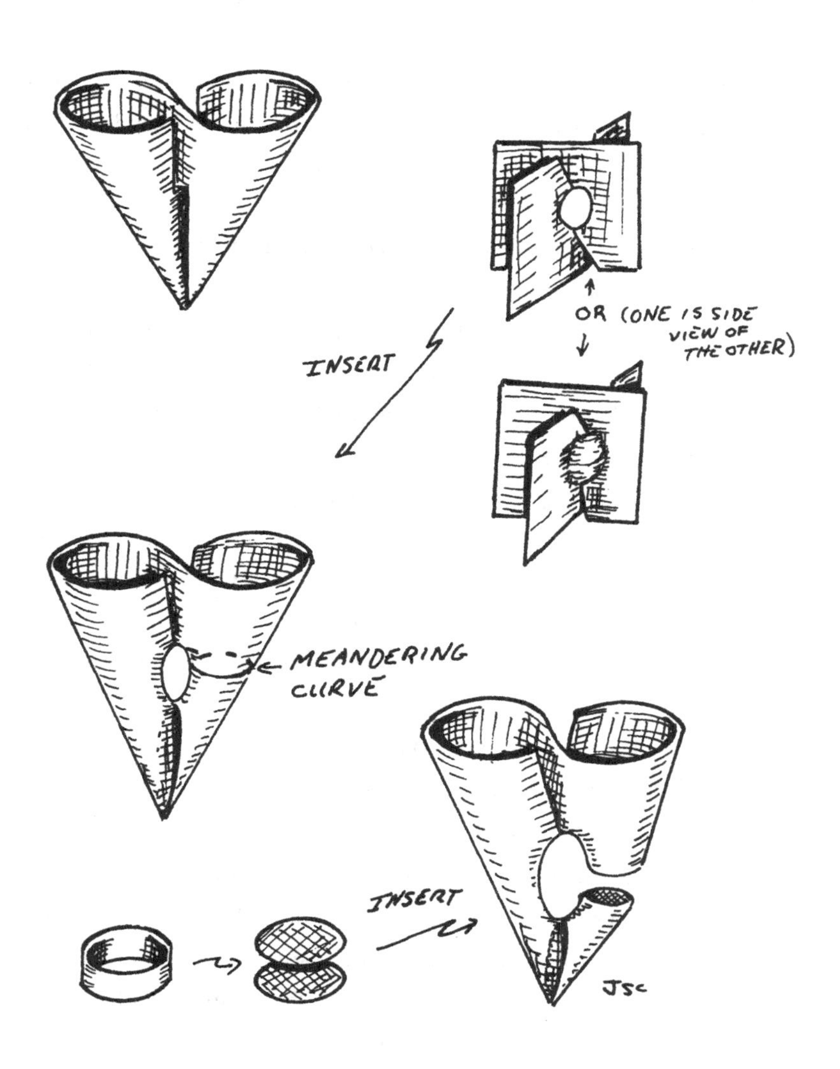

Figure 3.33: Removing the silly double point

center circle of the twisted annulus. The other substantial circle meanders around the surface.

On the cone picture, the meandering substantial circle is depicted. It has an untwisted annulus as its neighborhood in the surface, and it bounds a disk that does not intersect the rest of the surface. We cut out its annular neighborhood, and replace it with a pair of disks. The surface that results is a disk even though it is more complicated than the one that you thought of in the first place.

Now we know how to define silly. A double point is **silly** if it is like the double point on the cone on a figure 8. When this double point was replaced with an annulus, one of the two substantial circles that resulted had an untwisted annular neighborhood, and it bounded a disk that didn't intersect the rest of the surface. So this substantial circle could be removed, and a disk resulted.

Two things are measured in silliness. One is the shape of the annular neighborhood of the meandering curve; is it twisted or not? The other is the type of intersection that a surface bounded by the substantial circle has with the rest of the surface. If either the neighborhood is twisted or a surface bounded by the circle has an odd number of intersections with the rest of the surface (in 4-space), but not both, then the double point that results is not silly.

Figure 3.34 and Figure 3.35 illustrate that the double points on the disks bounded by the trefoil and the figure eight knot (the unique knot with four crossings) are not silly. This shows that they are knotted.

To conclude, let me summarize the definition of the Arf invariant. Given a knot, we find a self intersecting disk in 4-space that is bounded by the knot. Replace each intersection with an annulus. Then examine the pairs of substantial circles on the resulting surface. One such circle is the center of the twisted annulus. The other meanders around the surface. See if the meandering circle has a neighborhood that is not twisted and if it bounds a surface that doesn't intersect in 4-space the rest of the disk. If both these conditions hold, then that pair of substantial circles contributes nothing to the parity. If both conditions fail, that pair contributes nothing to the parity (but I didn't attempt to explain why). If either condition (but not both) fails,

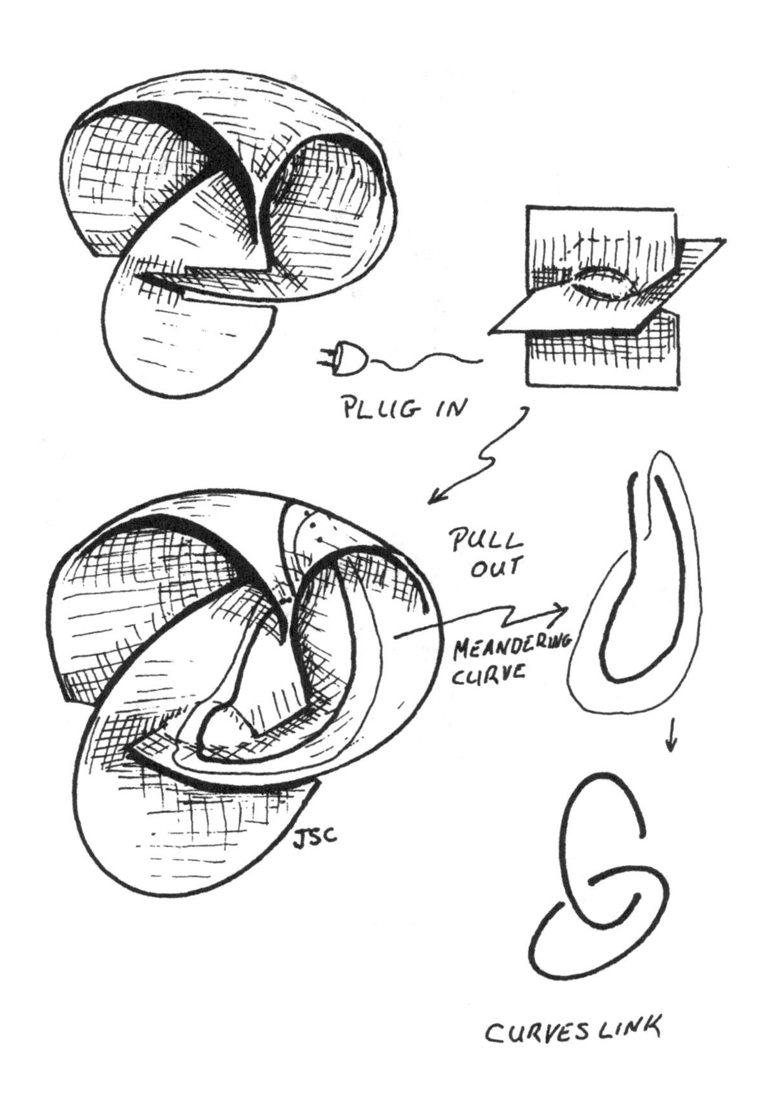

Figure 3.34: The trefoil is knotted

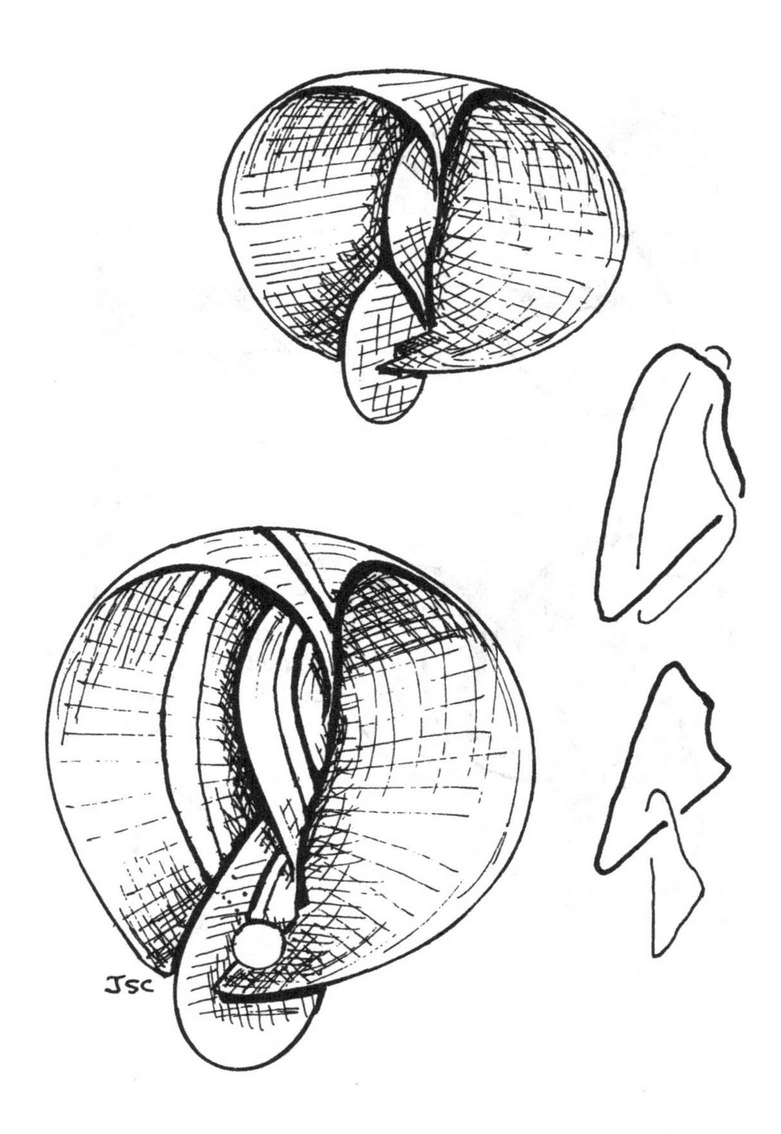

Figure 3.35: The Figure eight knot is knotted

then the substantial pair contributes 1 to the parity. Add up all of the contributions to the parity, and examine the evenness or oddness of the result. If the result is odd, then you know that the given knot is not slice, and in particular it is knotted.

Finally, I need to remark that the Arf invariant will remain unchanged under the Reidemeister moves. This is because the Reidemeister moves can be realized by embeddings. So that the Arf invariant is defined independently from the representative diagram.

Slice Genus

The **slice genus** of a knot is the smallest of all the genera of the surfaces in 4-space that are bounded by the knot. Thus each knot is bounded by a collection of embedded surfaces in 4-space, and among these surfaces there is (at least) one whose genus is no bigger than all the rest. This genus is an invariant of the knot but it is hard to compute. If the slice genus is zero, then the knot is a slice knot. Thus the slice genus is closely related to slice disks.

We have a method for finding an upper bound on the slice genus using Gauss words. Namely, the slice genus of a knot is always less than or equal to the genus of the surface that is constructed from the Gauss word of the represented knot. In other words, we take the Gauss word, construct a surface with boundary via the Gauss word, and cap off all but one of the boundary components with disks. This surface embeds in 4-space with the given knot on its boundary. Each crossing point contributes to the rank, and the genus is determined by the rank and the number of boundary components.

This upper bound does not give very much information. For example, the Gauss word genus for the stevedore's knot is 1, but the stevedore's knot is slice.

3.5 Notes

The definitions of the Arf invariant took some work, but it was worth it.

Michael Freedman's [36] theory of 4-dimensional spaces is built on the same ideas. He examines, via Andrew Casson's techniques, when and how to remove disks under the assumption that the geometric intersection data are trivial. Casson [30] found when there were flexible handles that could be placed in the 4-ball, and Freedman used the flexible handles to give a topological characterization of 4-dimensional space that depended only on the ability to shrink loops and the lack of substantial surfaces therein. This work earned Freedman the Fields Metal, the highest level of recognition that a mathematician can achieve.

Another recent Fields Metal was given to Vaughan Jones [50] for his discovery of a new invariant for knots. One way of looking at Jones's polynomial (and its generalizations) is to consider surfaces that are bounded by a given knot diagram. The Gauss word surfaces are among those considered. The others come from either smoothing or changing crossings in the knot diagrams. And to each smoothing or crossing change an index is assigned. In this way, a knot is thought of as a weighted sum of the surfaces bounded by it. The possible weighted sums yield the invariants.

The sketch of the Arf invariant that I have given was based on the definition given by Rochlin [65] . Other methods of defining the invariant that allow for easy computations are possible. Some of the techniques in this chapter are based on previously unpublished joint work with Masahico Saito.

For further reading about knot theory the texts [67], [52], and [51] are excellent. Since the first edition of the current text was printed Colin Adams's book [1] has appeared. This is another great source from which to learn about the culture of knots.

We have come to understand how surfaces in 4-space can be schematized and understood. In Chapter 6, we will learn how to manipulate surface diagrams via moves that are analogous to the Reidemeister moves. So given two diagrams of the same surface in 4-space, we know how to move from one diagram to another.

Chapter 4

Other Three Dimensional Spaces

In this chapter, we explore some new spaces and investigate a few of the surfaces that can be found within them. Mathematicians believe in spaces of every conceivable dimension and then some. There is a simple reason for our belief.

Your position in space can be specified by three numbers. For example, you might be found in suite 1231, 929 North 606th Street. This address indicates that your office is located on the 12th floor of the building that is just north of the intersection of 606th Street and 9th Avenue. The position of a particle in the plane is specified by two numbers. And the position of a point on the line is specified by a single number.

From a mathematicians point of view, there is no reason to stop at three. We think of n-dimensional space as a space in which locality can be specified by n different numbers. More precisely, n-dimensional space is the set of all possible n-fold sequences of numbers. Distance in such a space is measured via the Pythagorean Theorem (which the Scarecrow gets wrong when he gets his brains).

Inside any n-dimensional space there is a sphere that consists of points which are a unit distance away from the point with coordinates $(0, 0, \ldots, 0)$. For example, the 2-dimensional sphere is the set of points (x, y, z) such that $x^2 + y^2 + z^2 = 1$. The unit circle (some people slip and say 1-sphere) is the set of points (x, y) such that $x^2 + y^2 = 1$. We define the **0-sphere** to be the set of points x such that $x^2 = 1$. These are the points ± 1; they are a unit distance away from 0. By analogy, the **3-sphere** is defined to be the set of all points (x, y, z, w) such that $x^2 + y^2 + z^2 + w^2 = 1$. The

3-sphere is sometimes called the **solid sphere**. We'll leave the higher dimensional spheres for Chapter 7.

In this paragraph, you can safely assume that n is either $1, 2, 3$, or 4. The unit sphere in n-dimensional space is of dimension $(n-1)$. We call this the $(n-1)$-sphere. The **dimension of a space** is the number of local degrees of freedom in the space. There are two degrees of freedom on the 2-sphere, and these are known as longitude and latitude. The $(n-1)$-dimensional sphere is the boundary of an n-dimensional ball that consists of the points in space that are closer than a unit distance to the origin $(0, 0, \ldots, 0)$.

The terms **interval, disk,** and **ball** will be used to distinguish the set of points close to the origin in 1-, 2-, and 3-dimensional space, respectively. These are denoted by letters I, D, and B. The boundary of the interval is $\{-1, 1\}$. The boundary of the disk is the circle, C, and the boundary of the ball is the sphere, S. The pattern among spheres and the balls that they bound goes as follows. A pair of points is to an interval, as the circle is to a disk, as a 2-sphere is to a ball, as a 3-sphere is to a 4-ball.

4.1 The 3-Dimensional Sphere

We'll use three tricks to understand the 3-dimensional sphere. First, we'll look at it in the same way that Santa views the 2-sphere. Second, we'll look at the equators and meridians of the 3-sphere. Third, we'll think of the 3-sphere in terms of tropics and poles. The descriptions that I give, can be formulated algebraically, but we're not going to go much beyond the limits of perception, so you don't have to worry much about the algebra.

Equators and Meridians

Santa Claus thinks of the earth as a plane together with a point at infinity. This view of the world is called the **stereographic projection** of the sphere. The essence of this projection is given in Figure 4.1. A line from the south pole that passes through the sphere intersects the plane that is tangent to the sphere at the north pole. The point on the sphere through which the line passes is mapped to the intersection point of that line with the northern tangent plane. The only point on the sphere that is not mapped to the plane is the south pole.

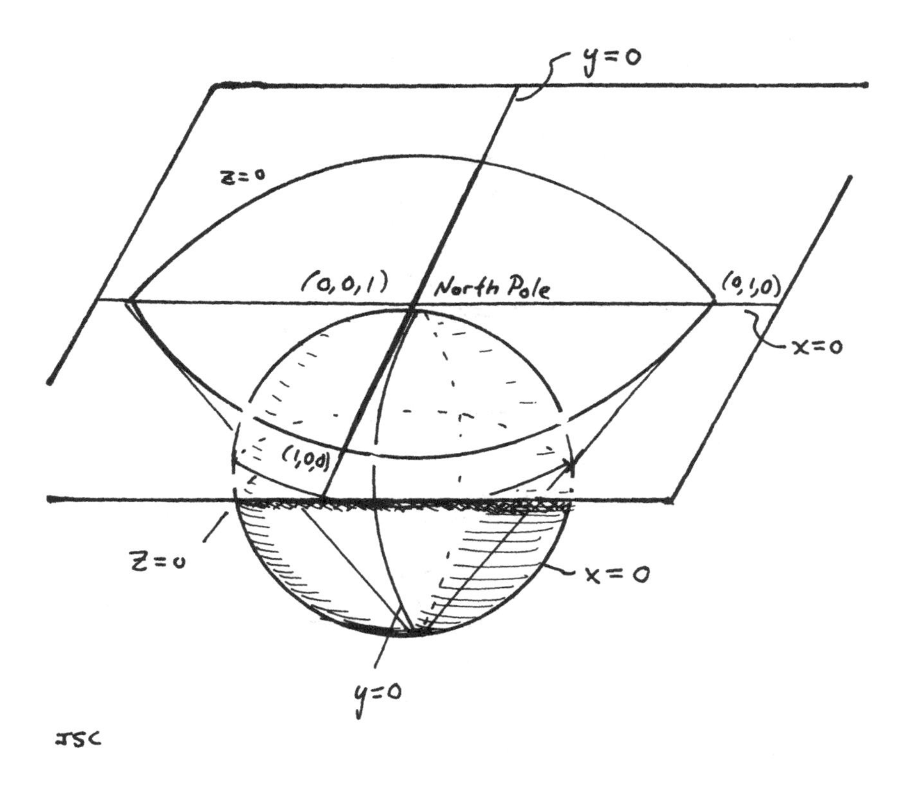

Figure 4.1: Stereographic projection or Santa's map

In the stereographic map, the north pole is mapped to the origin of the plane (0,0), the equator is mapped to the unit circle, and the intersecting meridians are mapped to the coordinate lines. To see all three circles on the plane, we choose a

plane that is not tangent to these circles and project from a diametrically opposite point. This views and another is indicated in Figure 4.2.

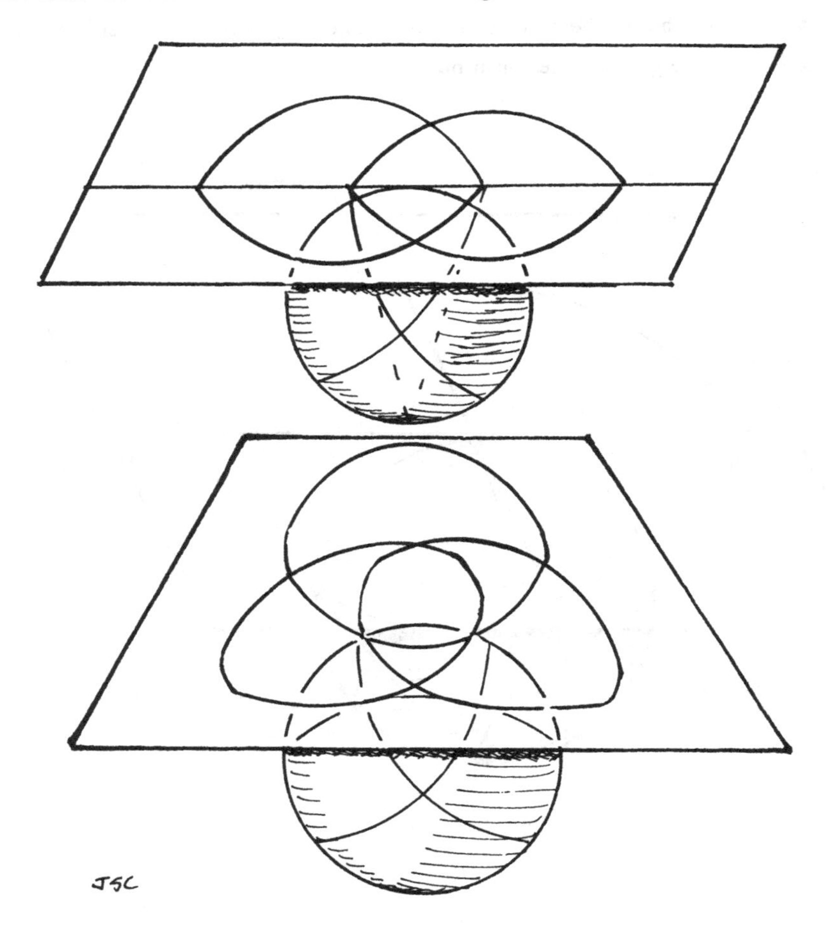

Figure 4.2: Projecting onto other tangency

The circle is the line with its ends sewn together, the sphere is the plane with a point at infinity, and the 3-sphere is ordinary 3-space with a point attached at infinity. In Figure 4.3, some views of the equatorial 2-spheres in the 3-sphere are shown. These are equatorial in the sense that they are defined by the equations $x = 0$, $y = 0$, $z = 0$, or $w = 0$, just as the equator and meridians of the 2-sphere are given by the first three of these equations. The double decker sets of the intersections among the equatorial 2-spheres in the solid sphere are the equatorial circles. The views that are depicted in Figure 4.3 are analogous to the planar views of the 2-sphere.

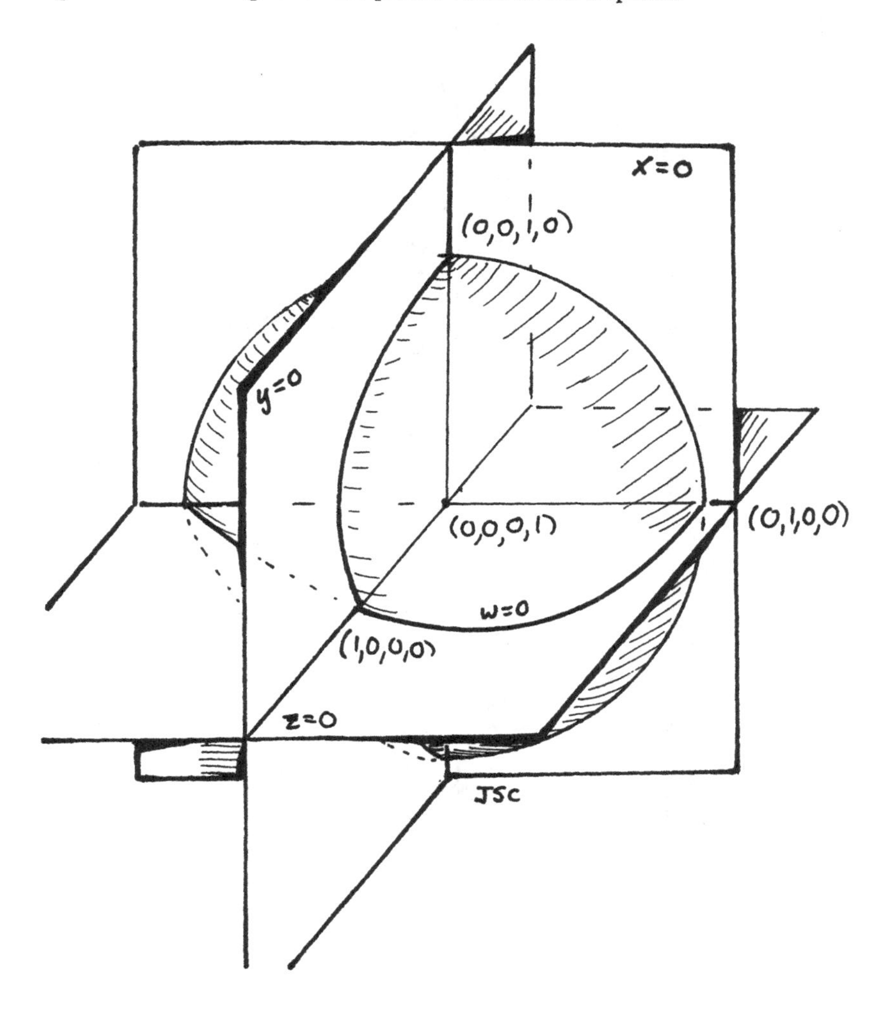

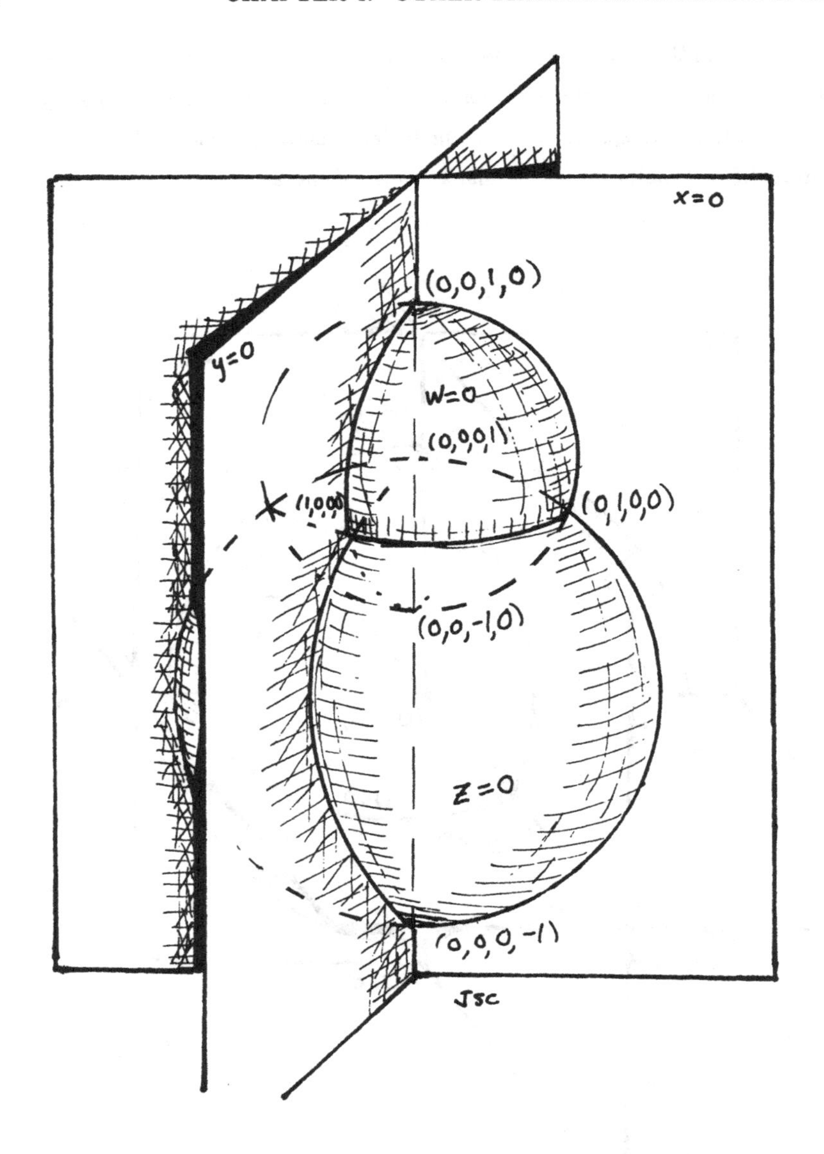
x=0
(0,0,1,0)
y=0
w=0
(0,0,0,1)
(1,0,0,0)
(0,1,0,0)
(0,0,-1,0)
z=0
(0,0,0,-1)
JSC

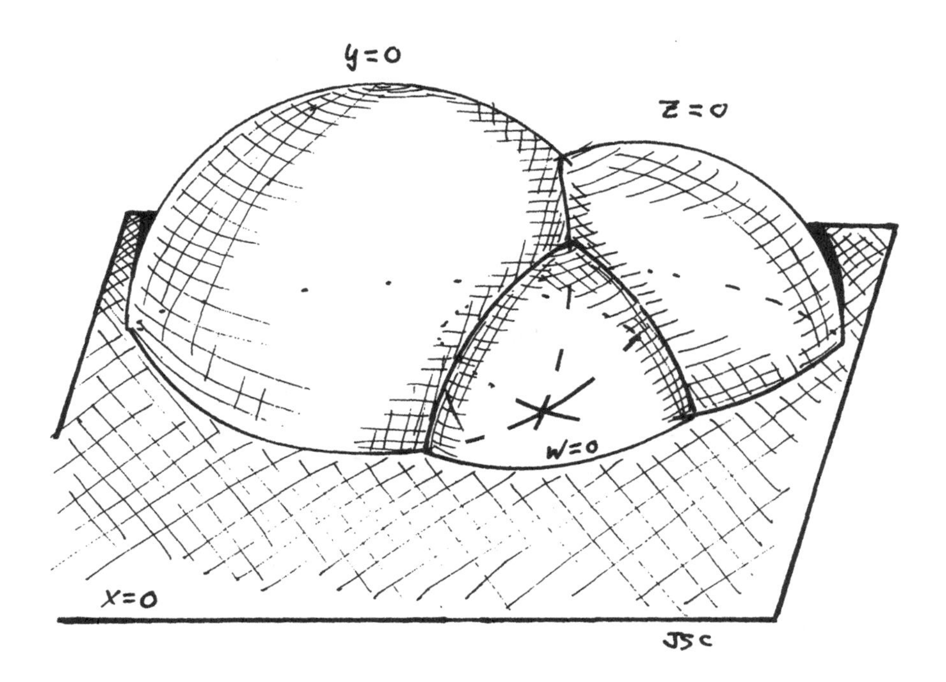
y=0
z=0
w=0
x=0
JSC

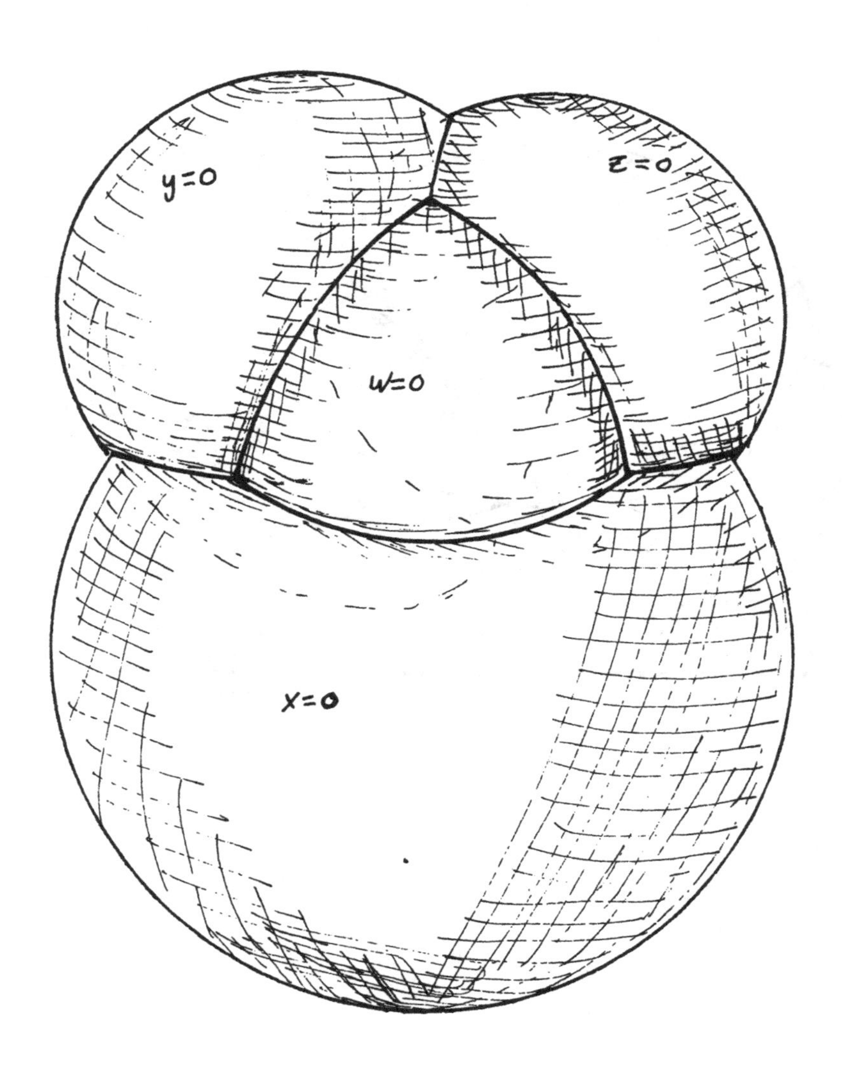

Figure 4.3: Four views of the 3-sphere

The Hopf link, by the way, is the link of the circles $x^2 + y^2 = 1$ and $z^2 + w^2 = 1$ in the solid sphere. It is not hard to see from the pictures that those linking circles are determined by these equations. The unit *disk* is the set of points for which $x^2+y^2 \leq 1$. The disks that are bounded by the Hopf link intersect in 4-dimensional space at the point $(0,0,0,0)$.

A sphere of any dimension has polar/tropical decompositions. Tropical and polar regions can each be defined as the Cartesian product of a disk and a sphere. As the dimension of the sphere gets larger, there are more decompositions. We begin with the two familiar cases.

In the 2-sphere, the polar regions consist of a pair of disks. One such disk, say the south pole, can be labeled as a negative disk, and the north pole can be labeled as a positive disk. The boundaries of these two disks are a pair of circles that are also the boundaries of the tropical annulus on the sphere. So the 2-sphere is the union of an annulus and a pair of disks. This decomposition of the 2-sphere is familiar: Go look at a can of soup.

In general, the **Cartesian product** of two spaces is the collection of pairs of points where one point in the pair comes from one space and the other point comes from the other space. The address of a point in such a space is given by specifying two addresses. For example, the north pole is specified by a (+)-sign and the origin in the unit disk; the sign is a coordinate in the 0-sphere. The address of a point in the tropics is given by a point in the interval (latitude) and a point on the circle (longitude).

On the circle, the polar regions are the Cartesian product of the an interval and a 0-sphere. The tropics are the Cartesian product of a 0-sphere and an interval. I changed the ordering on the two regions because of a certain mathematical convention. Namely, think of a square disk topologically as the product of 2-intervals, $I \times I$. Then the boundary of the disk is $(\{-1,1\} \times I) \cup (I \times \{-1,1\})$ where $\cup$ means the union of these sets. In this decomposition, the first set represents the poles; the second represents the tropics. (The method for computing boundaries of Cartesian product

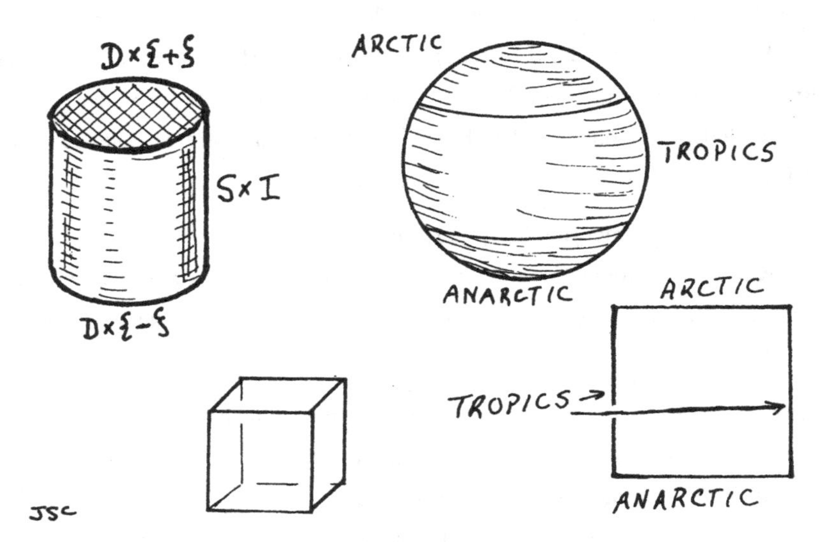

Figure 4.4: Tropics, poles, cylinders, and squares

spaces is the product rule from first term calculus: d(Hi Ho) = Ho d Hi + Hi d Ho; in the product spaces the operator d means to take the boundary. See Figure 4.4.)

The 3-sphere can be decomposed in two ways. In the first way, the polar regions are a pair of balls, and the tropics are a thickened 2-sphere. In the second decomposition, the poles and the tropics are both solid tori. These correspond, respectively, to writing the 4-ball as a product $B \times I$ or $D \times D$. In the decomposition as two solid tori, the Hopf link forms the core of these.

In the rest of this chapter, we will build 3-dimensional spaces out of three fundamental pieces. The ball will play the same role that the disk did in the construction of surfaces. There are two types of handles that we use: 1- and 2-handles are solid cylinders of the form $D \times I$ where, as before, D denotes the unit disk, and I denotes the interval. The 1-handles are attached to the ball along the disk ends of the cylinders, and the 2-handles are attached along the annular sides. In Figure 4.5, I have indicated a handle decomposition of the 3-sphere that uses two balls, a 1-handle, and a 2-handle. Attaching the outside ball is done so in the abstract sense, or we see the outside ball as the space that surrounds us and infinity.

When handles are attached to each other, the attaching surfaces melt away. A similar thing happens when handles are attached to surfaces. In the superficial case, the fabric of golden fleece fuses seamlessly. In the solid case, the walls between the solids evaporate in much the same way that a film of water disappears when two quantities of water are mixed.

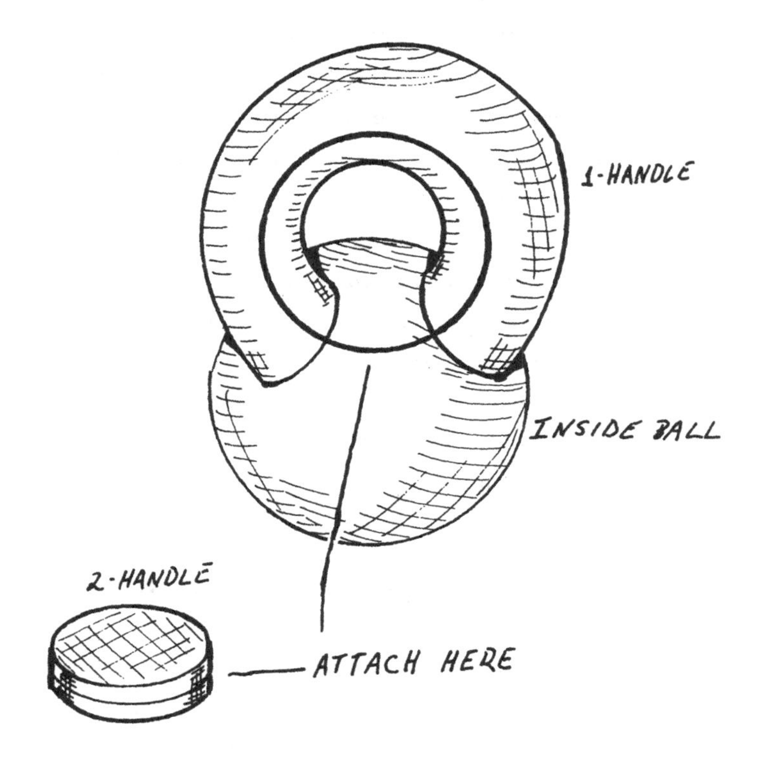

Figure 4.5: Handle decomposition of the 3-sphere

The solid 3-sphere has the property that every embedded 2-sphere separates. If the 2-spheres are embedded in a nice way, then either side is a ball. This property is analogous to the Jordan Curve Theorem of Chapter 1; the fact that nice embeddings bound balls is called the **Schoenflies Theorem.**

To understand 3-dimensional spaces, we'll need to consider surfaces and loops inside the spaces. The best known outstanding problem in 3-dimensional topology attempts to characterize the 3-ball in terms of loops inside it.

Poincaré Conjecture. *Consider a space that has a 2-sphere as its boundary. If every loop in that space bounds a disk, then the space is the 3-ball.*

The Poincaré conjecture was not conjectured by Poincaré; rather it was posed as a question. Analogous statements in higher dimensions have all been proven. Ten years ago, an approach to the Poincaré conjecture was proposed by William Thurston at Princeton University. Thurston translated the problem into a more general statement about geometric structures on 3-dimensional manifolds.

I have neglected any discussion about geometry in lower dimensional spaces. Jeff Weeks [74] already covered that topic quite well at an elementary level. In dimensions 2 and 3, there is a tremendous interplay between geometry and topology, and Thurston's program is to exploit that interplay. The goal of this book is not to develop the most general theory in low dimensional topology, but rather to survey the subject by means of a few examples.

4.2 A Space Not Separated by a Sphere

The annulus contains a substantial arc. A torus can be formed as the union of two annuli along their boundary circles as in Figure 4.6. The substantial arcs in the annuli glue together to form a substantial circle. The torus can also be decomposed as the quotient of an annulus where the two boundary circles are glued together. In this section, a 3-dimensional analog will be constructed.

A **solid torus** is the Cartesian product of a disk and a circle. In this product space, there is a **substantial disk:** a disk that has its boundary on the boundary

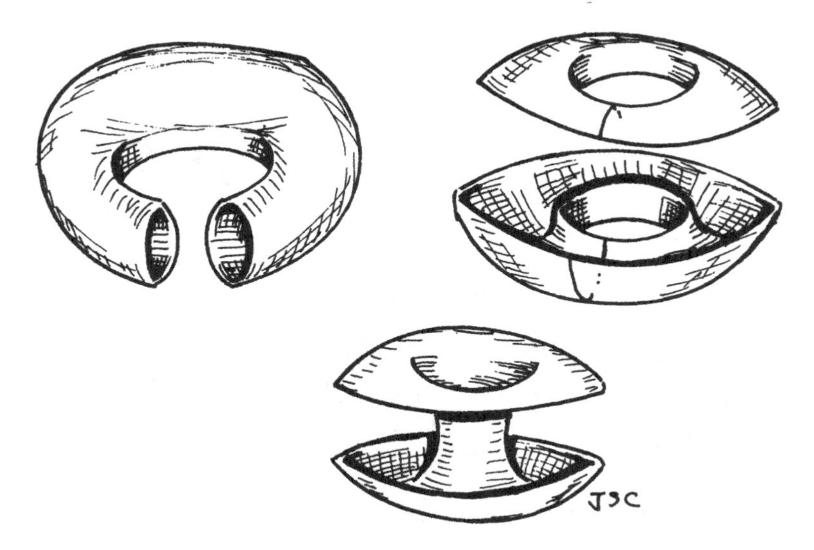

Figure 4.6: The torus as quotients of annuli

torus and that does not separate. We can see that this disk is substantial because a core circle in the solid torus intersects the disk in one point. This substantial disk is like the substantial arc in the annulus: The annulus is the Cartesian product of an interval and a circle, and an arc between the boundary circles is pierced once by the center circle. The solid torus is the union of a ball and a 1-handle. The solid torus is the topological space represented by a solid coffee cup or a doughnut.

Consider a solid torus with a 2-handle attached along the outside at a meridional curve. The **meridian of a solid torus** is the boundary of a substantial disk. The outside 2-handle can't be embedded in the 3-sphere, so you should only conceive of this space as an abstract object. The boundary surface of the resulting space is a 2-sphere. Figure 4.8 depicts the resulting space.

Let's see that the boundary is spherical by running briefly through the relevant part of the Classification Theorem of Surfaces. The boundary torus has an annulus removed and two disks attached. When an annular neighborhood of a substantial circle is removed from a torus, the remaining surface is an annulus. When two disks

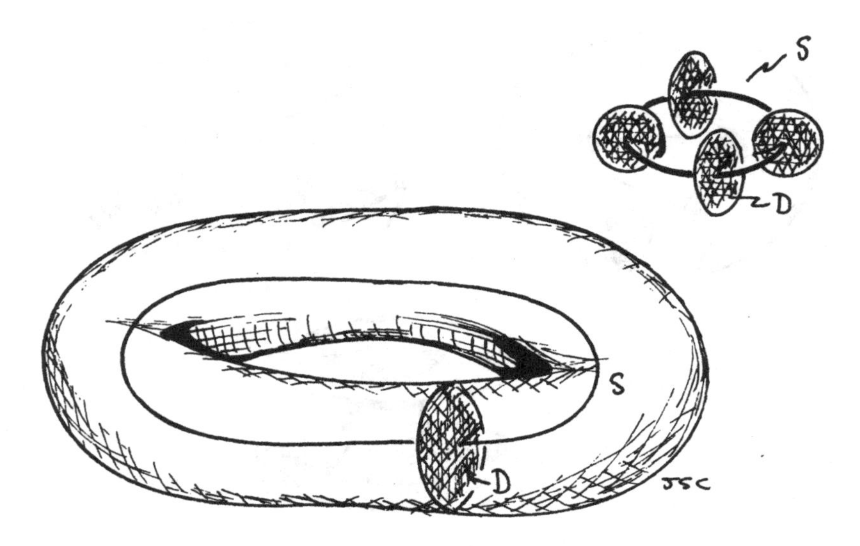

Figure 4.7: The solid torus

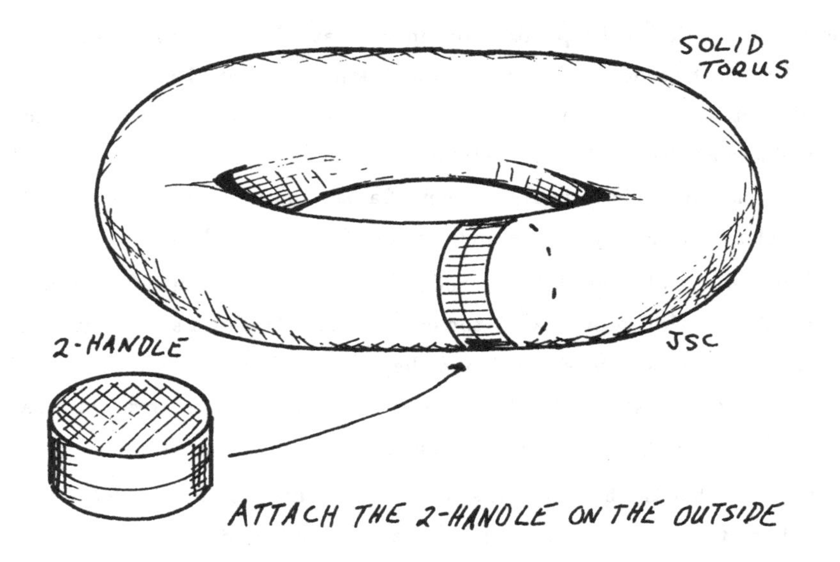

Figure 4.8: Adding a 2-handle

are attached to an annulus, a sphere results. In the process of adding a 2-handle to the solid torus (or to any space for that matter) an annulus is removed from the boundary, and replaced with two disks. These disks and the remaining annulus on the boundary form a sphere.

A ball is glued to the outside of this bounding sphere and a closed solid results (**a closed solid** is a 3-dimensional space that has no boundary). The solid is the Cartesian product of a circle and a sphere. We give two more ways to describe it.

In the next description, we begin with a thickened sphere in space. This is the Cartesian product of the sphere, S, and the interval, I. Imagine that the inner boundary sphere is glued to the outer boundary sphere. The gluing process is akin to the gluing that occurs when the boundary circles of an annulus are glued together to form a torus. The sphere is substantial in this space because cutting along the sphere will turn the space back into the product space $S \times I$. In Figure 4.9, an arc is depicted that pierces the sphere once.

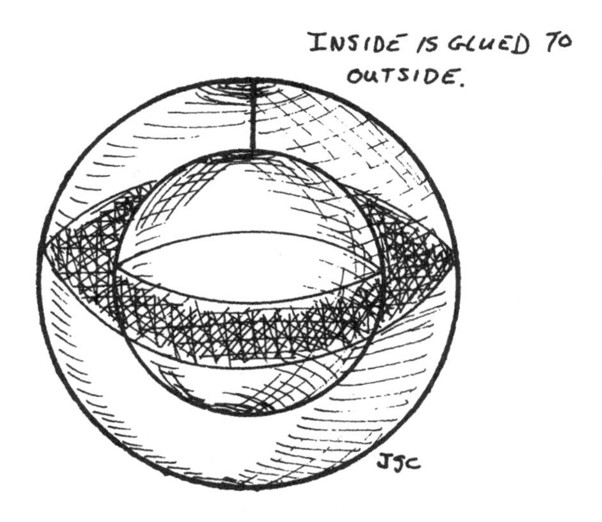

Figure 4.9: A thickened sphere

The final description is akin to the description of the torus as the union of two annuli along their boundaries. In one higher dimension, we glue two solid tori together along their bounding tori so that the substantial disks in the solids glue together to form a sphere. This description is not much different that the very first one, the center disk of the 2-handle glues to the substantial disk of the solid torus to form this very same non-separating sphere.

The space that results from any of these constructions is the Cartesian product of a circle and a sphere. The sphere factor is the union of the substantial disks in the solid tori. The circle factor is the center core of either of these tori. We can find a point in this space by specifying its longitude and latitude on the sphere and an angle that measures displacement from a fixed sphere. A coordinate circle intersects a coordinate sphere at a single point.

Figure 4.10 depicts an $\alpha\gamma$-curve on a 2-sphere that we think of as the boundary of a ball which has been removed from the space $S \times C$. The curve is the boundary of a punctured torus inside this solid. The standard disk that is bounded by the $\alpha\gamma$-curve has a triple point inside the 3-ball. If this disk and this ball are sewn back into the space, then we have constructed a substantial torus in $S \times C$ that has one triple point. The torus is substantial because there is a thread in the space that intersects the disk attached on the outside exactly once.

What's the deal with that example? Banchoff's Theorem [5] says that in ordinary space the only surfaces that have an odd number of triple points are the surfaces of odd rank. The torus constructed from the $\alpha\gamma$-curve torus has even rank and one triple point. The space $S \times C$ is extraordinary. In fact, Figure 4.11 shows you a surface of rank 3 that fits into this space without triple points. Indeed, this non-orientable surface is substantial in $S \times C$ even if it is not embedded. The rank 3 surface in $S \times C$ looks a bit like the figure 8 map of the Klein bottle. The difference is that the current figure 8 has a quarter twist in it and the Klein bottle has a half twist in it.

Ki Hyoung Ko and I used this last example and other examples in [55] to give a generalization of Banchoff's triple point result. Banchoff's result says that the parity of the number of triple points of a closed surface in 3-space (with no branch points)

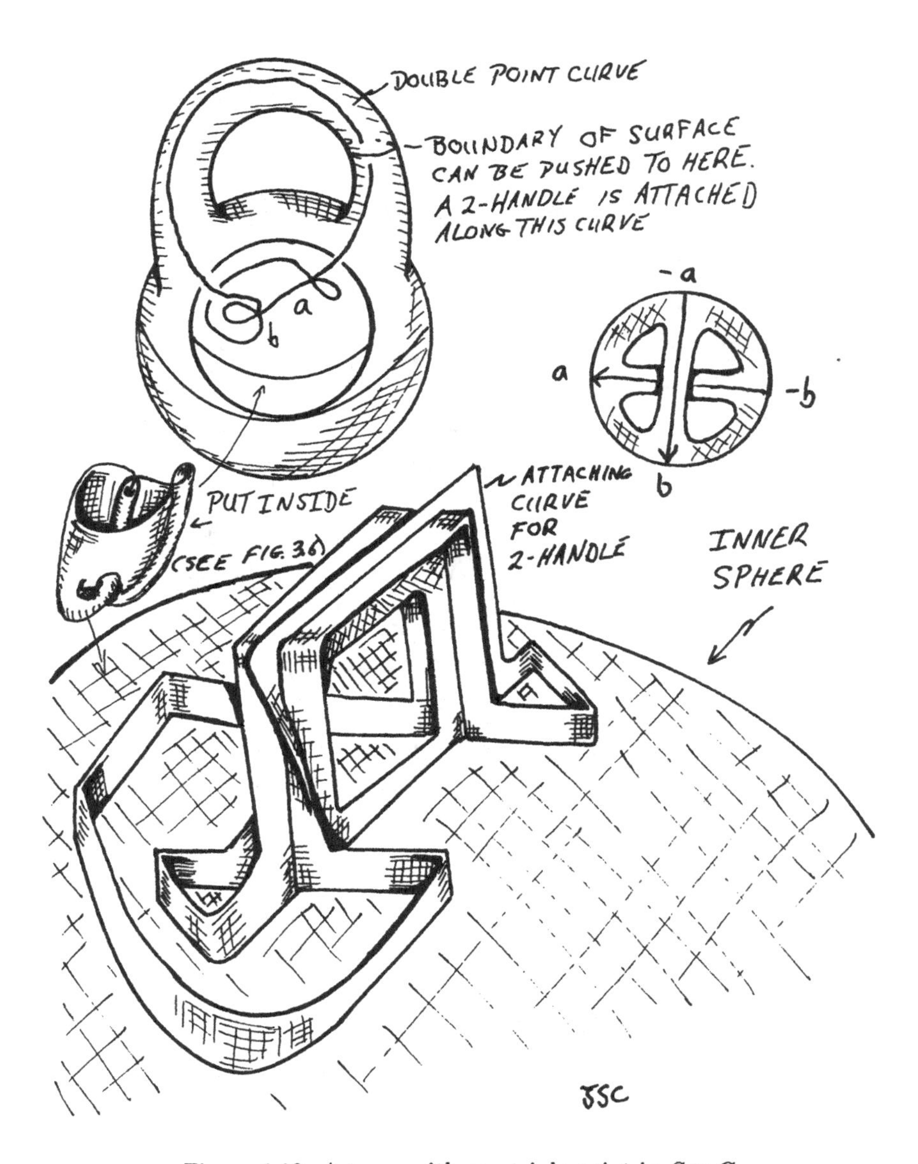

Figure 4.10: A torus with one triple point in $S \times C$

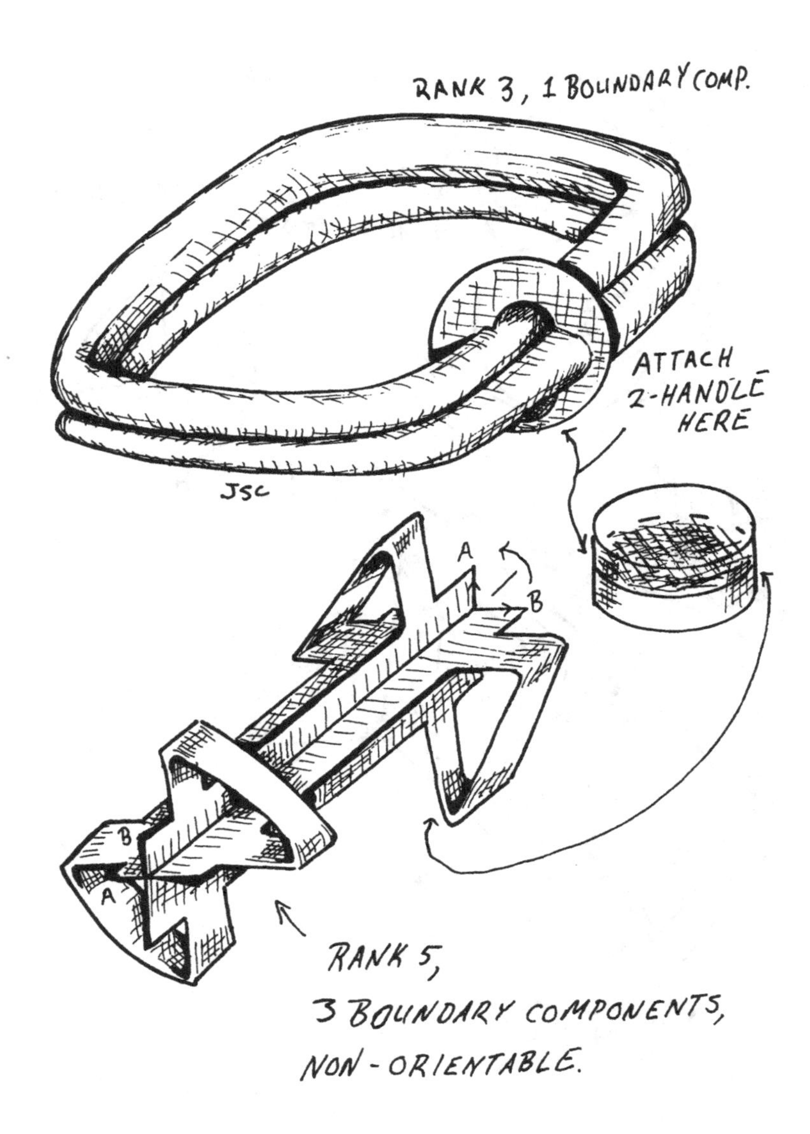

Figure 4.11: A rank 3 surface in $S \times C$

is the same as the parity of the rank of the surface. Our formula contains a simple correction factor for surfaces in any 3-dimensional space. The central idea in the proof is found in Banchoff's proof. A triple point formula relating to rank that is true for embeddings, for certain examples, and remains unchanged under a certain set of moves is true for all surfaces with no branch points.

The correction term that we put into Banchoff's formula only measured things up to parity. In Chapter 5, we will show how this term can be used to measure the difference in the number of triple points and the rank. The proof of the more general result (which is due to Izumiya and Marar) is presented in Chapter 5.

4.3 Three Dimensional Projective Space

The projective plane is the result of gluing a disk to the boundary of a Möbius band. It is also described as the result of identifying diametrically opposite points on the 2-sphere. In this section, we construct an analogous 3-dimensional space in which the projective plane is embedded as a substantial surface.

Revisiting the Projective Plane

Since the projective plane doesn't embed in ordinary space, we will find a space in which the projective plane embeds naturally. The Möbius band fits nicely inside a solid torus as in Figure 4.12. The boundary of the Möbius band wraps twice in the longitudinal direction and once in the meridional direction. The meridional wrapping is a bit tricky to see because it is occurring gradually.

Recall that a solid torus is the union of a ball and a 1-handle. In the current discussion, the 1-handle is attached to the ball with a half twist. The equatorial disk in the ball is glued to the equatorial disk in the handle; the latter is twisted, and the two glue together to form the Möbius band. Meanwhile, the core of a 2-handle is a flat disk. The core disk will be attached to the Möbius band as the 2-handle is attached to the torus along a neighborhood of the boundary circle.

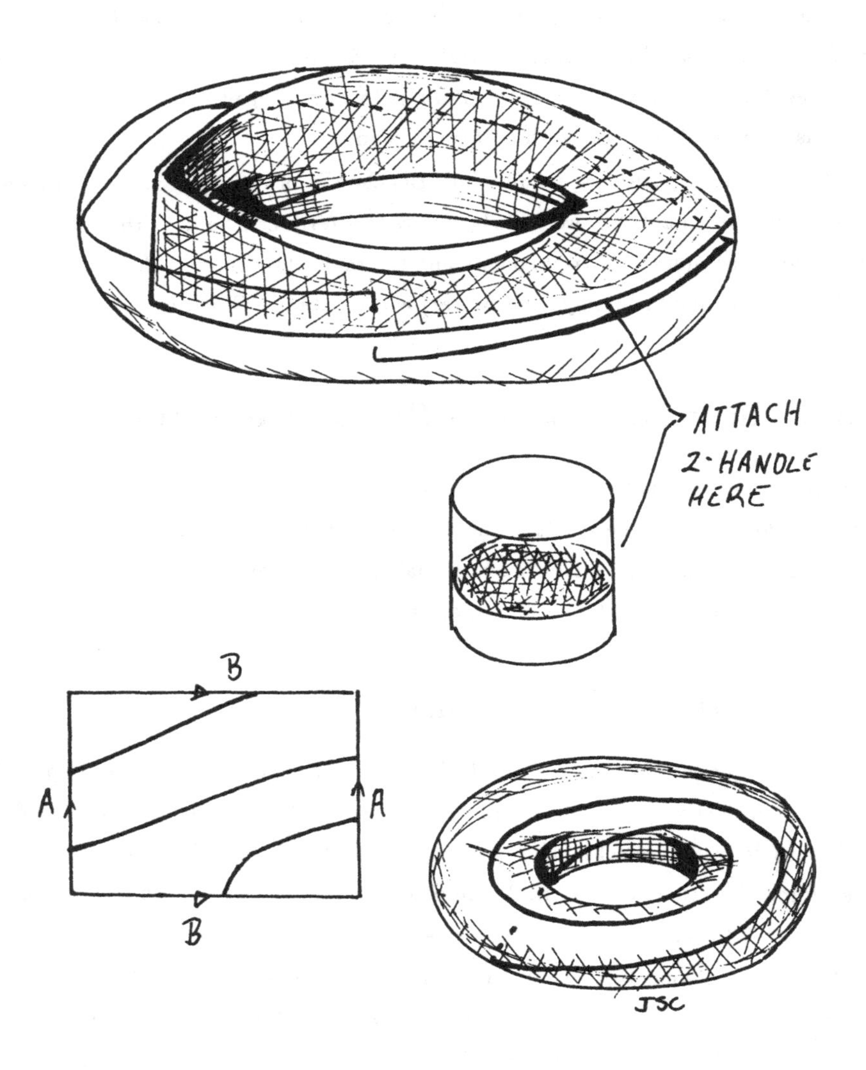

Figure 4.12: A Möbius band in a solid torus is attached to a disk in a 2-handle

We also attached 2-handles to solid tori in the previous two examples. In the 3-sphere, the 2-handle was attached along a longitudinal curve thereby filling the hole in the doughnut. In the space $S \times C$, the 2-handle was attached on the outside along a meridional curve. The resulting space had a substantial 2-sphere that was the union of the core disk in the 2-handle and the substantial disk in the solid torus. In general, a 2-handle can be attached to any simple closed curve on the boundary torus. The more interesting attaching regions are the ones that are substantial on the boundary.

After the 2-handle is attached, the boundary of the resulting solid is a sphere because the attaching curve is substantial in the torus as indicated in Figure 4.13. We glue a ball along this sphere to get a closed solid 3-dimensional space called **projective 3-space**.

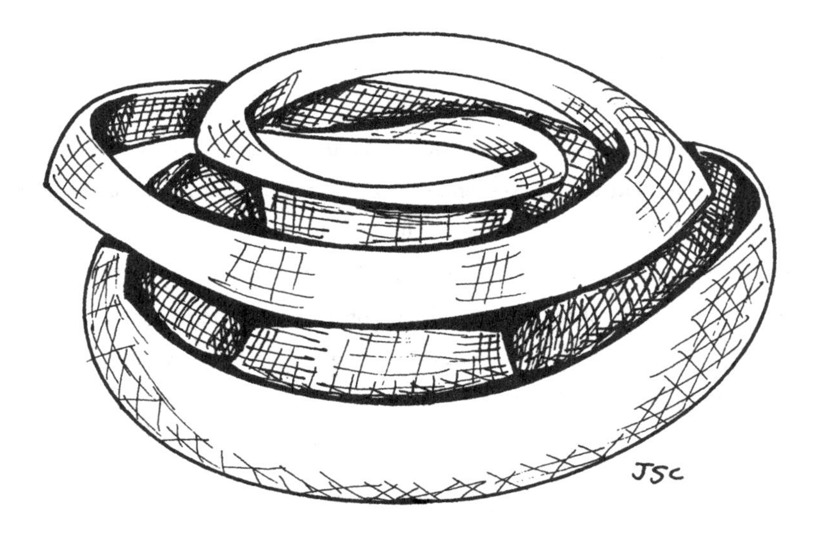

Figure 4.13: The attaching curve on the torus is substantial

The projective *plane* is substantial in projective space because the core disk of the 2-handle is pierced by a transverse arc in a point. This arc can be closed to a loop through the outside 3-ball. Similarly, the Möbius band is substantial in the solid torus. Figure 4.12 shows the once intersecting loop.

Identifying Points on the 3-sphere

Consider the possibility of identifying diametrically opposite points on the 3-sphere. As in the 2-dimensional case, we can think of the quotient map in two pieces. The polar regions of the 3-sphere are two balls; we can see one of these, and the other one surrounds infinity (the south pole). The polar balls are identified with each other, and a single ball results. The tropical region is the Cartesian product, $S \times I$, of a sphere and an interval.

To see the result of the identification on the tropical regions, we cut the equatorial 2-sphere into two polar disks and a tropical annulus. The polar *disks* on the sphere are identified to each other, and their neighborhoods in the 3-sphere get mapped to the 2-handle in the previous description.

The tropical annulus is mapped to a Möbius band. A neighborhood of the annulus in the 3-sphere is a solid torus. Remember that to map the annulus to the Möbius band, we put a full twist in the former. The annulus with a full twist fits inside a solid torus as indicated in Figure 4.14, and this figure indicates the quotient map.

The equator of the 3-sphere is mapped to a projective plane in the 3-dimensional projective space, the poles of the 3-sphere are mapped to the outside ball, and the projective plane is substantial there.

Projective Space Is the Set of Rigid Motions

A standard exercise in graduate school is to show that projective space is the same as the set of rigid motions of 3-dimensional space. The solution to the exercise is quite easy, so easy, in fact, that most instructors can't wait for students to discover the

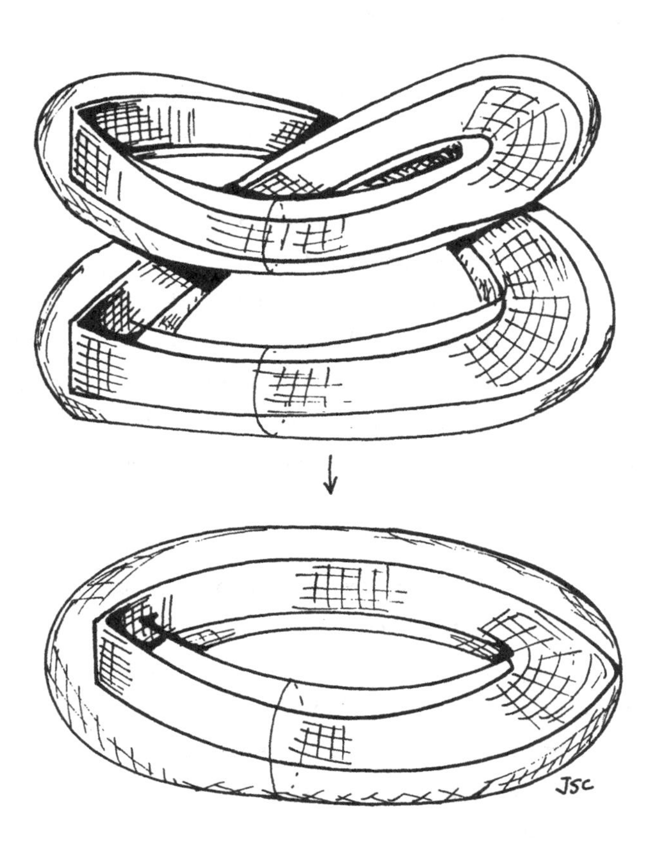

Figure 4.14: A neighborhood of a twisted annulus maps to a neighborhood of the Möbius band

answer for themselves. You have had all of a paragraph to think about it, so now I'll tell you the answer.

A **rigid motion of 3-space** that keeps a point in space fixed is determined by an axis and an angle of rotation about that axis. The axis runs through the fixed point, and the angle of rotation can be any number between $-180°$ and $180°$. But the result of a rotation through an angle of $-180°$ is the same as the result of rotation through an angle of $180°$.

Projective space is the quotient of the 3-dimensional sphere where diametrically opposite points are identified. Alternately, think of a ball in which the diametric points on the bounding sphere are glued together: All the points in the quotient space have a preimage on the inside ball of the 3-sphere, and the boundary points are the only ones in the ball that are mapped two-to-one. Thus projective space is a ball on which antipodal points on the boundary are identified. (Similarly, the projective plane is the quotient of the disk where opposite points on the boundary are identified.)

A point in the ball can be specified by an axis through the origin and a distance along that axis. (To specify the axis we need two numbers; these determine a direction vector in 3-space.) The correspondence between the set of rigid motions and the projective space now is easy. Given a rotation of, say, x degrees about an axis we go out a distance $x/180$ along the axis of rotation. If x is a negative number we go backward along the direction axis. The points $-180°$ and $180°$ are mapped to points on the unit sphere that are joined by a diameter of the sphere. These two angles determine the same rotation, but they also determine the same point in projective space. Figure 4.15 depicts the correspondence.

Loops in Projective Space

A fascinating aspect of the projective plane and of projective space concerns loops within these spaces. Put yourself into the mind of a Massey bug walking on a Möbius band in projective space. (Got a virtual reality machine handy?) On a journey once around the Möbius band left and right are switched, but so are top and bottom. So

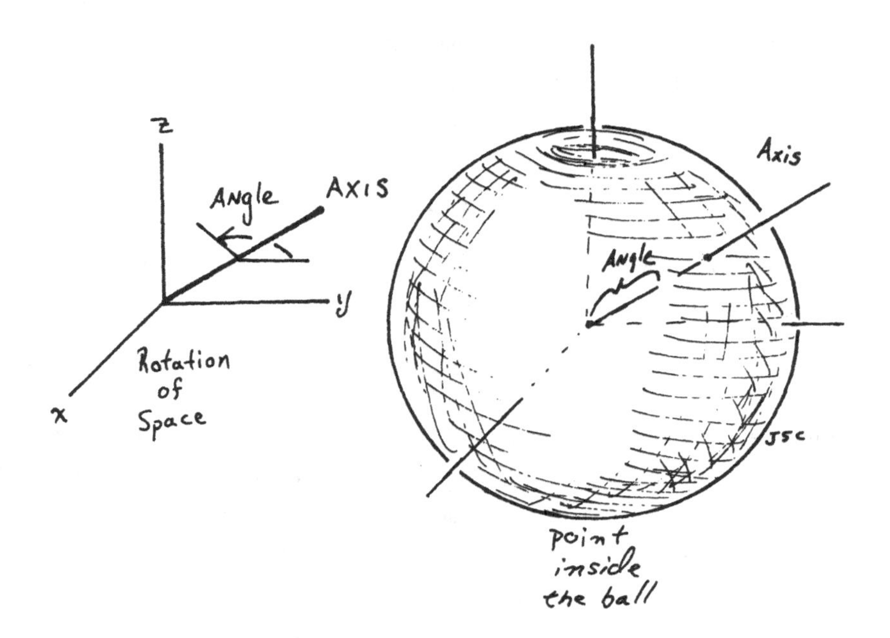

Figure 4.15: Rotations of space correspond to projective space

miraculously, this loop is not orientation reversing in the larger space. If on your journey you nail a piece of string down at your starting point and if you nail the other end down upon completing a second walk around the Möbius band, then the resulting string bounds a disk in projective space.

In other words, the core circle of the Möbius band does not bound a disk, but the boundary circle does, and a walk around the boundary circle is tantamount to walking twice around the core. In Lou Kauffman's book [52] this contractible loop is discussed in the context of rotations as the Dirac string trick and the belt trick.

⋆⋆⋆

This discussion of loops in the set of rotations is summarized with some high powered notation in the following equation:

$$\pi_1(SO(3)) = \mathbf{Z}/2.$$

The letters have the following meanings: π_1 denotes the set of loops in a space that are considered to be the same if they are the image of an annulus mapped into the space. $SO(3)$ is the proper name for the set of rotations of 3-space; the S is used to denote orientation preserving (special), and the O denotes orthogonal. $\mathbf{Z}/2$ is the set even/odd with addition table: even + even = even; even + odd = odd; odd + even = odd; and odd + odd = even. That is, loops in the set of rotations can be added, and their addition structure is just the same as even/odd. Once around the Möbius band represents odd, and twice represents even. The equation,"pi one of ess oh three is zee two" isn't as profound as $E = mc^2$, but it is as poetic, and it does have physical consequences.

4.4 Lens Spaces

In this section we discuss the general class of spaces that the projective space, the 3-sphere, and the product space $S \times C$ fit within. These are the **lens spaces;** they are the result of attaching a 2-handle and a 3-ball to a solid torus. The important datum that is used to define them is the attaching curve along the boundary of the torus.

Lens spaces can be completely classified, but we won't go into that classification in this book. Certain descriptions of lens spaces involve polyhedral lens shaped objects; that's where they got their name; the advanced text [70] contains that description. My purposes for describing them here are three-fold. First, they are pretty examples. Second, in many of these there are substantial self intersecting projective planes. Third, the triple points that the projective planes can have are related one of the movie moves of Chapter 4.

Substantial Circles on the Torus

The torus is usually described as the quotient of a square in which opposite edges are identified. Figure 4.16 shows how to get from that description to the basket shaped thingy and back.

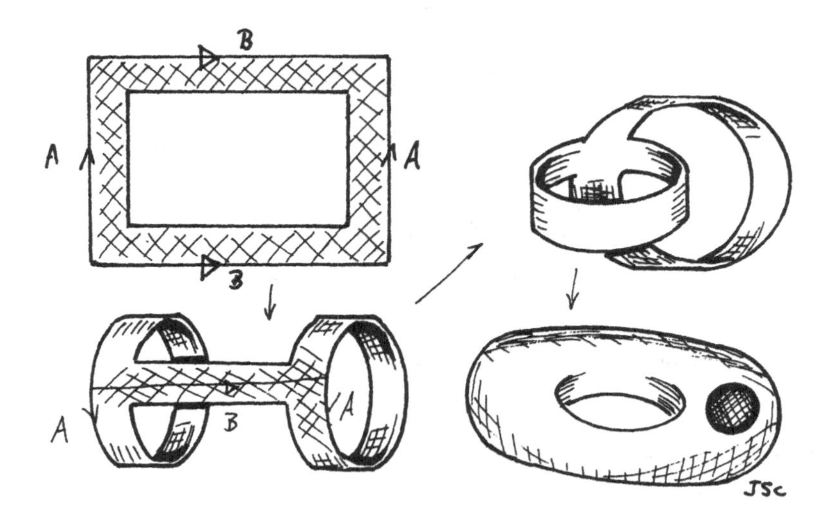

Figure 4.16: The torus as the quotient of the square

Suppose that p and q are a pair of whole numbers that have no common divisors except for 1. For example, the pair 5 and 3 are used in Figure 4.17. In that figure, 3 lines were drawn horizontally and 5 were drawn vertically along the square. These yield 3 longitudinal curves and 5 meridional curves on the torus. Replace each crossing point by a northeast detour as indicated. In the example depicted, a (5,3) curve results on the torus. In general, a (p, q) curve will result; this is a curve that wraps p times around the longitude and q times around the meridian.

Any simple closed curve that is substantial on the torus can be fitted to a (p, q) curve where p and q have no common divisors except 1. The indices only make sense within the context of a fixed meridian and a fixed longitude.

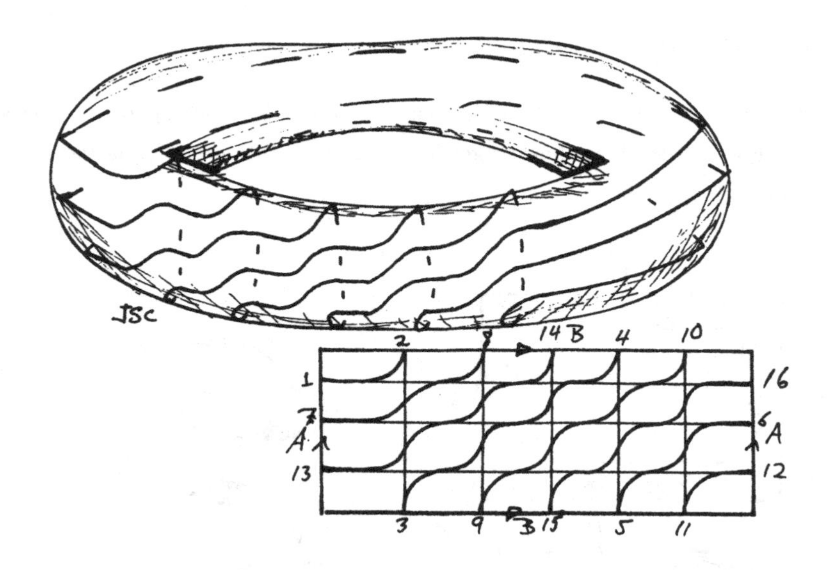

Figure 4.17: Constructing a (p, q) curve on the torus

The (p, q) lens space is constructed by attaching a 2-handle along the (p, q) curve and closing the boundary by attaching a 3-ball on the outside. In this way, the 3-sphere is a lens space of type (1,0). The product, $S \times C$, of a sphere and a circle is a lens space of type (0,1). And projective space is a lens space of type (2,1).

I encourage you to try the following experiment. Consider a pair of whole numbers, such as 4 and 6, that have a common divisor. Follow the procedure to draw a (4,6) curve on the torus. You should have found a pair of parallel (2,3) curves. In general, the number of curves that you find will be the common divisor of the whole numbers with which you started.

In the lens spaces, the 2-handle and the ball that are attached on the outside also form a solid torus: The top and bottom of the 2-handle are attached to two disks on the ball. All of the lens spaces, then, are the result of gluing together two solid tori along their boundary. Some distinctions between these spaces can be made by means of the indices (p, q). But according to the classification of lens spaces, not every distinct pair of indices gives rise to a new lens space.

(When we think of the outside piece of the lens space as the inside, we are performing a technical feat that is called turning a handle decomposition upside down. Believe it or not, the upside down move can be accomplished by turning this space over in some higher dimensional space. Chapter 7 says more about this. When a handle decomposition of a 3-dimensional space is turned upside down the 2-handles become 1-handles and the 1-handles become 2-handles.)

Projective Planes in Lens Spaces of Type $(2k, q)$

Lens spaces of type $(2k, q)$ contain embedded substantial surfaces that are non-orientable. Glen Bredon and John Wood [9] found a way of constructing a surface that has the smallest rank among all the embedded surfaces that are substantial in these spaces. On the other hand, there are always projective planes in these lens spaces that are substantial even if they are not always embedded.

To construct these projective planes, we'll leave the world of general position for a while. Let k be fixed. For the examples here, we'll use $k = 4$, but as usual it can be any number. Draw $2k$ equally spaced points on the boundary of a circle and join these in pairs by k diameters of the circle. Now take the Cartesian product of this asterisk and an interval. The result is k disks intersecting inside a 1-handle. Twist one end of this handle through an angle of $360°/q$. The figure has caduceus -like arcs

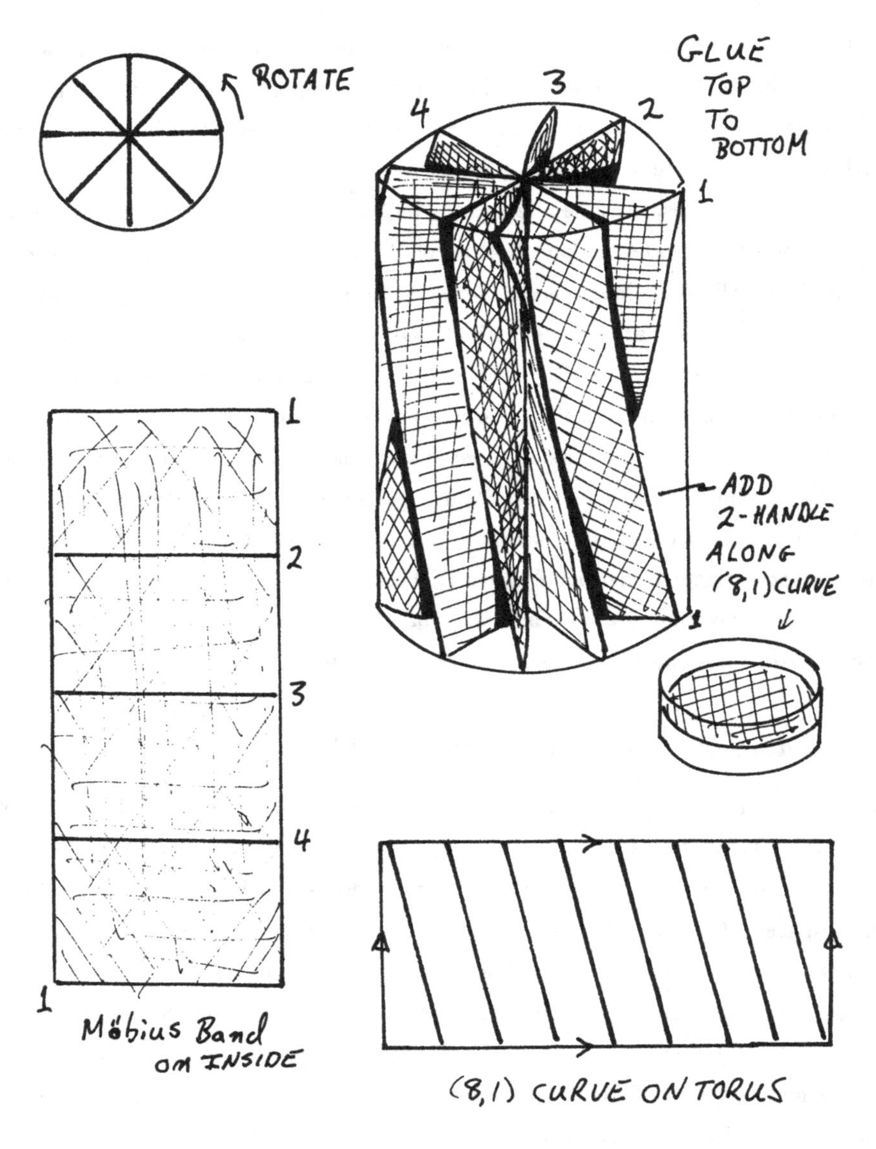

Figure 4.18: A projective plane in the lens space $L(8,1)$

spiraling up the sides of the cylinder. When the ends of the cylinder are identified, a $(2k, q)$ curve results on the boundary torus.

The inside intersecting disks glue end-to-end to create a Möbius band. To get the substantial projective plane, a 2-handle is glued to the outside. Figure 4.18 depicts the details.

Since we understand the embedded disk that is the core of the outside 2-handle, let's look more closely at the self intersecting Möbius band that appears on the inside. It is not in general position: Four sheets are intersecting along an arc. One way try to push the Möbius band into general position is to push the arcs in the asterisk into general position. In Figure 4.19, a movie shows that as the cross sectional disk is rotated, the direction in which these arcs are pushed changes. Consequently, the pushed-off arcs at the top and the arcs at the bottom don't fit tightly.

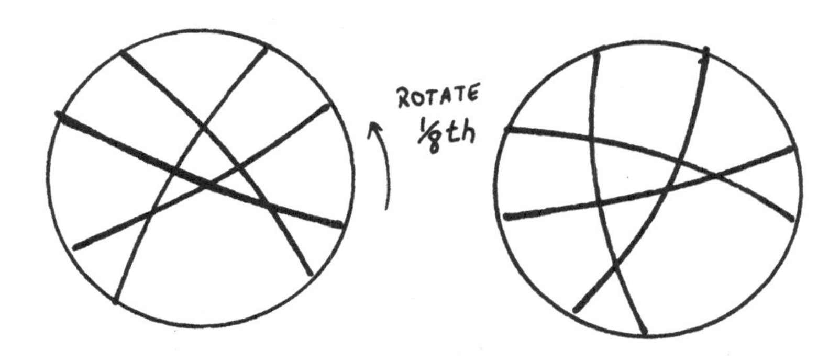

Figure 4.19: Pushing into general position without compensating for twisting

The situation is a bit like when square metal plate is screwed into place. If all of the screws on the left hand side are tightened before the screws on the right are inserted, then the screw holes on the right will be out of alignment. The intersecting disks are pushed too far in one direction at the top of the can, and so the bottom of the can is out of alignment.

To compensate for the misalignment, we allow the disks to intersect at three triple points. This will push the bottom back into place. Figure 4.20 indicates how this works.

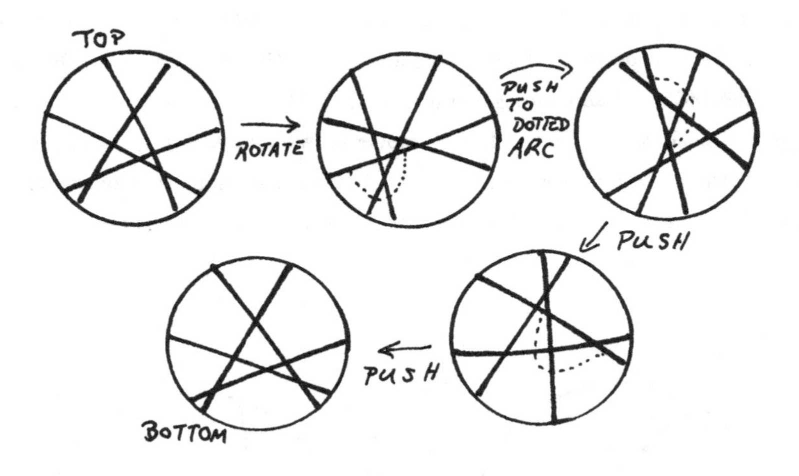

Figure 4.20: Three triple points causes realignment

⋆⋆⋆

There are two reasons that I find these projective planes interesting.

First, the paper by Bredon and Wood [9] describes substantial embedded surfaces in lens spaces of this type. (Such surfaces are necessarily non-orientable, but I won't tell you why we know that.) Among all the embedded substantial surfaces there is one of smallest rank. This smallest rank depends on a linear function of the index k. And so the rank of the surface does not grow very fast. On the other hand, I showed in [17] that the smallest number of triple points of a substantial projective plane depends on a quadratic function of k, and consequently must grow much faster.

Thurston has defined a way of measuring the size of a 3-dimensional space in terms of the rank of substantial orientable surfaces. The size of a space is a nebulous concept in topology, but Thurston finds ways to avoid such technicalities. David Gabai [38] proved that there are embedded orientable surfaces achieving the smallest rank among the substantial surfaces.

In the lens spaces, the substantial surfaces are non-orientable, and the projective plane is always of smallest rank. So a measurement of size in these cases is the minimal number of triple points that can be achieved. This notion of size is by no means as rigorous as is Thurston's notion. But it is a good rough estimate. Spaces in which there must be a lot of triple points seem small to me.

Second, one of the moves that is necessary to move surfaces around in 4-dimensional space involves diagrams similar to the cross sectional disks on the inside torus. There are higher dimensional analogues to this move and they are mimicked in the lens spaces. A similarity between diagrammatic formulations should not be ignored! There is usually a deeper connection between the ideas that are expressed in diagrams.

Substantial Subsets in Other Lens Spaces

Why did we need the first index of the lens space to be $2k$ when we constructed substantial projective planes? In the odd indexed lens spaces there are substantial sets, but they are not represented by *closed* surfaces. Let's construct these.

In a lens space of type $(5, 1)$ we'll construct a substantial set from the core disk on the outside 2-handle. On the inside the construction is similar to the asterisk. Take five equally spaced points on the circle and connect them to the center of a disk by radii. Then twist this star one fifth of a rotation gradually up a cylinder. When the ends of the cylinder are identified, a (5,1) curve along the boundary torus is constructed, and this bounds a substantial spiraling object inside the solid. The spiral is glued to the core of the outside 2-handle to form a substantial set in the lens space.

The substantial set that arises is the continuous image of a disk, but it is not the continuous image of a *closed* surface. Each closed surface is the quotient space of a disk where arcs on the boundary are identified in pairs. For example, the projective plane results from the upper and lower semi-circles being identified. So that in the lens space, $L(8,1)$, when an octagonal disk is mapped to the substantial set, the antipodal arcs on the disk are on the same diameters of the disk of the inside torus. In contrast, the substantial set in the lens space of type (5,1) has an odd number of arcs glued together along the core of the inside torus. So no matter how these arcs are identified, there is always one that is not mated. Figure 4.21 depicts the difference in these cases.

The construction outlined for $L(5,1)$ works for any lens space. As a result in a lens space of type (p,q), a loop that wraps p times around the center circle always bounds a disk. The disk can be thought of as a regular polygon that has p sides: a triangle, square, pentagon, hexagon, *etc.*

⋆⋆⋆

Consider the result of rotating a pentagon through an angle of 72°. Mark one of the vertices before rotating. Five such rotations will bring the polygon back to its original position, and this position is determined by the fact that the marked vertex is back at its original position. Any rotation that is a whole multiple of 72° will send the pentagon back to itself in a way that the vertices are in the correct cyclic order, but the marked vertex might be any of five places. Any whole multiple of 72° can be represented by a whole number in the set $\{0,1,2,3,4\}$. And the elements in this set can be added as if they were rotations. For example, $4 \oplus 3 = 2$ because a rotation through an angle of 288° plus a rotation through 216° is a rotation through 504° which is the same a rotation of 144°.

There is nothing special about the number five in the previous paragraph. And so the set of rotations of any regular polygon through a fundamental angle can be added and subtracted.

The set of loops (that start and end at the same point) in a lens space of type (p,q) can be added by first tracing one loop and then the next. Loops are considered to be the same if they are the boundary of the continuous image of an annulus. The

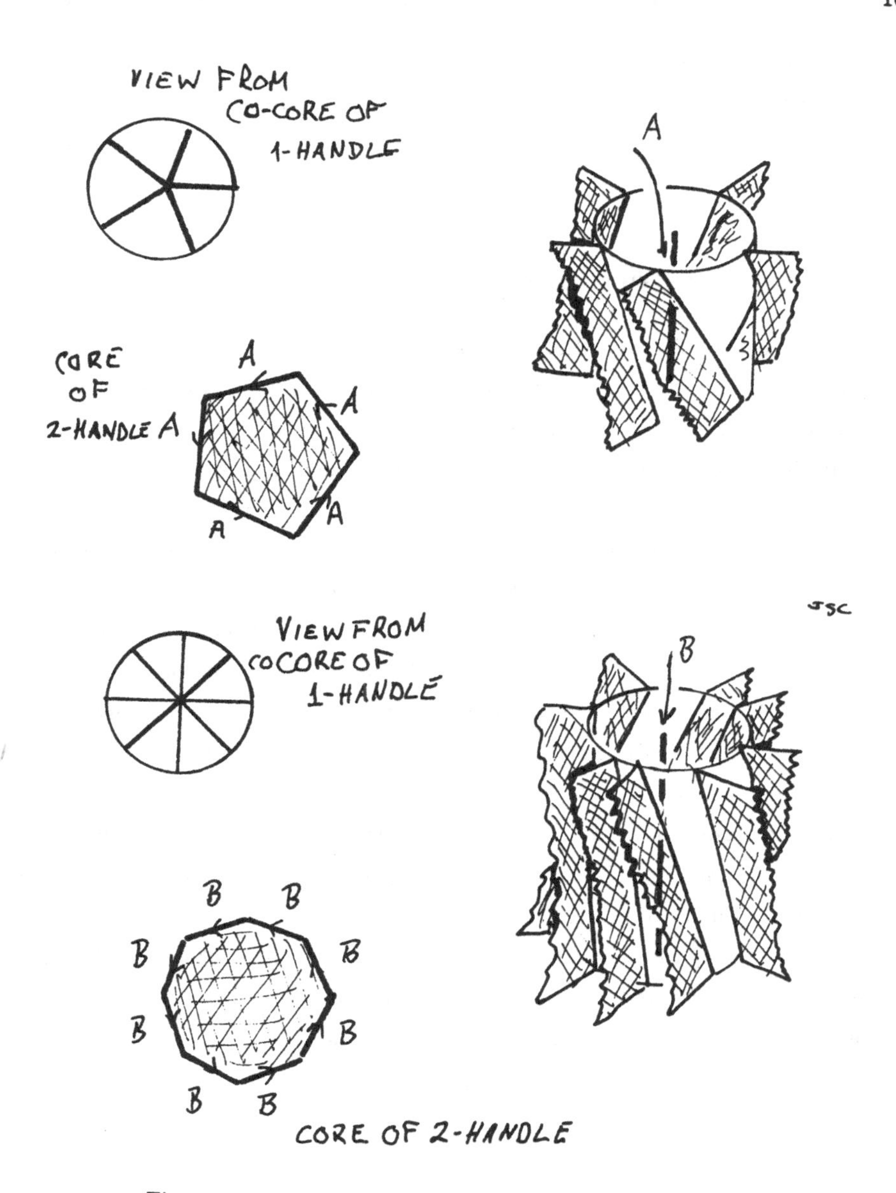

Figure 4.21: The substantial sets in $L(8,1)$ and $L(5,1)$

addition structure is the same as the set of rotations of a regular p-sided polygon. The similarity should be pretty clear. A generating loop is represented by the center circle of the inside torus, and p times this loop is the boundary of the substantial disk.

4.5 The Quaternions

Early in this century, Professor Hamilton came up with a new algebraic system called the quaternionic numbers. They were the result of a failed experiment. Hamilton thought he could get points in 3-dimensional space to multiply just as points on the line multiply as real numbers and points in the plane multiply as complex numbers. His experiment failed because it could not succeed. There is no such arithmetic structure on 3-dimensional space. However, there is almost such a structure in 4-dimensional space, and this is what Hamilton discovered.

The "almost" part of the discovery is this. The multiplication that he found was not commutative. That is, there are points A and B in 4-dimensional space such that $AB \neq BA$. But even though the operation of multiplication is not commutative, there is a notion of division, and the ability to divide by points in 4-space made Hamilton happy enough.

We will use Hamilton's quaternions to find a space that is intimately related to the 3-sphere. In fact, the space that we construct will be a quotient space of the 3-sphere that is similar to projective 3-space.

Reexamine the figure of four 2-spheres sitting in the 3-sphere where any three can intersect. The arcs of intersection are now labeled with letters **i**, **j** or **k**. These **unit quaternions** satisfy the following rules of multiplication:

$$\mathbf{i}^2 = \mathbf{j}^2 = \mathbf{k}^2 = -1,$$

$$\mathbf{ij} = \mathbf{k} = -\mathbf{ji},$$

$$\mathbf{jk} = \mathbf{i} = -\mathbf{kj},$$

$$\mathbf{ki} = \mathbf{j} = -\mathbf{ik}.$$

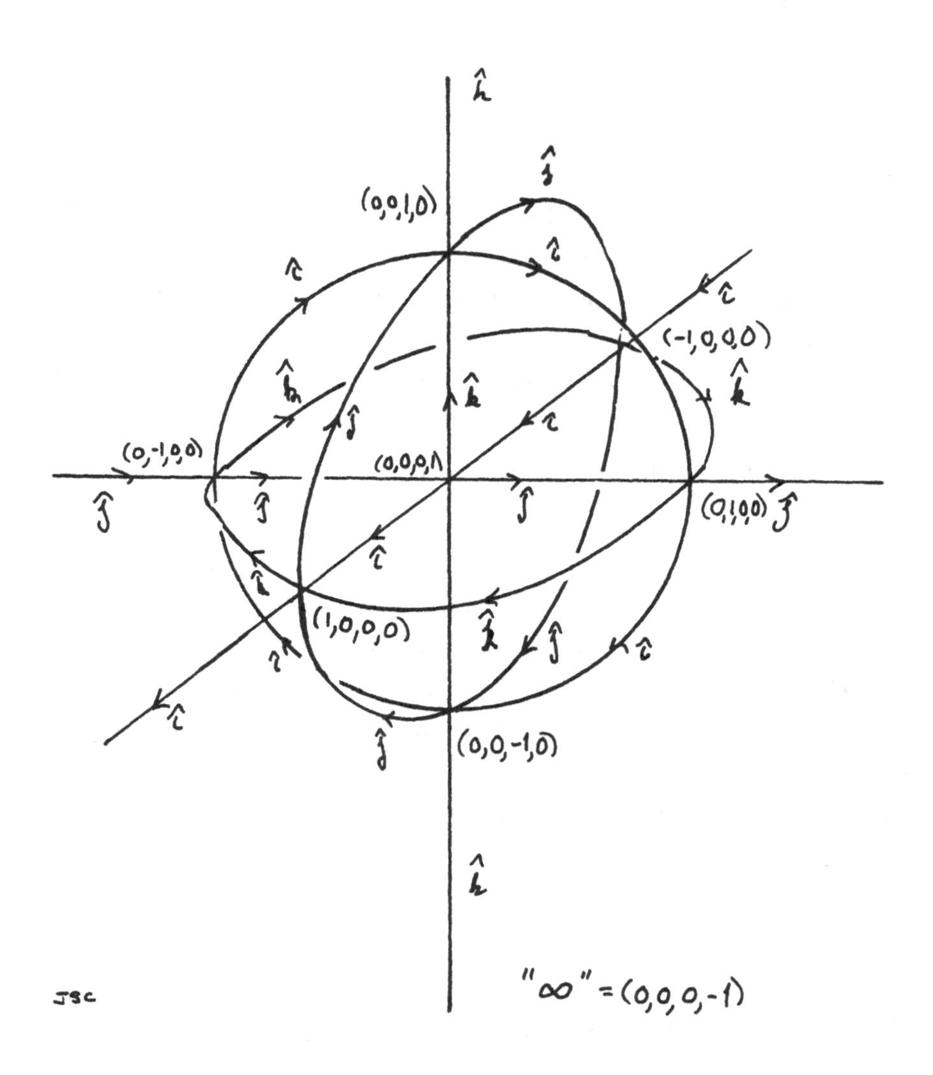

Figure 4.22: The quaternions in the sphere

Now observe that the multiplication rules are encoded by the labels that are on the double arcs. Specifically, if you read the boundary of any triangle in the 3-sphere you'll see one of the relations above. The point at infinity has coordinates $(0,0,0,-1)$, and any point with a single negative coordinate is reached from a positive coordinate by traveling along two arcs with the same label.

In the quotient space all eight triple points are identified to a single point. There are three 1-handles and these will have at their core, arcs with labels $\mathbf{i}$, $\mathbf{j}$, and $\mathbf{k}$. Inside, each such handle there are a pair of intersecting disks. The result of gluing these 1-handles to the ball is depicted in Figure 4.23. The closed curves on the boundary indicate the attaching regions of the 2-handles that contain the triangles. Finally, two balls of dimension 3 are added to the outside to create the solid. These two balls correspond to any two adjacent tetrahedral regions in the 3-sphere.

In Figure 4.24 the surface that is intersecting itself is shown to be a projective plane.

The set of loops in the quaternionic space can be added together as in the lens spaces. Actually, in this space we say that the loops are multiplied because the composition operation is not commutative ($AB \neq BA$). (For some reason, we are more likely to assume that addition is commutative than to assume that multiplication is. For example, matrix addition is commutative, but matrix multiplication is not.) The multiplication table for the set of loops is the same as above on the set $\{\pm 1, \pm\mathbf{i}, \pm\mathbf{j}, \pm\mathbf{k}\}$. because each of the relations above is represented by a triangular disk in the space.

4.6 Orientable 3-Dimensional Spaces

In this section, two general methods for constructing arbitrary 3-dimensional spaces are sketched.

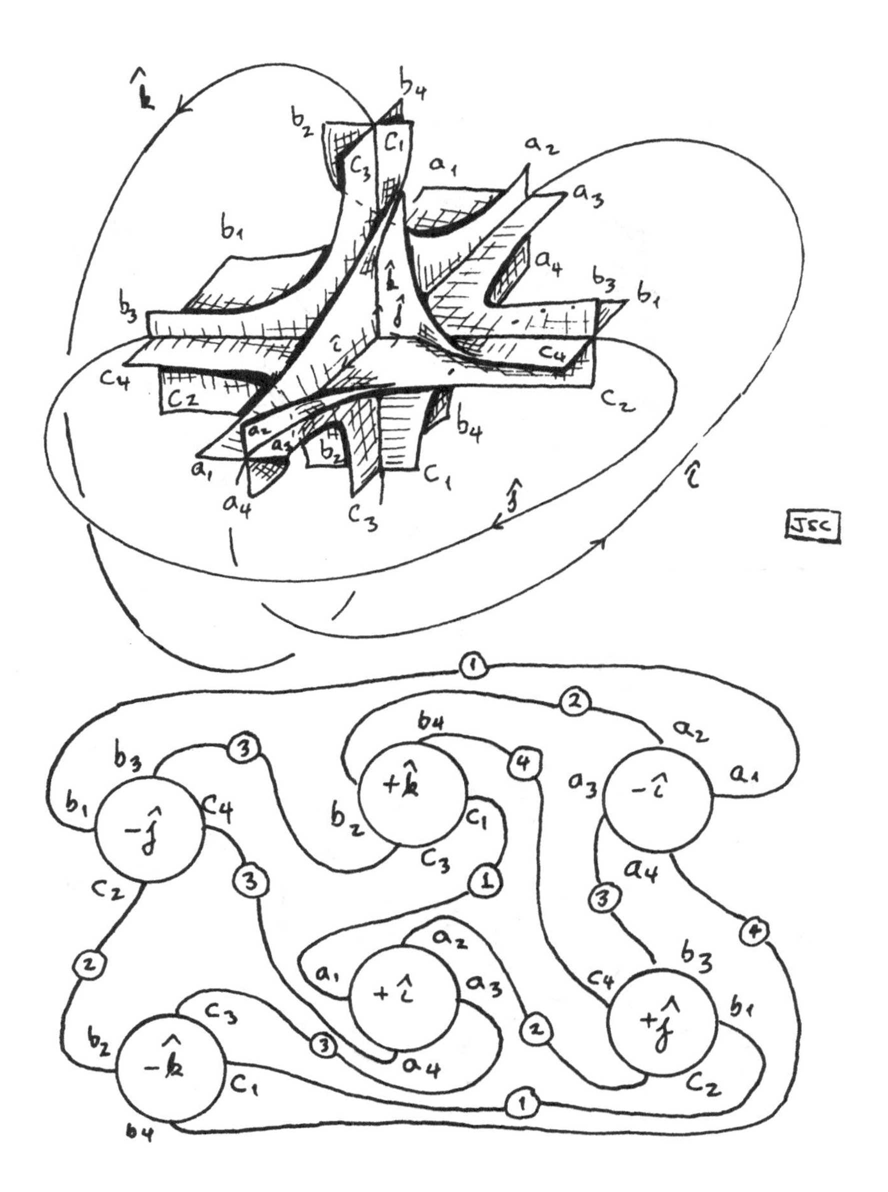

Figure 4.23: The 1-handles for the quaternionic space

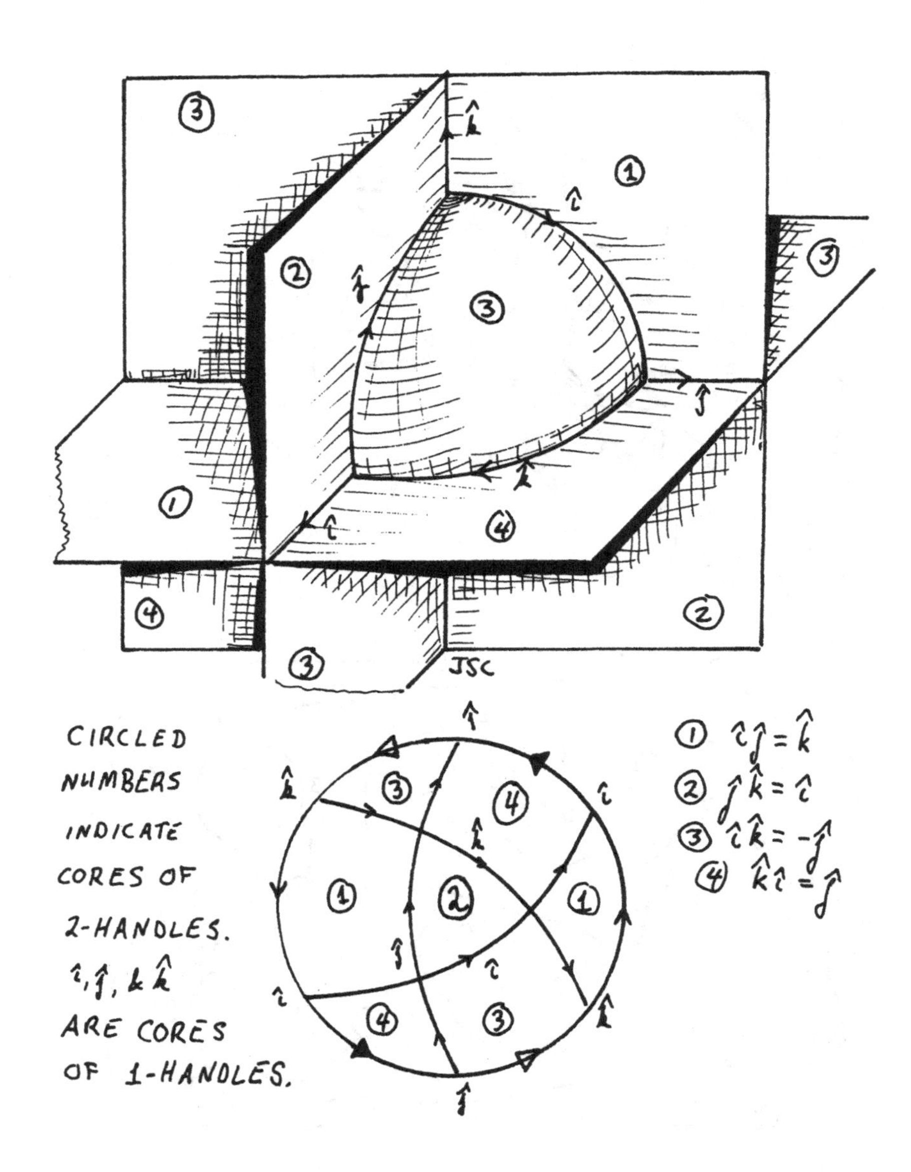

Figure 4.24: The map of the projective plane into the quaternionic space

Heegaard Diagrams

Consider a rank $2k$ closed surface. This surface is the boundary of a solid that has k holes. As an example consider the surface of a chair. The standard solid bounded by a genus k surface is called a **cube-with-handles**. It is the result of attaching k different 1-handles to a 3-ball. On the boundary of this surface draw a collection of non-intersecting simple curves, and consider these as the attaching regions of a collection of 2-handles.

Recall that a simple closed curve determines the attaching region of a core of a 2-handle, and the annular side of a 2-handle is glued to an annular neighborhood of the attaching curve on the cube-with-handles. The boundary of the solid that results is likely to have several components each of which is a closed orientable surface. Often when such a diagram is given, the spherical boundary components are capped with 3-balls, and the closing of the spherical portion of the boundary is completed tacitly.

If the number of attaching curves is the same as the genus of the surface and each curve is substantial, then the bounding surface that results is a 2-sphere. In this case especially, we attach a 3-ball along the sphere to get a closed solid.

A **Heegaard diagram** is the diagram that consists of a cube-with-handles together with a non-intersecting set of simple closed curves on the boundary. *Every orientable 3-dimensional solid can be described by means of a Heegaard diagram.* Even though every 3-dimensional solid has such a description, this result is a far cry from a classification theorem.

The Heegaard diagram is isn't used in a classification scheme for three main reasons. First, it is possible to give very complicated diagrams of very high genus of rather simple manifolds. Although it's really not too complicated, Figure 4.25 indicates a high genus Heegaard diagram of the 3-sphere. Second, although there are many experts in the calculus of sliding handles, there isn't really a standard way of diagraming every solid. Third, even though there is agreement on the meaning of a minimal genus Heegaard diagram, a given solid can have more than one miminal genus Heegaard splitting.

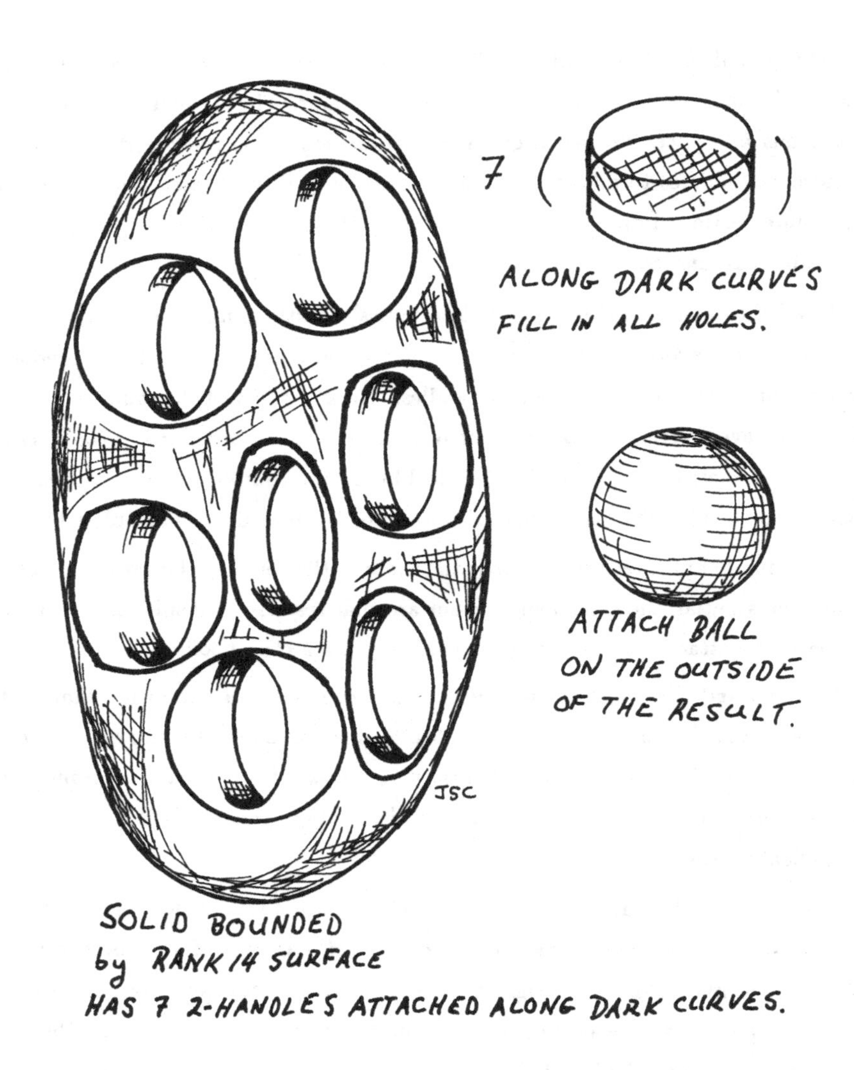

Figure 4.25: A Heegaard splitting of the 3-sphere

Gluing Together Knot Spaces

A knot in the 3-sphere is a simple loop in that space. The boundary of a small neighborhood of that loop is a torus. The torus, then, is the boundary of the complement of the knot, or the boundary of what is not the knot. Two knot complements can be glued together along their boundaries, and a closed solid results. This technique can be extended to allowing links glued together on a variety of toriodal fronts. Yet still no classification theorem results from such descriptions.

The hope of the Thurston program is to cut a given solid open along tori such as these and to recognize the individual pieces by means of their geometric properties. The key property that a toriodal knot boundary has is that loops in the torus do not bound disks in the complement of the knot. Furthermore, in a technical sense most knots have a hyperbolic geometric structure (constant negative curvature). It turns out that solids with a hyperbolic structure can be classified.

A Non-orientable Solid

The Klein bottle is the boundary of a solid that has a Heegaard-like description. Namely, we attach a 2-handle along the meridional curve on the Klein bottle. This curve is not orientation reversing in the Klein bottle. A 3-ball is attached along the boundary of the result. The 3-dimensional solid that results is called the **solid Klein bottle**. It has a Klein bottle as its boundary. To get a closed solid, two of these are glued along their boundaries in the same way that two solid tori were glued to get $S \times C$. The resulting space is called the twisted sphere bundle over the circle.

This space is mentioned only to indicate that there are non-orientable solids as well surfaces.

Constructing Spaces Equipped with Surfaces

Given a self intersecting surface in a space, there is a natural way to decompose the larger space in terms of handles. Each triple point appears inside a ball. An arc of double points forms the core of a 1-handle in which a pair of perpendicular sheets intersect. The core of a 2-handle is an embedded disk. And a neighborhood of a branch point is a 3-ball.

If there are no triple points, then a neighborhood of a double point can be considered to be a 3-ball rather than a 1-handle. If the surface has rank outside of the intersecting set, then the core of a 1-handle can be chosen to be a substantial arc in the surface without its self intersections. Finally, if there are no intersections, as in the case of the projective plane in projective 3-space, then a 3-ball can be found that contains an embedded disk. These situations are all depicted in Figure 4.26.

Therefore, a self intersecting surface in a 3-dimensional solid can be used to give a handle decomposition of that solid. Furthermore, a handle decomposition is the same as a Heegaard diagram. If the surface has a lot of self intersections, then the genus of the Heegaard splitting will be very large, at least as large as the number of loops in the double point set. For example, Boy's surface gives rise to a genus 3 Heegaard diagram in the 3-sphere, as does the projective plane inside the quaternionic space.

The Heegaard splittings that come from these surfaces are not intrinsically useful. But I find them pleasant because I can think of the surfaces inside the solid as walls, floors, and ceilings. They provide me with the comforts of home in unfamiliar surroundings.

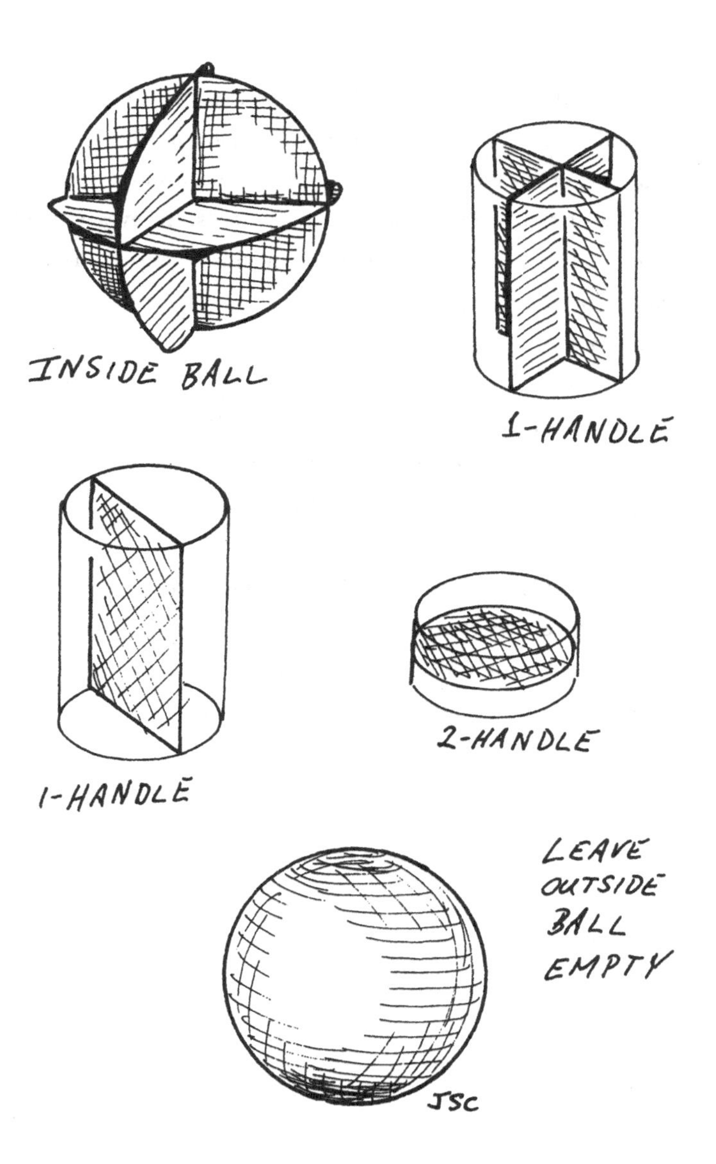

Figure 4.26: Handles and surfaces within

4.7 Notes

We have barely scratched the surface of 3-dimensional spaces. These are properly known as **3-manifolds**. An n-dimensional manifold is space in which each point has a neighborhood that looks like ordinary n-space. In dimensions four and greater it is necessary to specify what "looks like" means. In other words, the eyes with which you see these spaces might be the eyes of differential calculus where the neighborhoods have tangent planes, or they might be the eyes of topology where all notions of smoothness are irrelevant.

The point of these examples is mainly to indicate that there are realistic worlds other than our own, and in these worlds the ordinary ideas of separation that you might intuit do not hold true. If you are looking for a mystical reality in which you can hang your hat, you don't have to look much further than the examples presented here.

Jeff Weeks gives the following exercise: Pick a 3-dimensional example that you have read about, and write a short story about life in this space.

Further readings should include Milnor's book [59] , Armstrong's book [4], Hempel's book [44], Jaco's book [49], and Thurston's lecture notes [72].

Chapter 5

Relationships

In this chapter, I will show you some results of Cromwell, Izumiya, and Marar. Izumiya and Marar [48] discovered a formula that relates the rank of a surface to the numbers of triple points and branch points when the surface intersects itself in any 3-dimensional space. Cromwell and Marar [32] gave a list of all the fundamentally different surfaces in 3-space that have one triple point and six branch points occurring at the ends of the arcs that intersect at the triple point.

The triple point formula of Marar and Izumiya escaped me for many years. In fact, Ki Hyoung Ko and I [55] came very close to proving their result in our paper, but we only realized that the formula was true up to parity. The **parity** of a number is its evenness or oddness. We showed that certain numbers when added together would give an even number. Marar and Izumiya showed that these same numbers when added (and subtracted) appropriately would give 0.

The result of Cromwell and Marar has to do with the surfaces in ordinary 3-space that have exactly one triple point. Remember that three arcs of double points intersect at a triple point. Cromwell and Marar ask about the situation in which each double arc has a branch point at both of its ends. They give a complete list of these surfaces.

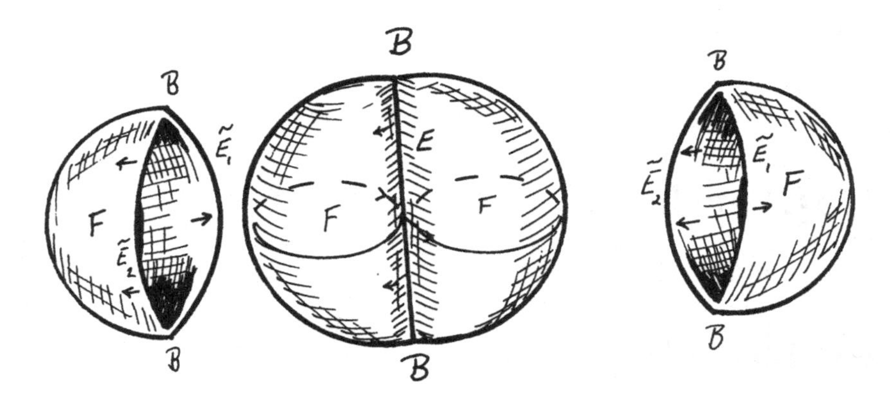

Figure 5.1: A sphere in 3-space with two branch points

5.1 A Triple Point Formula

Before the Izumiya-Marar triple point formula is stated, some new definitions and explanations are needed. The formula relates the rank of a surface to the number of triple points and branch points of a generic map. One might suspect that there is such a formula because Boy's surface (which is the generic image of a projective plane) has a single triple point, and the cross-cap map has none but does have branch points. The formula must have terms to take care of the map of a sphere with two branch points that is illustrated in Figure 5.1.

Rank Revisited

Remember from Chapters 1 and 2, that the rank of a closed surface (a surface that has no pin-holes, no boundary, and that can fit in a bounded region of some space) is the number of non-intersecting substantial arcs that are found when a disk is removed from the surface. It will be convenient in this chapter to define **the Euler number** of the surface to be two minus the rank of the surface. The Euler number is denoted by the Greek letter χ, and we have for a closed surface S,

$$\chi(S) = 2 - \text{Rank}(S).$$

The Euler number can be thought of in terms of handles as follows. A closed surface can be made from a disk, by attaching a collection of handles and then by adding disks to the resulting boundary. You might be tempted to think that the rank of the surface is the number of handles attached, but this thinking is not quite right because we want the rank of the closed surface. The disk with handles might have more than one boundary circle, and as the number of boundary circles goes up so does the rank of the surface with boundary. But handles that add to the number of boundary circles do not contribute to the rank of the **closed** surface since these new circles are removed by adding disks. We compute the Euler number by counting all of the handles (whether or not they contribute to the rank of the closed surface), and we subtract this number from the total number of disks added. The disks include the initial disk (onto which the handles are attached) and the disks added "outside" to close the boundary. In this way if a handle does not contribute to the rank, then its arithmetic contribution to the Euler number is canceled by the contribution of the disk that kills its boundary.

Another way to conceive of the surface is to consider it constructed as a (golden) thread with fabric (of golden fleece) draped over and attached to the thread. A physical model of the thread is the wire at the center of a twist-tie. The plastic or paper surrounding this wire is like the fabric that is draped over the thread. On the abstract surface, we take the cores of the handles to be like the thread and extend them by arcs to the center of the inside disk (Figure 5.2). The fabric corresponds to the disks that are added to the boundary of the collection of handles attached to a single disk. In this way, every surface can be constructed from a single vertex, a collection of arcs (corresponding to each handle), and a collection of golden fleece disks. The Euler number is the number of vertices, V, minus the number of arcs (edges), E, plus the number of disks (faces), F.

If we are given a closed surface (without a specific decomposition in mind), we can compute the Euler number as follows. We mark a set of points (vertices) on the

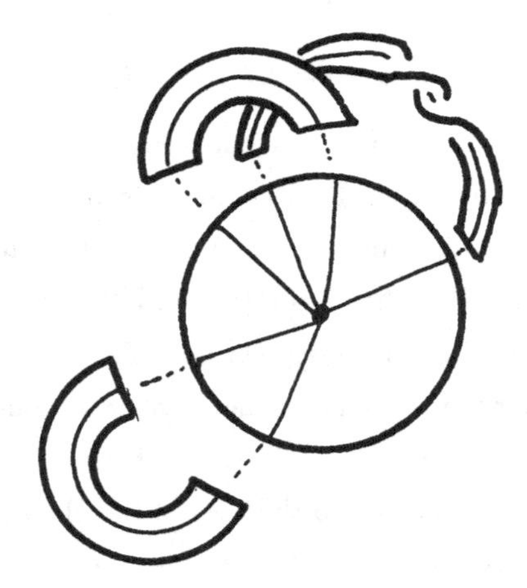

Figure 5.2: Making a surface from thread and fleece

connected surface (one is enough, but we may choose many), and connect these by arcs (edges) in such a way that the surface that results from removing this network is a collection of disks (faces). The Euler number is

$$\chi(S) = V - E + F.$$

The Euler number does not depend on the decomposition of the surface into vertices, edges, and faces. We'll prove this important fact by showing that the number computed from such a decomposition is always two minus the rank.

First, we find a large subset in the network that functions as the disk of the surface onto which handles are attached. The collection of vertices and edges form a connected network. Starting from a vertex, we include edges in such a way that if we include any more edges a loop arises. In this way we construct **a maximal tree**: a collection of vertices connected by edges such that there are no closed loops in the total set. It may be that the network in the surface consists of loops all joined to a single vertex. If so, then the maximal tree is that single vertex. If there is an edge that begins at the starting vertex and ends at another vertex (that is not already part of the tree), then this edge and its other vertex are added to the tree. Edges and their other ends are added to the tree until no more can be added; that's what makes the tree maximal. It's a "tree" because it has no branches with both ends on the trunk — *i.e.* it has no loops.

Now each time an edge is added to the tree, a vertex is also added. So the Euler number of a tree, $(V - E)$, is always 1. A neighborhood of the maximal tree in the surface is a disk since the surface can be thought of as being draped over the network in question. This is the disk onto which the remaining handles are attached. Also every vertex of the network is already in the tree because if a vertex weren't in the tree, we could join it to the tree by adding edges without introducing loops.

The remaining edges in the network act as cores of handles since their ending vertices are found in the maximal tree. So the Euler number of a network of vertices and edges in a surface is 1 minus the number of handles or equivalently one minus the number of edges not found in the maximal tree. The number of handles is the rank of the surface with boundary that is a neighborhood of the network in the surface.

On the other hand, the rank of a closed surface is the rank of the surface that results when all but one boundary component is closed by adding disks. So the rank of a closed surface that results when the multiple boundary components are sewn away is the number of handles minus the disks added plus 1. Let's do that computation explicitly:
Rank = # of handles − (# of outside disks −1) = # of handles − # of outside disks +1). Moving terms around in this last equation, we get
# of handles = Rank +# of outside disks −1.

The Euler number of the closed surface is the Euler number of the network plus the number of outside disks:

$$\chi(F) = 1 - \#\text{of handles} + \#\text{of faces}$$

$$= 1 - (\text{rank} + \#\text{of faces} - 1) + \#\text{of faces}$$

$$= 1 - (\text{rank} - 1) = 2 - \text{rank}.$$

In conclusion, the Euler number of the surface is 2 minus the rank of the surface, and this can be computed as $V - E + F$ for any decomposition of the surface into vertices edges and faces. The $V - E + F$ formula will be used subsequently to define the Euler number for the image of a self-intersecting surface.

Just as the rank and orientability can be used to classify closed surfaces, so can the Euler number and orientability. Observe that the largest possible value for the Euler number is 2, and this value is achieved on a sphere which can be constructed without handles. The Euler number of the projective plane is 1. The Euler number of a Klein bottle or a torus is 0. If the Euler number is odd, then the surface is non-orientable.

The Euler Number of a Generic Intersecting Surface

The Euler number of a surface is an intrinsically defined quantity. What I mean by that sentence is that the Euler number does not depend on how the surface sits inside any particular space, but it only depends on the topological nature of the surface. In this section, we will consider generically intersecting surfaces in an arbitrary 3-dimensional space, and we will assign an Euler number to the image surface. Hence, there will be two distinct Euler numbers considered here: The first is the Euler number of the intrinsic surface, the second is the Euler number of the image surface. The second quantity measures some aspects of the intersecting surface. The exact measurements are given by the Izumiya-Marar formula.

For the surface to be intersecting generically, each point must have a neighborhood in the image in which the surface is either embedded, looks like the intersection of 2 or 3 coordinate planes, or looks like a cone on a figure 8. Figure 1.29 contains illustrations of these situations.

In Chapter 4, many examples of 3-dimensional spaces that contained self-intersecting surface were presented. Usually these were given in general position. In the case of the lens spaces, I showed self-intersecting projective planes with a single arc of multiple points. These too were put into general position by pushing certain diameters off of each other and compensating for mis-alignment by introducing triple points.

A generically self-intersecting surface is the image of a certain continuous function f defined with its domain the surface S and its range some 3-dimensional space N. We can consider the image surface, $f(S)$, to be constructed from a fabric with vertices (like buttons or snaps), edges (zippers or seams), and disks (swatches of golden fleece). The intrinsic surface is constructed in a similar fashion, but on the intersecting surface we may sew the disks multiply around the edges that are arcs of double points. Similarly branch points and triple points can be thought of as being among the buttons.

The **Euler number of** $f(S)$ is again defined as $V - E + F$ where V is the number of vertices, E is the number of edges, and F is the number of faces (or disks) used in the construction. The difference between the Euler number of S and $f(S)$ is that the image surface $f(S)$ may employ fewer edges and vertices in its construction since the triple points and arcs of double points can function multiply.

Let's review some examples before we continue. If f is an embedding (so $f(S)$ has no double points, branch points, or triple points), then the Euler number of S and of $f(S)$ are the same. The Euler number of a sphere is 2 since the sphere is made from two disks (or one vertex, one edge, and two disks). Meanwhile the Euler number of the self-intersecting sphere in Figure 5.1 is $2 - 1 + 2 = 3$: There are two branch points that function as vertices, one arc of double points that functions as an edge, and two disks — one on the left and one on the right of the double arc. In the cross-cap map of the projective plane, there are 2 branch points, one double arc, and one disk attached to the double arc. So the Euler number of the cross-cap is $2 - 1 + 1 = 2$. The Euler number of Boy's surface is $1 - 3 + 4 = 2$: The one vertex is the triple point, three arcs of double points end there, and four disks are attached.

The Result

Theorem [48]. *Let $f : S \to N$ denote a generic map from a closed surface S to a 3-dimensional manifold N. Let $T(f)$ denote the number of triple points of f, let $B(f)$ denote the number of branch points of f. The numbers $\chi(S)$ and $\chi(f(S))$ denote the Euler numbers of S and of the image $f(S)$ respectively. Then*

$$\chi(f(S)) = \chi(S) + T(f) + B(f)/2.$$

For the examples given before the statement of the Theorem, we see that the Theorem holds. By the way, whenever you are presented with the statement of a theorem, you should check that the theorem holds on the examples with which you are familiar. There is nothing so satisfying as finding a simple counter-example (even when you or one of your friends claims to have proven the theorem), nor is there anything so frustrating as a proof without a guiding example. The Izumiya-Marar Theorem holds; we just checked some examples, now let's check the proof.

In the case that f is an embedding the formula holds trivially. A decomposition of the image surface into vertices, edges, and faces yields the same decomposition of the intrinsic surface. So this case is taken care of.

Let's examine a case that is quite the opposite. Namely, consider a generic map, f, of S for which the set of vertices is the same as the set of triple points plus the set of branch points, the set of edges are the double point arcs, and the set of faces, are the remaining sheets of the surface. Each of the surfaces we reviewed above can serve as an example, and for a surface with both triple points and branch points, consider the Roman surface constructed in Chapter 2.

From the decomposition of the image surface, $f(S)$, we construct a decomposition of the intrinsic surface S. Each branch point in the image, can be considered to be a vertex on the intrinsic surface since the lift of a branch to the double decker set is a single point. Each triple point of the image, $f(S)$, lifts to three distinct triple points in the double decker set (Figure 2.8 contains an illustration.) Each double arc of $f(S)$ lifts to a pair of edges in S. Meanwhile, the faces of $f(S)$ are the faces of S.

At the risk of introducing an alphabet soup of notation, I will call $V(S)$, $E(S)$, and Face(S) the number of vertices, edges, and faces of S, respectively, and $V(f(S))$, $E(f(S))$, and Face($f(S)$) the respective quantities of the image surface, $f(S)$. Now we have to do some arithmetic. From the previous paragraph we know:

$$V(S) = 3T(f) + B(f)$$

(the vertices of S are the preimages of the triple points and the branch points),

$$V(f(S)) = T(f) + B(f),$$

(the vertices of $f(S)$ are the triple points and branch points),

$$E(S) = 2E(f(S))$$

(the edges of S are arcs of double decker points), and

$$\text{Face}(S) = \text{Face}(f(S))$$

(the faces of the intrinsic and image surfaces coincide).

Let's measure the difference between $\chi(f(S))$ and $\chi(S)$.

$$\begin{aligned}\chi(f(S)) - \chi(S) &= V(f(S)) - V(S) - (E(f(S)) - E(S)) + \text{Face}(f(S)) - \text{Face}(S) \\ &= T(f) + B(f) - (3T(f) + B(f)) - (E(f(S)) - 2E(f(S))) \\ &= E(f(S)) - 2(T(f)).\end{aligned}$$

Each edge of $f(S)$ has two ends that can be found among the triple points and branch points. Six edges end at a given triple point, and one edge ends at a branch point. This means that $E(f(S))$ is half of the number of its ends, and this number is $3T(f) + B(f)/2$. Therefore, we have

$$\chi(f(S)) - \chi(S) = E(f(S)) - 2(T(f)) = T(f) + B(f)/2.$$

We have demonstrated the Izumiya-Marar result in two important special cases: (1) in case the surface S is embedded, and (2) the case in which the vertices and edges of the image $f(S)$ can all be found among the double, triple, and branch points of the surface.

The final case to consider is the case in which the double, triple, and branch points do not give a subdivision of the surface or its image. We can include the multiple points among the vertices edges and faces, but we may have to add some more to this set.

A good example to understand is the projective plane in a lens space $L(4,1)$. This example is one constructed in Chapter 4. It has a simple loop of double points. Label one of the double points a vertex. Then there is a single edge that starts and ends at that vertex. Also there is a face wrapped four times around the edge. (OK, the face is wrapped around the outside of the solid torus that contains an immersed Möbius band, but we can gradually add fabric to the outside disk until it is adjacent to the core of the handle.) The vertex on the image surface lifts to two vertices in the preimage, and the single edge lifts to two edges that form (with the lifted vertices) a circle in the projective plane. The Euler number of the image surface is $1 - 1 + 1 = 1$ because it has one vertex, one edge, and one face. The Euler number of the projective plane is also 1, and we have divided it into two vertices, two edges, and one face.

More generally, for each simple loop of double points, one adds a vertex on that loop. Then there is one edge on the image surface, and the net contribution, one vertex minus one edge, to the Euler number of the image surface is zero. On the surface that is being mapped, the one vertex in the image is lifted to two vertices, and the one edge on the double point set is also lifted to two edges on the abstract surface. But these lifted vertices and edges do not contribute to the Euler number because twice zero is still zero.

In fact, if it becomes necessary to insert an extra vertex anywhere along a double curve, then the contribution of added vertices and edges in both the image and the abstract surface is zero. Why? Because the vertex lifts to two vertices in the intrinsic surface, but there are also two new arcs in the intrinsic surface. So on both the intrinsic surface and the image surface, the contribution of vertices and edges cancels.

So we start once again with a generic surface in any 3-dimensional space, and we begin to subdivide the surface into vertices, edges, and faces in such a way that the branch points and triple points are among the vertices, and arcs of double points are among the edges. If necessary, vertices are added on the double point set, and these do not affect the Izumiya-Marar formula. Finally, we may need to add some more vertices and edges outside of the set of double points so that the rest of the surface consists of disks. For example on Boy's coffee cup, the cup handle must be further subdivided. But a vertex or edge added outside of the double points on the image is a single vertex or edge on the intrinsic surface just as a face away from the double and triple points is a face on both image and intrinsic surfaces. The net contribution of these extra vertices and edges to the difference $\chi(f(S)) - \chi(S)$ is 0.

Let's summarize the proof. In the extreme case of an embedding, the formula is trivial because vertices, edges, and faces on the image are the corresponding quantities on the intrinsic surface. In the other extreme when the vertices and edges are found among the multiple points, the formula holds by simple arithmetic — we count the number of vertices and edges in the intrinsic surface based on the number on the image surface. In the intermediate case, the multiple and branch points can be included among the vertices and edges but other vertices and edges might need to be added. Those that are outside of the multiple points don't affect the formula because they are akin to the first case. Those vertices in the double point set don't affect the formulas because they add the same number of vertices and edges.

So we have the Izumiya-Marar formula in all cases. Even though the formula is not very difficult to prove, it was remarkably difficult to discover. In 1974 Banchoff [5], proved a "folk theorem" that says the parity of the triple points of an immersed surface in 3-space is the same as the parity of the Euler number. Since Banchoff calls it a folk theorem, we may assume the result had been known for a while before 1974. Many people, including me, spent some time trying to generalize that result with marginal success. Banchoff's proof also applied to certain surfaces that were not in ordinary 3-space, but in some other space, provided the surface was not substantial in that space. My proof with Ki Hyoung Ko considered surfaces immersed in any possible 3-space, but our result only counted things up to parity and we did not allow branch points.

5.2 Surfaces with One Triple Point and Six Branch Points

One thing that impresses me with the Cromwell-Marar [32] surfaces that will be presented in this section is their inherent simplicity and beauty. In their paper, they

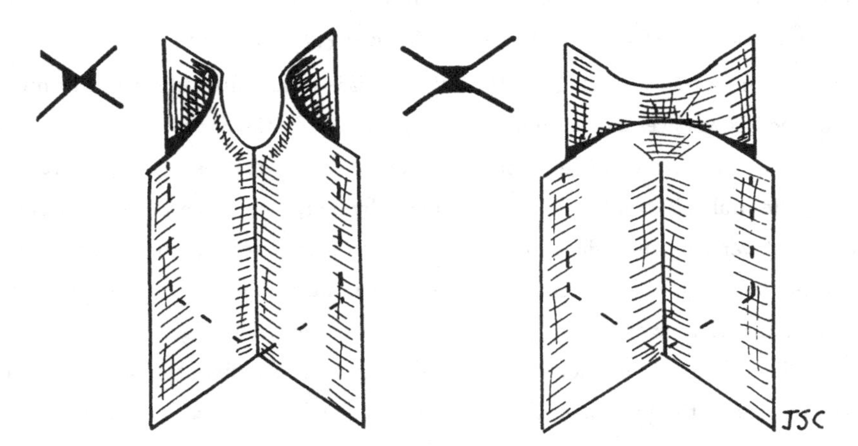

Figure 5.3: Arcs of double points ending at branch points

had asked (among other things), "What are the possible surfaces in 3-space that have one triple point, and six branch points?"

The triple point is the intersection of three arcs of double points that can be taken to be arcs along the coordinate axes. To fix our ideas here, each such arc is the set of points in which either the x, y or z coordinate ranges between plus and minus 1. So there are six end points to this set of arcs, and each arc is to be a branch point of a surface that has no further singularities.

Figure 5.3 illustrates an arc of double points and two possible ways of ending the top of the arc with branch points. Of course these two ways are specified in reference to some fixed coordinates, but the arcs that we want to end at branch points are fixed arcs along coordinate axes. Figure 5.4 illustrates that each arc ends in a pair of branch points in such a way that a neighborhood of the double decker arc in the intrinsic surface is either a Möbius band or an annulus. There are two ways to get a Möbius band and two ways to get an annulus.

Since there are 3 coordinate arcs, each has two ends, and each end can end in branch points in one of 2 ways, we should consider $2^6 = 64$ possible configurations of branch points. Some of these configurations will result in the same surface, and

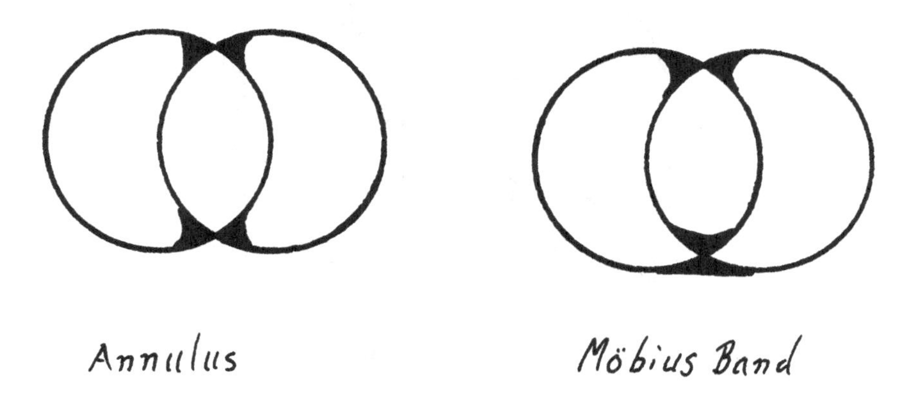

Figure 5.4: A Möbius band and an annular neighborhood of a branch point

we will need a method of detecting when these surfaces are the same. Meanwhile, we should find a way of making sure that no further double and triple points are necessary to complete the surface.

Look at the boundary of a neighborhood of a triple point. As mentioned before, since the triple point can be thought of as the intersection of round disks, the boundary is three great circles on the sphere. One circle corresponds to the prime meridian joined to the "ideal" international dateline, another is the equator, and a third is a meridian 90° west of the prime meridian. (The international dateline is approximately 180^0 from the prime meridian, but it is topographically not a great semi-circle because of political considerations. It is a great semi-circle topologically.)

For the current discussion, we will flatten these three great circles to three circles in the plane via stereographic projection, as in Chapter 4. Then the circles become the projection of Borromean rings on the plane. The branch points that are added can be thought of as smoothings of the double points on the boundary as in Section 3.3. Each double point can be smoothed in one of two ways, and these will be diagrammed in a planar picture as indicated in Figure 5.3.

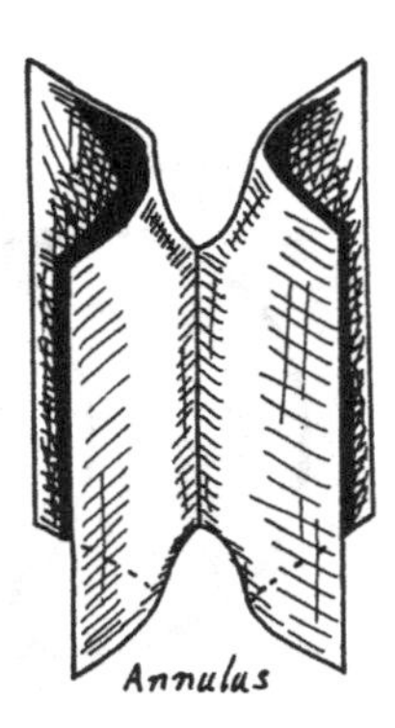

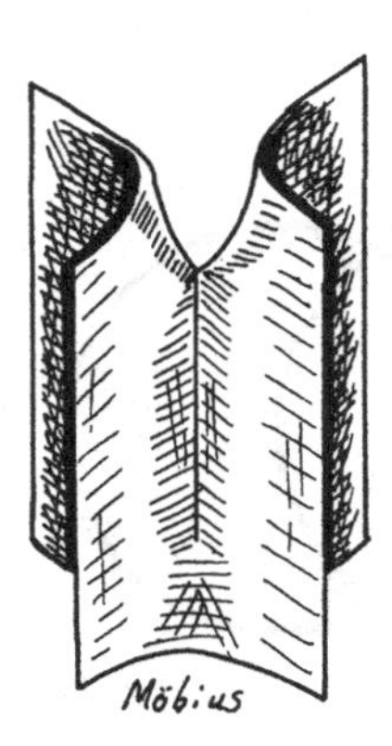

Figure 5.5: Schematic diagram for annular and Möbius arcs

Consider a given coordinate arc. The ends can be smoothed to give a Möbius band or an annulus as indicated in Figure 5.4. On the stereographic projection, these smoothings are indicated in Figure 5.5.

In order to construct the Cromwell-Marar surfaces, it is sufficient to list all the 64 possible configurations of smoothings by planar diagrams. Listing 64 possibilities is sort of drastic especially when there will only be 7 different surfaces that result. I want to get by with listing 32 of these. Even 32 is a lot to list, but the figures will only take up 4 pages, and I hope that this smaller exhaustive list will also be informative.

The argument that only 32 possibilities are needed goes as follows. Flatten out (Figure 5.6) one of the boundary circles of a triple point so that one of the Borromean circles becomes a horizontal line and the other two intersect each other in two points and intersect the horizontal in a total of four points circles. The horizontal line will represent the $z = 0$ equator of the boundary. The circle on the left represents the $x = 0$ equator, and the circle on the right represents the $y = 0$ equator. We will concentrate on the 4 possible smoothings of the z-axis. Two of these yield a self-intersecting annulus, and two yield a self-intersecting Möbius band.

First consider the smoothings that yield annuli. At the double point at the top of the diagram, there are four triangles. By blackening antipodal triangles, we are indicating that the smoothings are done on that side of the double arc. Since the arcs

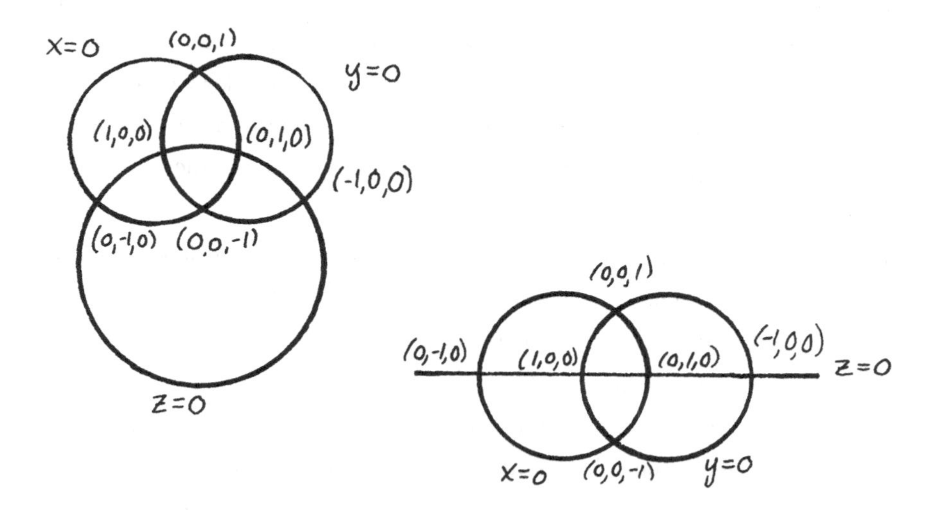

Figure 5.6: Flattening the $z = 0$ boundary circle

that are intersecting run southeast to northwest and southwest to northeast, we can call the triangles either east-west triangles or north-south triangles. Suppose that we smooth by blackening the east-west triangles. Choose a smoothing of the remaining four double points (the intersections between the $z = 0$ circle and the $x = 0$ or $y = 0$ circle). Now remember that the planar figure schematizes the intersection among coordinate planes. Rotate the $x = 0$ plane to the $y = 0$ plane by a 90° rotation about the vertical axis. This rotation will take the smoothing of the double points along the $z = 0$ plane to possibly another smoothing, and it will take the east-west triangles to the north-south triangles in the planar diagram.

Any smoothing along the horizon with the east-west triangles blackened gives a smoothing along the horizon with the north-south triangles blackened. So for the annulus intersecting along the z-axis, we only need to think about the east-west smoothing at the north and south poles.

In the case of a Möbius band intersecting along the z-axis, the argument is similar. However, instead of rotating the picture 90° about a vertical axis, we will rotate 180°

about a horizontal axis. You will see from the figures that we only consider an east-west smoothing at the north-pole. If we had a Möbius band intersecting along the z-axis and a north-south smoothing at the north pole, then there would be an east-west smoothing at the south pole. So by rotating south-to-north, we obtain an east-west smoothing at the north. Again the smoothings along the horizon are interchanged.

So if all the possible ways of smoothing the horizontals are considered, then we need only consider one annular and one Möbius smoothing along the vertical, and these will be codified by the east-west blackening at the north pole.

Actually, Cromwell and Marar take this argument as far as it can go. They examine the set of symmetries of the three intersecting coordinate arcs, and determine which possible figures are interchanged under these symmetries. Then they are sure that they have found a complete list of surfaces with 6 branch points and one triple point. These surfaces will be pursued from a less sophisticated point of view here. We will list all the possibilities and observe which of the surfaces are the same after the fact.

What should we look for in our list? First we should examine each double arc to see if it is a Möbius band or an annulus. Since there are three double arcs, and since each double arc is either an annulus or a Möbius band, we have the possibilities three annuli on the double arcs (denoted $\{A, A, A\}$), two annuli and a Möbius band (denoted $\{A, M, A\}$), two Möbius bands and an annulus (denoted $\{M, A, M\}$), or three Möbius bands (denoted $\{M, M, M\}$). In addition, when the ends of the double arcs are smoothed a certain number of circles will result on the boundary of the surface. We will denote this number by B.

The surfaces constructed will be classified by these data: the number of annular double curves and the number of circles on the boundary. In regards to the boundary, we need to remove these by adding disks. Our final illustrations of the surface show where the disks can be added. It turns out that there may be more than one way to add these disks, but once the intersection pattern is chosen among the annuli and Möbius bands the intrinsic topology of the surface is determined. Furthermore, if different disks closing the boundaries are chosen then the resulting surfaces in 3-space are inside-out versions of each other.

The illustrations in Figures 5.7, 5.8, 5.9, & 5.10 gives the complete list of surfaces constructed by stereographic projection of the boundary of a neighborhood of a triple point together with smoothings of the double points. In each of these 32 figures, the numbers of Möbius bands, annuli, and closed curves on the boundary are given.

The figures were generated as follows. The east-west smoothing is chosen at the north pole. Then there are 16 surfaces with east-west smoothings at the south pole, and 16 with north-south smoothings there. These are divided between Figures 5.7,5.8 and Figures 5.9,5.10. Figures 5.7,5.9 use a specific smoothing along the positive y-axis and Figures 5.8,5.10 use the opposite smoothing. The figures in the left-hand column of each page use a certain smoothing of the negative y-axis and this is chosen such that in Figures 5.75.9 the y-axis is an annular double curve in the left column and an annular double arc in the right column in Figures 5.8,5.10. The positive x-axis is consistently smoothed in 2-by-2 blocks on each page, and the negative x-axis is consistently smoothed on each row of every page. Again the choice of smoothing is alternating here.

The table below shows how many surfaces of each type there are. Illustrations of each of the surfaces are given in Figures 5.12 through 5.17. It is an easy exercise to classify the resulting closed surfaces.

Double arc type \ $B=$	1	2	3	4
$\{A, A, A\}$	2		2	
$\{A, M, A\}$		12		
$\{M, A, M\}$	6		6	
$\{M, M, M\}$		3		1

Actually, Marar has constructed a nice model for all of these surfaces out of cardboard, adhesive tape, and paper clips. The model uses 12 sturdy paper clips, 3

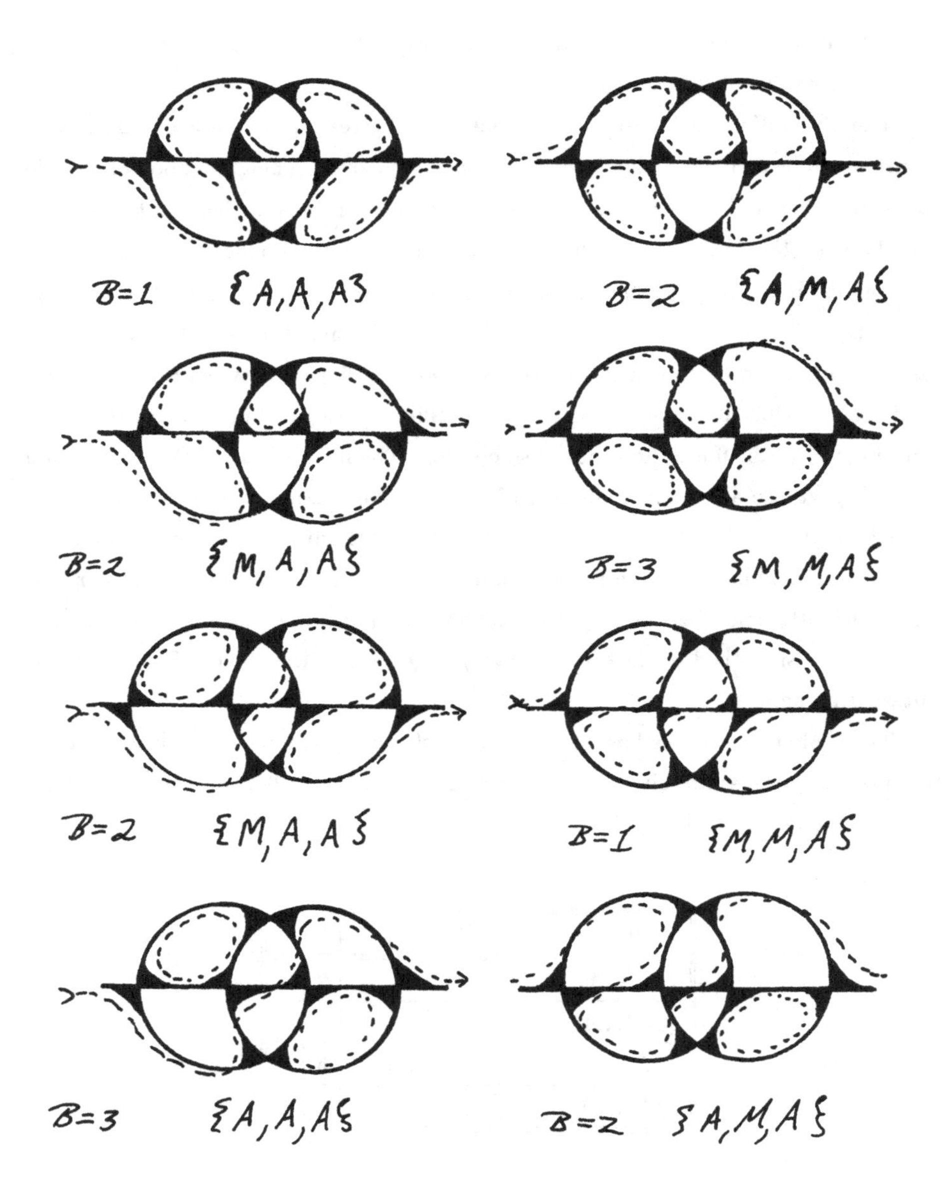

Figure 5.7: The Cromwell-Marar Surfaces, part 1

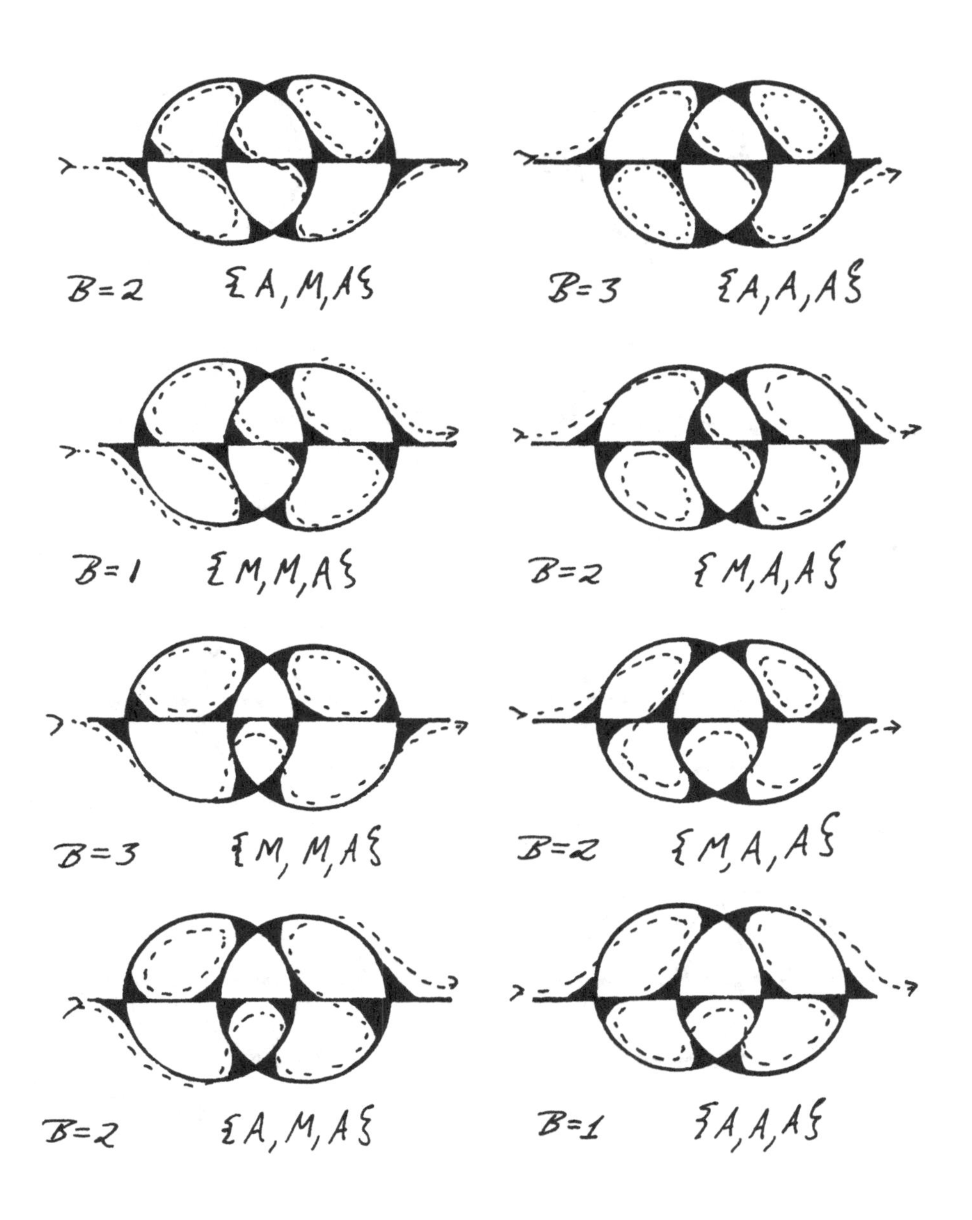

Figure 5.8: The Cromwell-Marar Surfaces, part 2

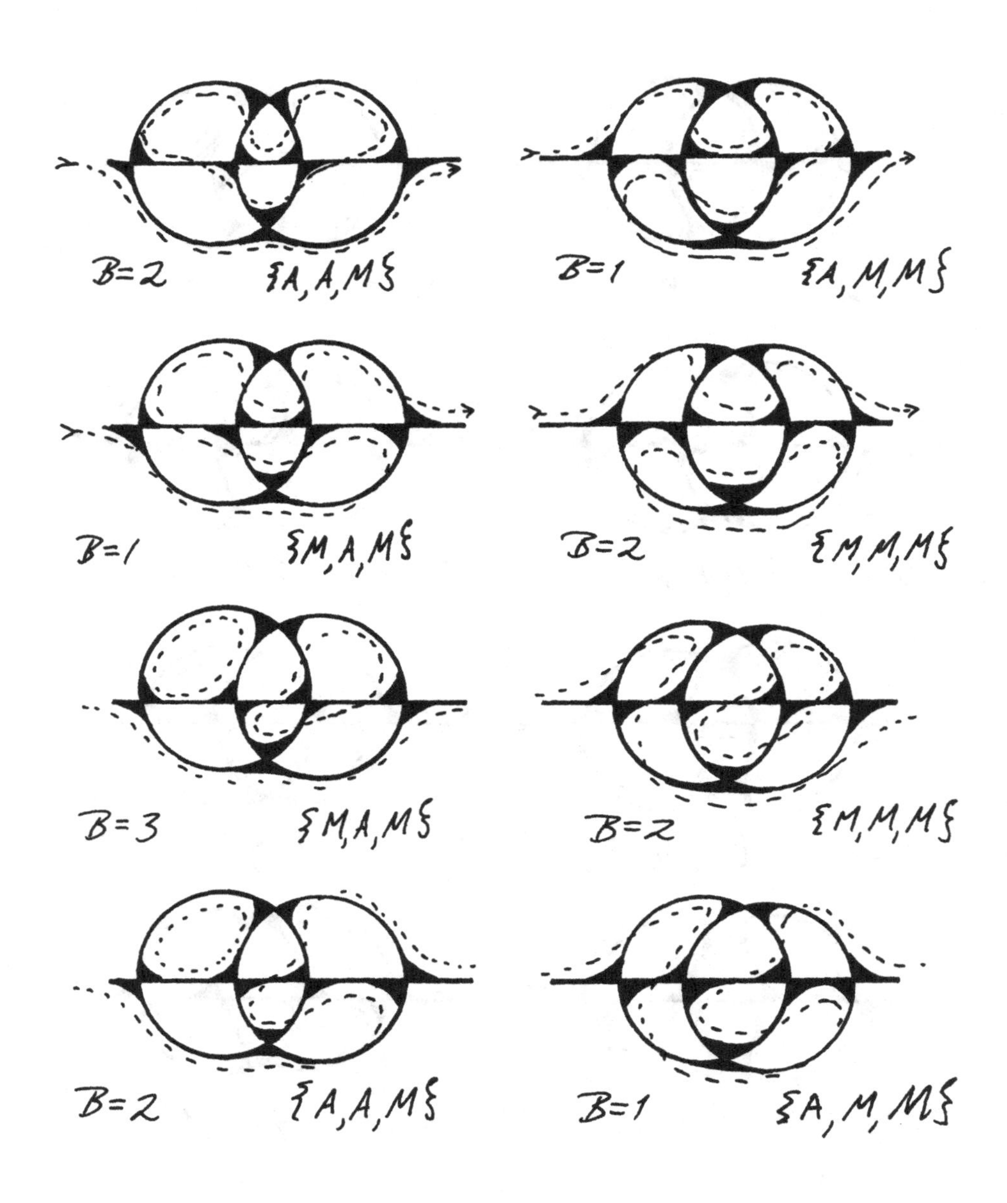

Figure 5.9: The Cromwell-Marar Surfaces, part 3

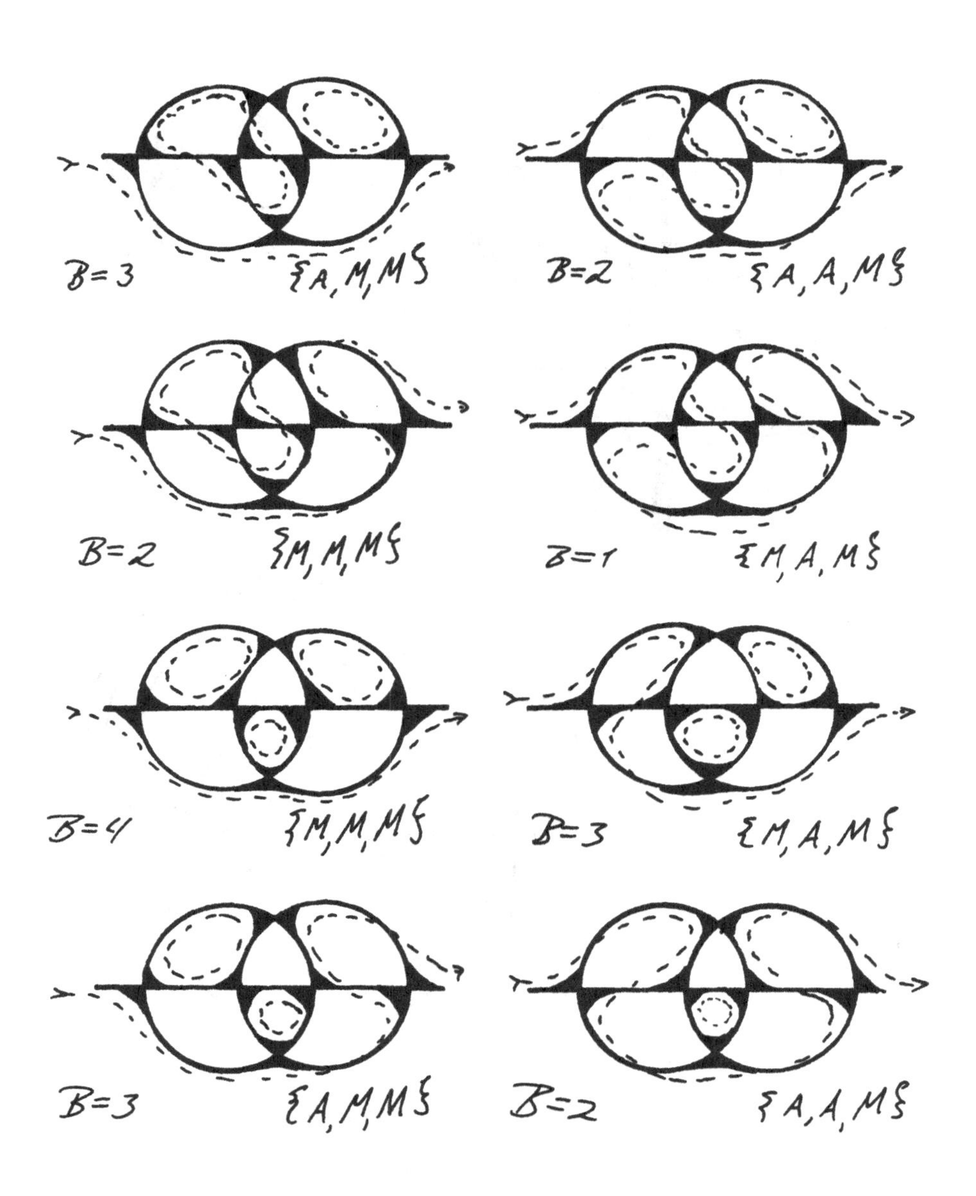

Figure 5.10: The Cromwell-Marar Surfaces, part 4

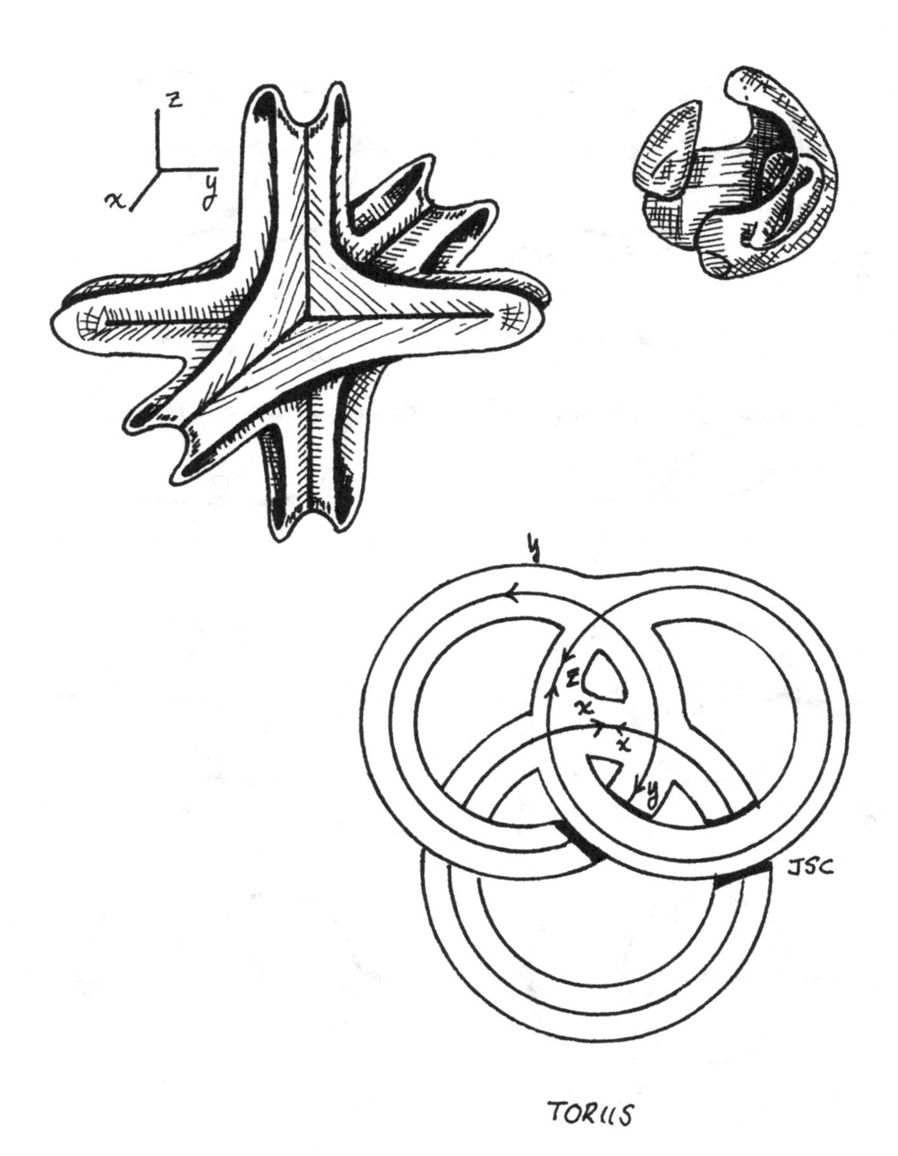

Figure 5.11: The $\{A, A, A\}, B = 3$ surface

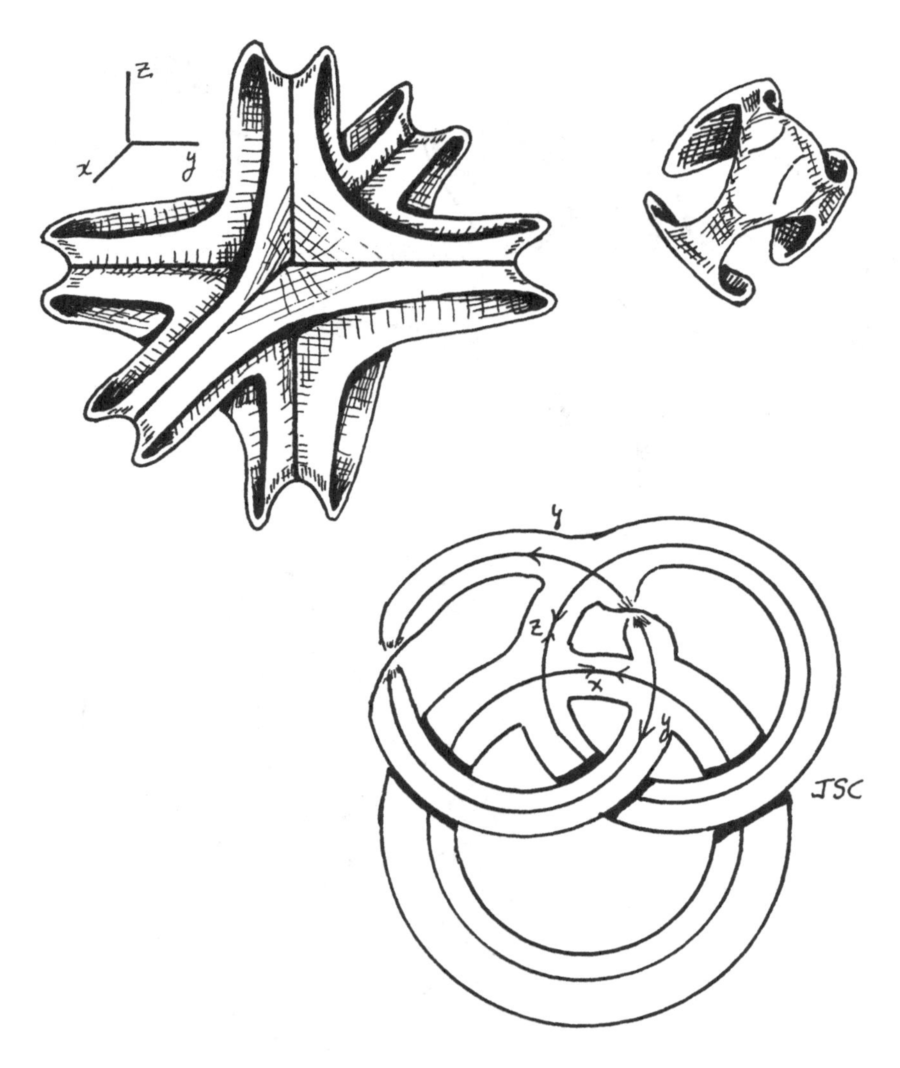

Figure 5.12: The $\{A, A, A\}, B = 1$ surface

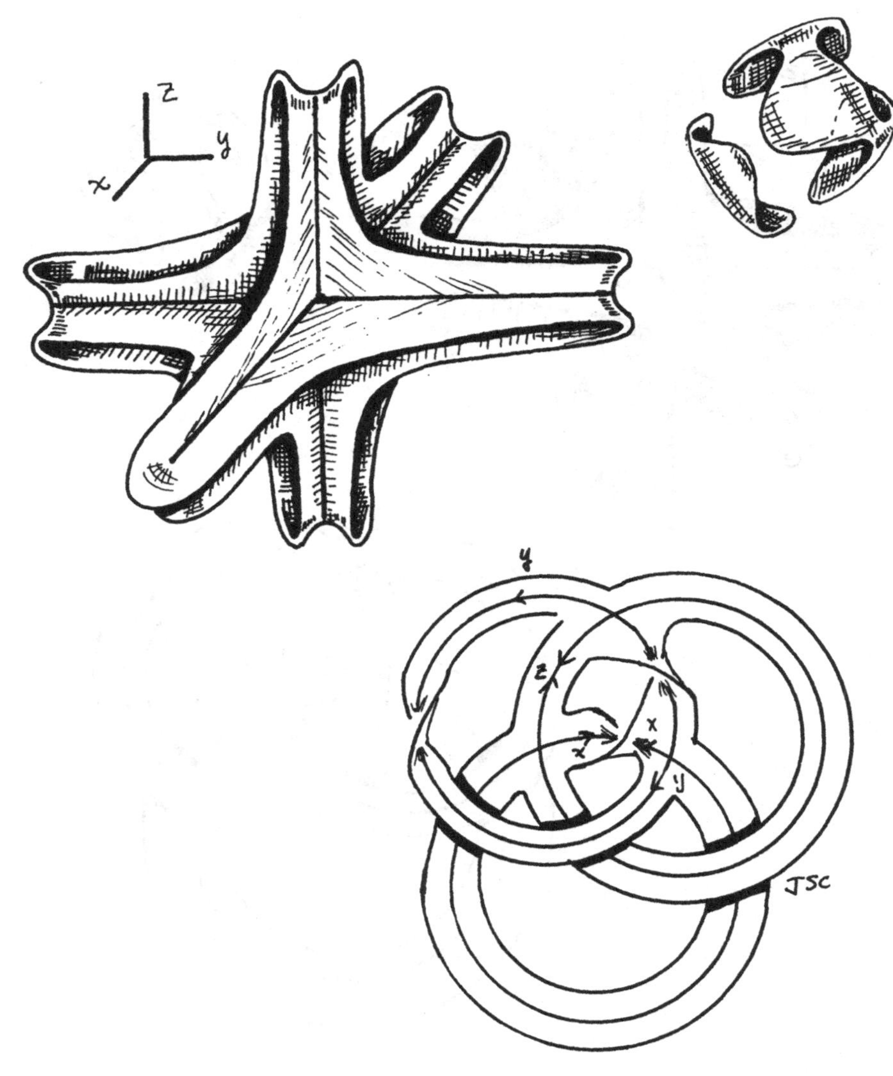

Figure 5.13: The $\{A, M, A\}, B = 2$ surface

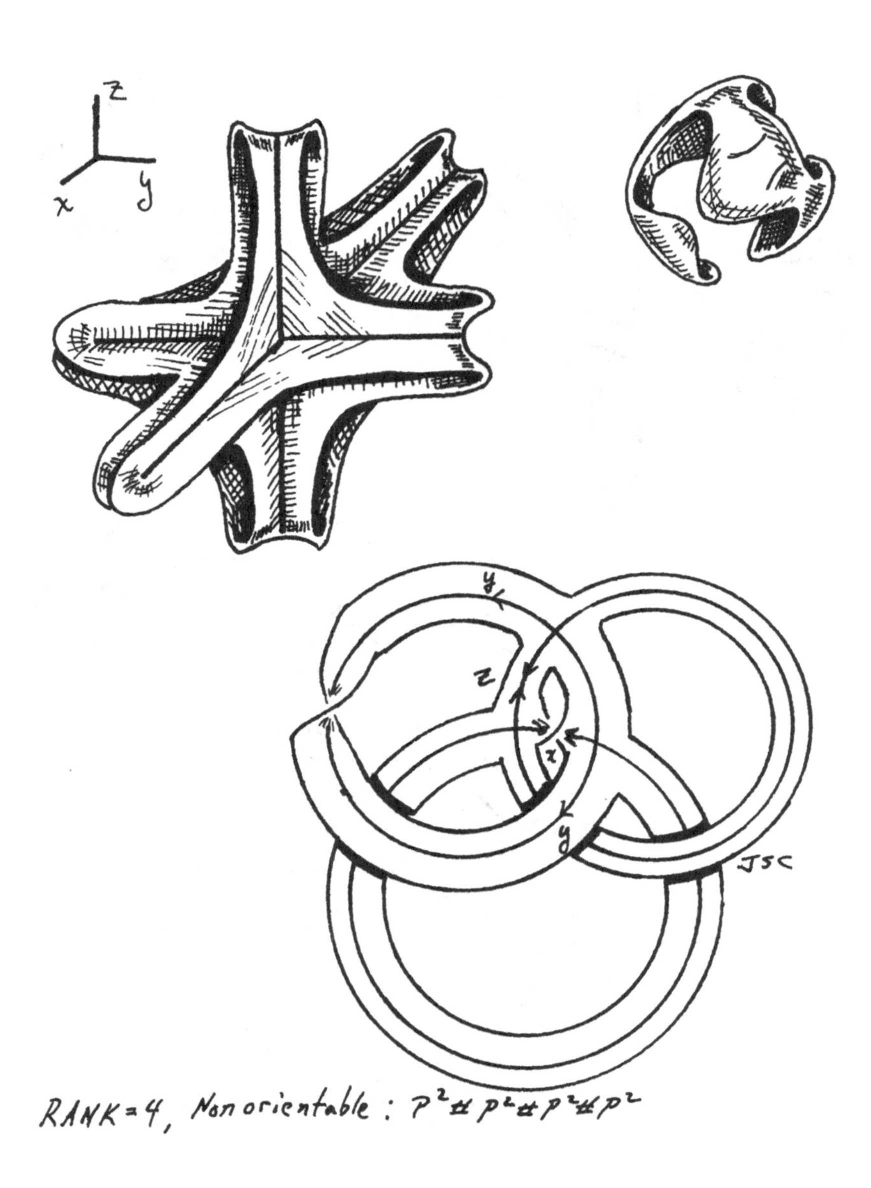

Figure 5.14: The $\{M, A, M\}, B = 1$ surface

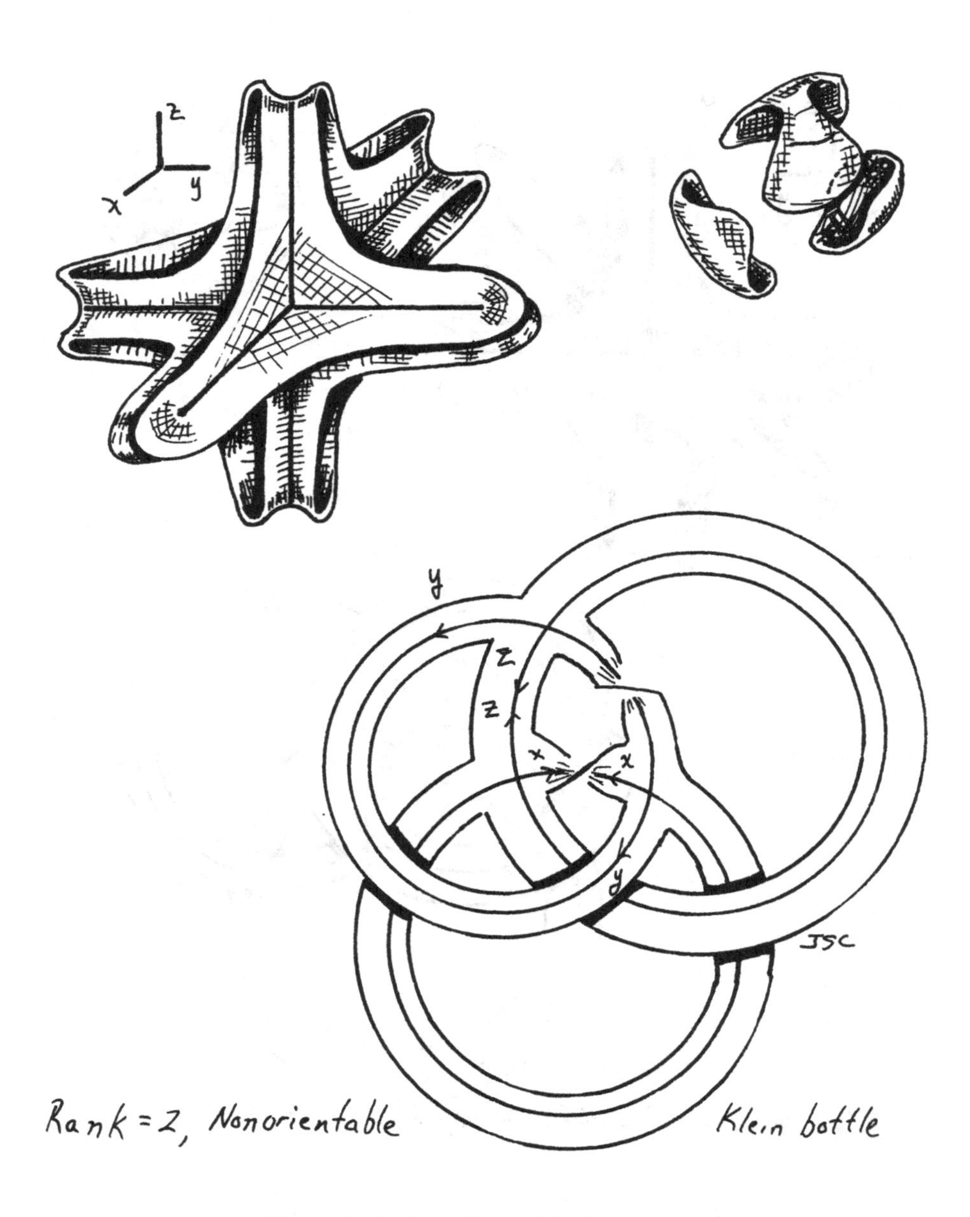

Figure 5.15: The $\{M, A, M\}, B = 3$ surface

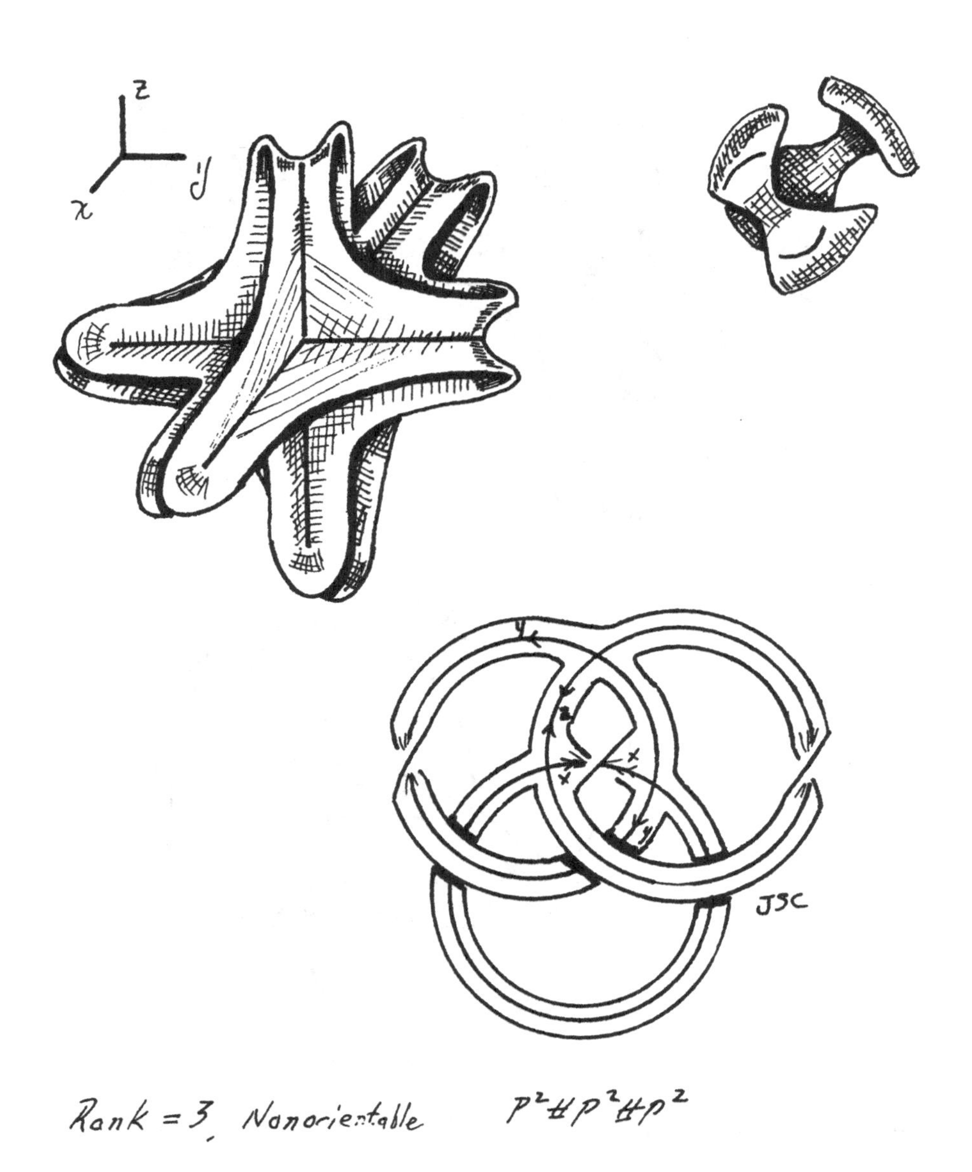

Figure 5.16: The $\{M, M, M\}, B = 2$ surface

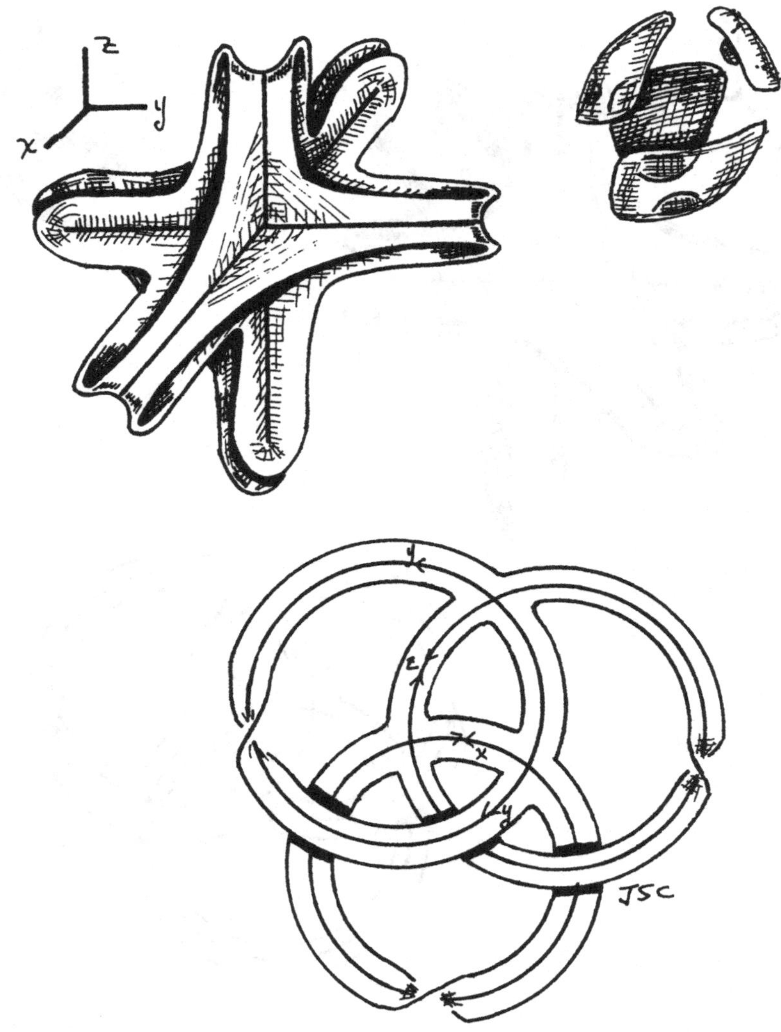

Figure 5.17: The $\{M, M, M\}, B = 4$ surface

pieces of sturdy (uncorrugated) cardboard of size 4 inches by 4 inches, and 12 strips of flexible cardboard (lighter than shirt cardboard — perhaps the kind used to package Christmas presents); these strips are roughly 11 inches by 2 inches and their front and back sides should be differently colored.

Pair the long strips together as in the figure and place them long edge to long edge with one having one color up and the other having the other color up. Use the illustration in Figure 5.18 as a guide. Tape these together at two points about on either side of the center about an inch from the center. You should have 6 pieces of that form. Now fold these with out a crease at the center, and bend the two strips so that along there common edge they meet at a 90° angle away from the center. Cut slits in the 3 square pieces as indicated in the figure, and attach these to form a neighborhood of a triple point. Finally use the paper clips to fix the strips (which now form models of double arcs ending in branch points) to the triple point. Any pattern will do. See if you can reattach all or some of these strips to get models for all the seven surfaces. Observe that the coloring can be used to determine which of the surfaces is orientable.

5.3 Notes

The notes on this chapter are necessarily brief. Obviously, one should refer to the primary sources for further information. I will mention that W. L. Marar's area of expertise is a branch of mathematics called singularity theory. The primary objects of study are certain equivalence classes of maps between Euclidean spaces. The maps are perturbed into general position and the intersection patterns among the perturbations are studied.

Professor Marar has told me that he has recently considered the case of surfaces with two triple points and ten branch points and he has a complete list of these.

One interesting note should be made here. Thanks to the generosity of the Institute of Mathematical Sciences of San Paulo in San Carlos (and my home institution), I was able to visit San Carlos, Brazil to attend a conference hosted by Professor

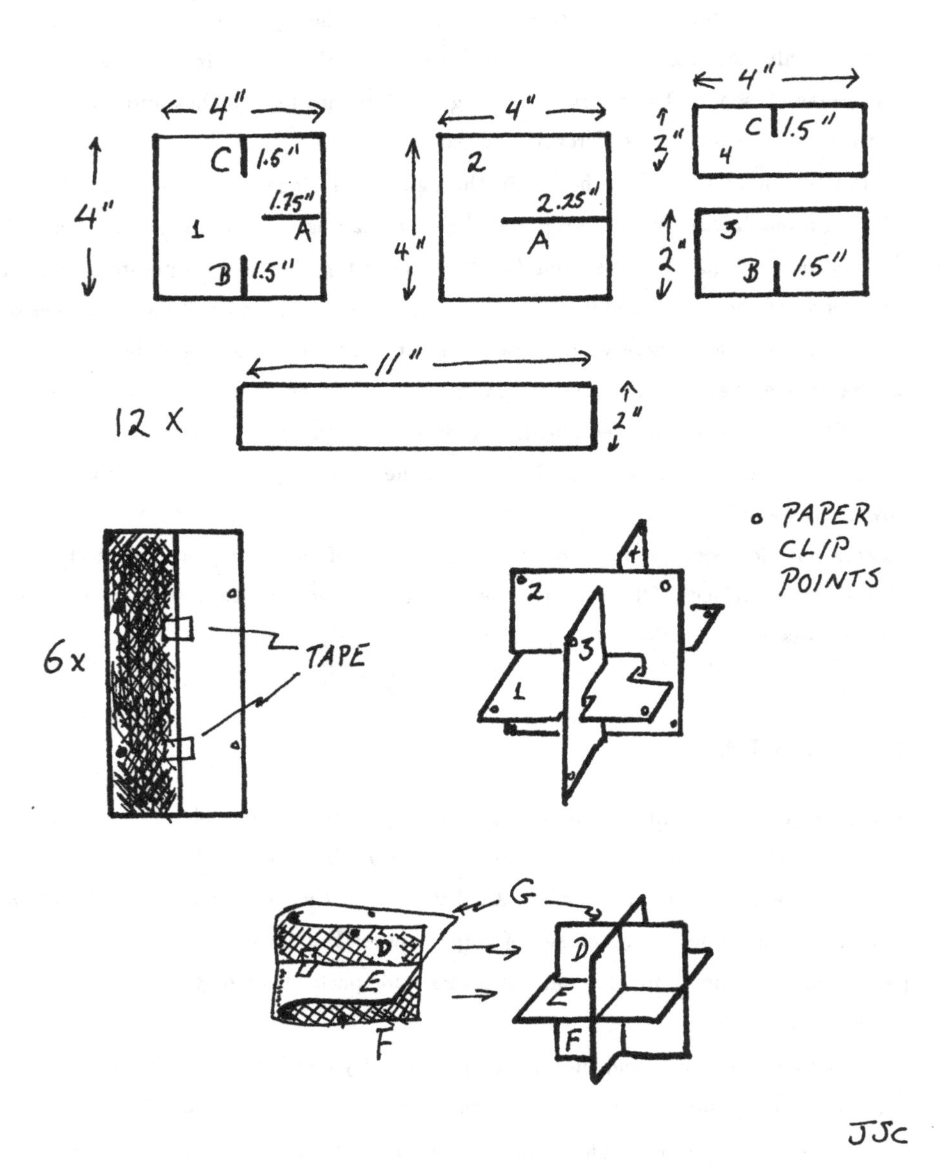

Figure 5.18: Marar's models of the surface

Marar. Imagine my shock when I saw in the garden at the math department a sculpture that resembled the cover of this book! In fact, the disk drawn on the cover is a certain subset of that sculpture. The considerations that led to the commission of the sculpture and the cover illustration were completely independent. Once again, we see that similar ideas are being pursued simultaneously all around the world.

Chapter 6

Surfaces in 4-Dimensions

This chapter shows how to move surfaces around in 3- and 4-dimensional space. In particular, the various pictures of the Klein bottle that have been given are all the same in 4-dimensions. We study the cross cap, Boy's Surface, and the Steiner surface in 4-dimensions. And we show how to turn a sphere inside out in 3-space.

6.1 Moving Surfaces in 4-Dimensions

Another Look at the Reidemeister Moves

Before we delve straight into the 4-dimensional world, let's recall the Reidemeister Theorem on knot diagrams. The Reidemeister Theorem says that two knot diagrams that represent the same knot can be transformed into each other by a sequence of moves of type I, type II, and type III. In his book [64], Reidemeister also mentions two other moves that he calls Δ-moves (delta). All five of these moves are depicted in Figure 6.1 together with their 4-dimensional interpretations.

The Δ-moves can be understood as follows. If we are given two knot diagrams in which the relative heights of the crossings, maxima, and minima play an essential role, then one diagram can be transformed into another by means of these five moves. There is a move that is made tacitly as well. The levels of two distant crossings can be interchanged.

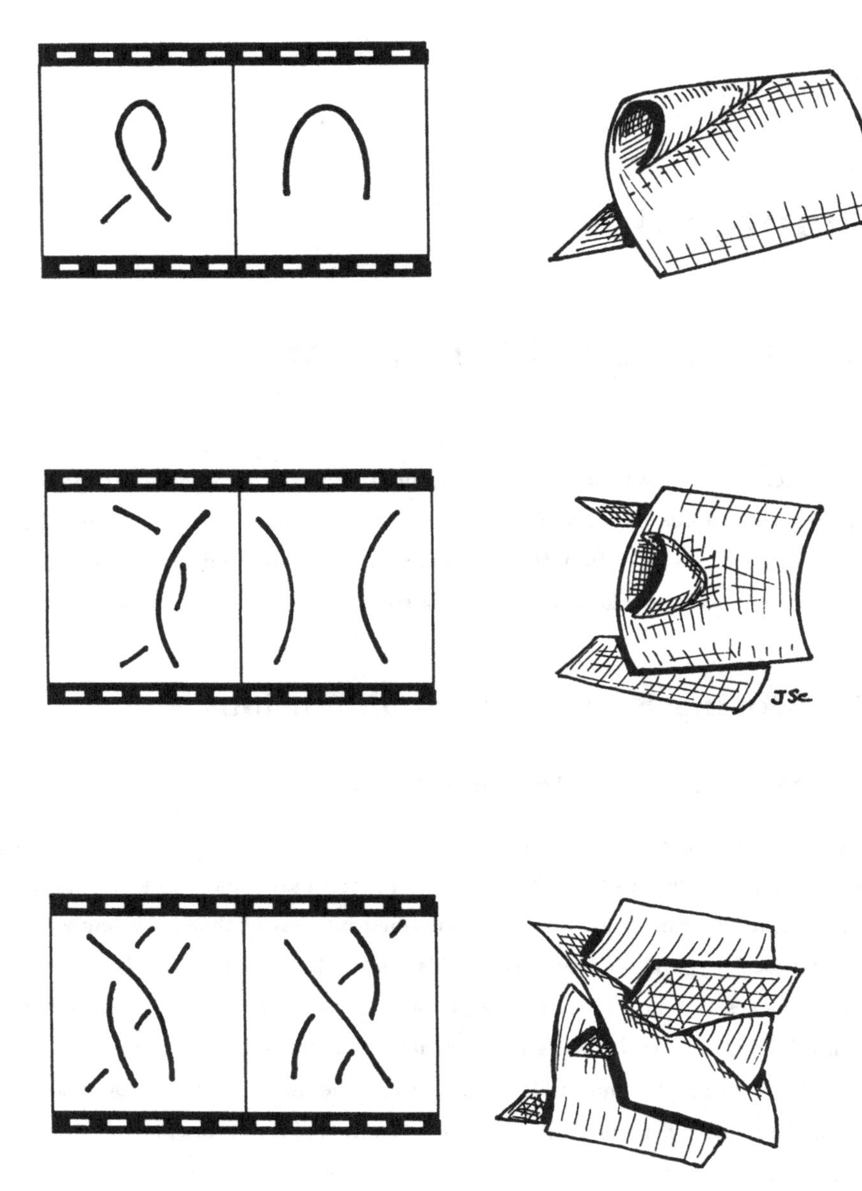

Figure 6.1: The Reidemeister moves revisited

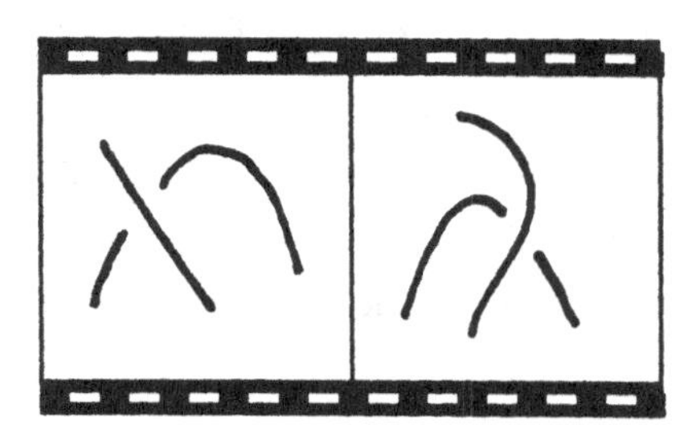

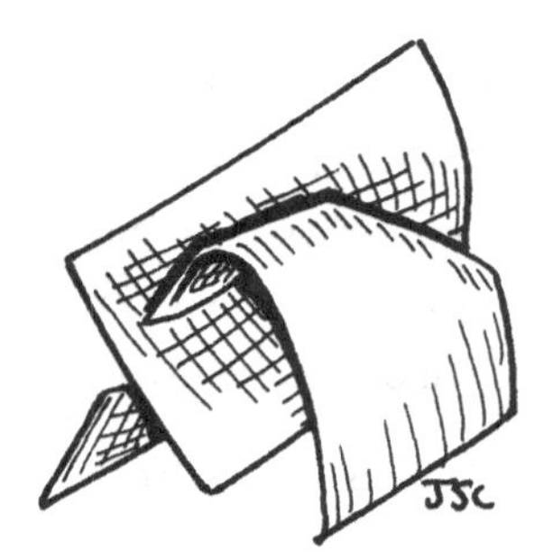

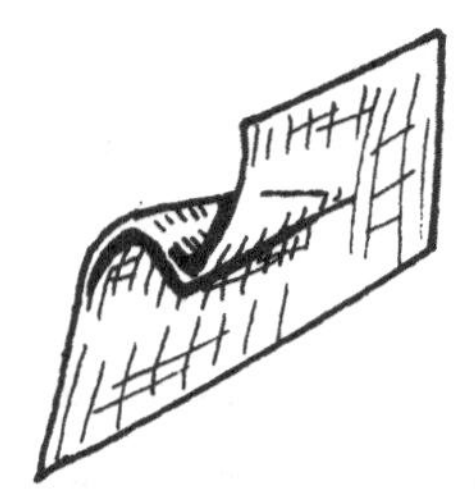

Figure 6.1b: More Reidemeister moves

In modern knot theory (modern = after Jones [50]) it became apparent that keeping track of relative height might be desirable. Some of the newer invariants are defined by means of a chosen height function, and for them to remain invariant requires that they remain unchanged under the Δ-moves.

The need for keeping track of heights of surfaces in 4-space comes about, in part, because one wants to find invariants that are analogous to Jones's. But more specifically, the movie description of a surface requires that a height function be specified.

Elementary String Interactions

There are five elementary string interactions. These are the three main Reidemeister moves (type I, II, and III), the birth/death of a simple closed unknotted curve, and the operation of fusing two components into one or fissuring one component into two. In all of the movies that are used to depict embedded surfaces, successive stills differ, at most, by an elementary string interaction. Birth corresponds to a (local) maximum of the height of a surface, and death corresponds to a local minimum. A fusing or a fission corresponds to a saddle point.

The Roseman, Homma-Nagase Moves

Dennis Roseman [68] and Tatsuo Homma with Teruo Nagase [46] showed that seven moves are sufficient to move embedded surfaces around in 4-dimensional space. These moves are like the Reidemeister moves for knot diagrams. If two surfaces can be deformed in 4-space into one another and they have given 3-dimensional diagrams that are different, then their diagrams can be transformed into one another by a sequence of the seven moves shown in Figures 6.2 through 6.8.

These figure are depicted with a given choice of over and under sheets (except in the case of the quadruple point move where such depictions become unwieldy). The broken surface indicates which sheet is above from the point of view of the projection into 3-space. Different choices of breaks are possible as long as the breaks match up. On the abstract surfaces, the preimage of the double points are indicated. Movie versions of these moves are also depicted, and these correspond to cutting the surfaces by hyperplanes that are parallel to the horizon.

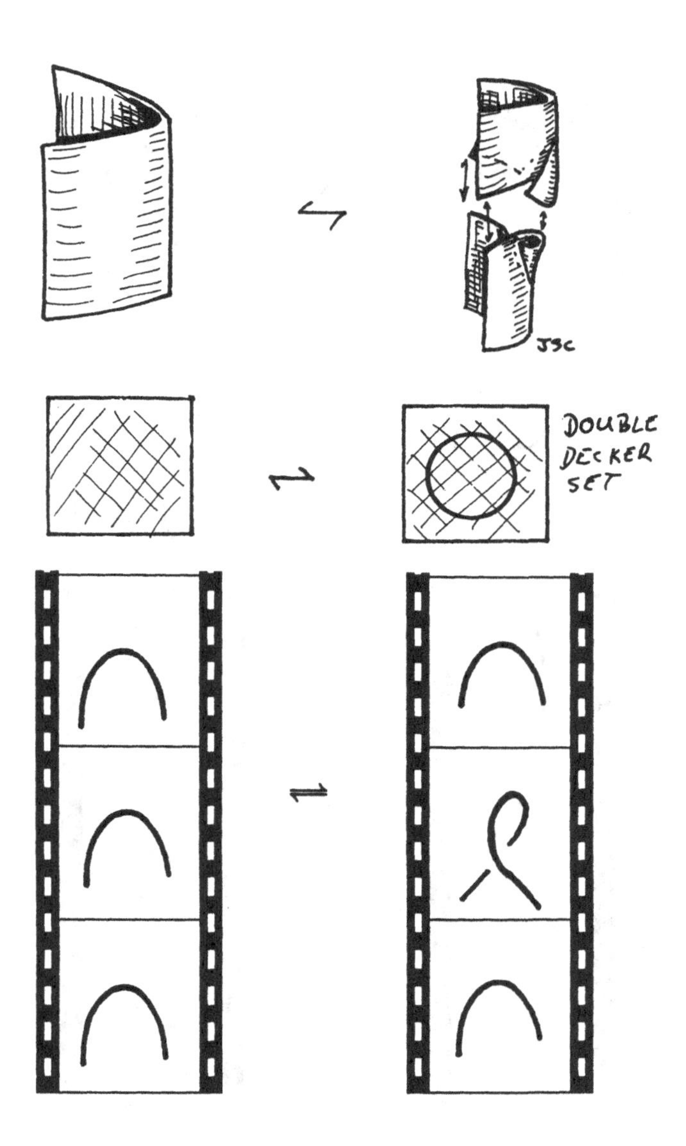

Figure 6.2: Canceling or adding branch points through a blister

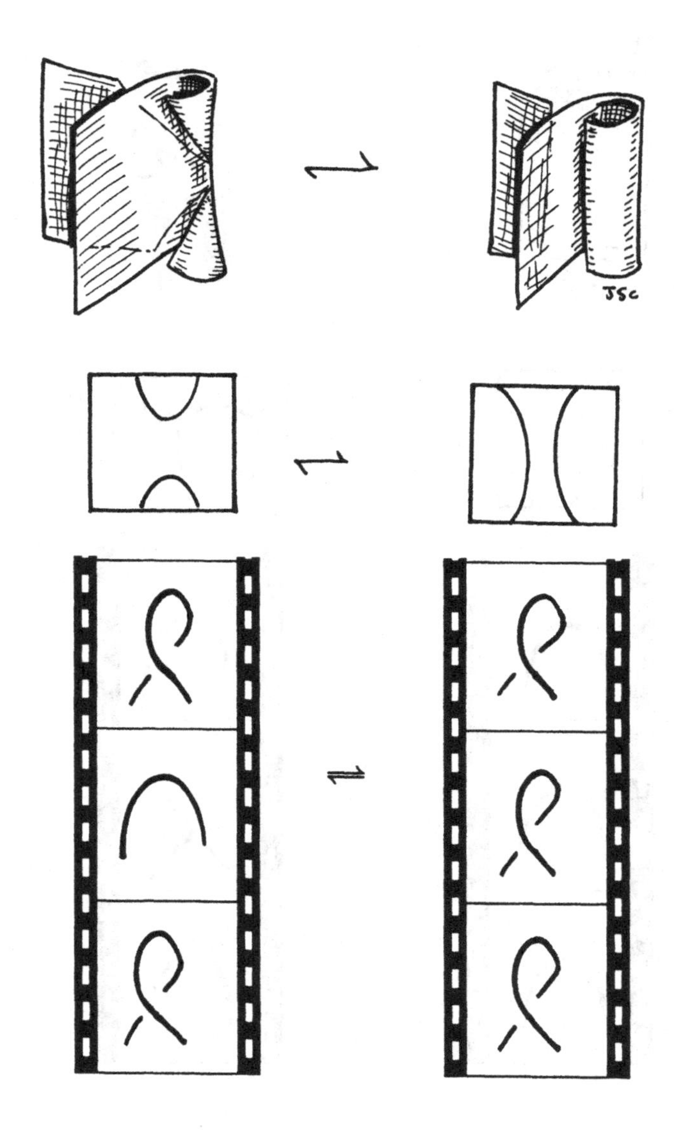

Figure 6.3: Canceling or adding branch points through a saddle

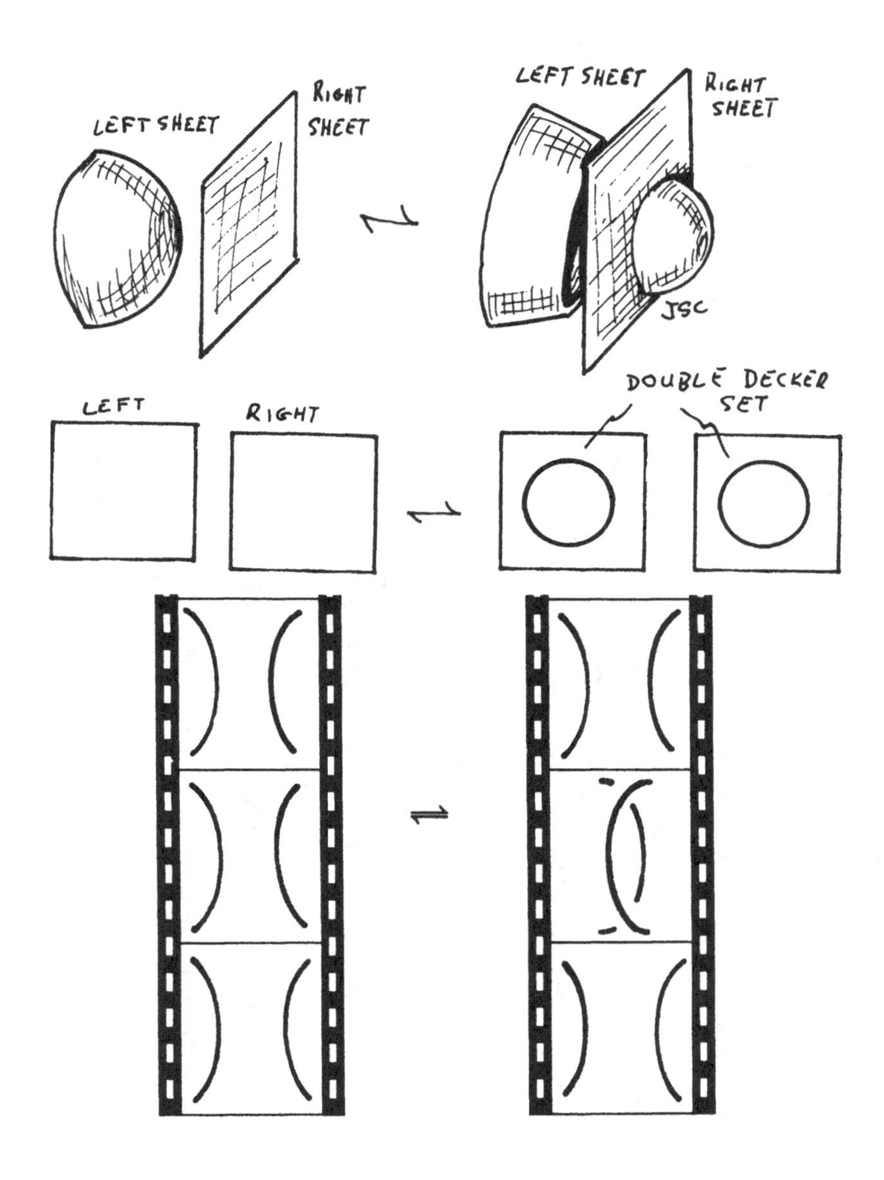

Figure 6.4: The bubble move

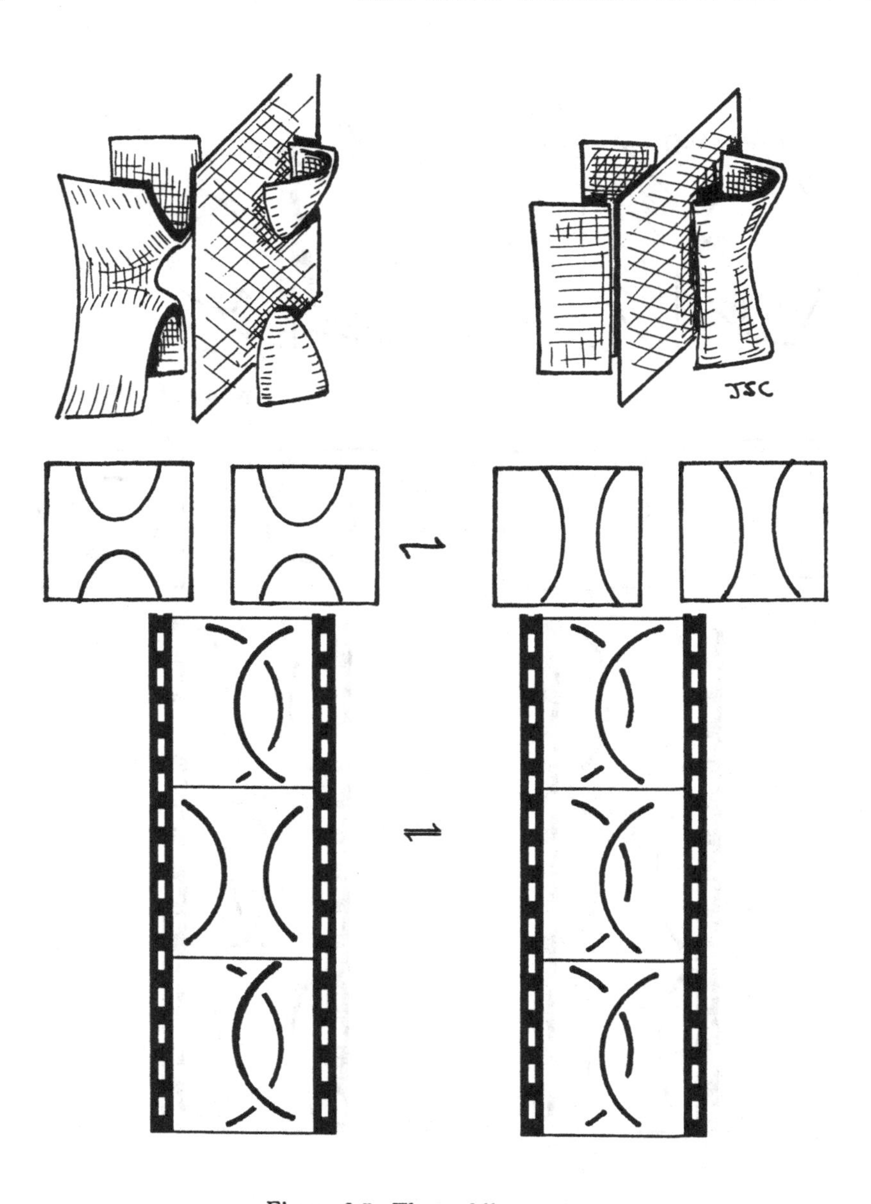

Figure 6.5: The saddle move

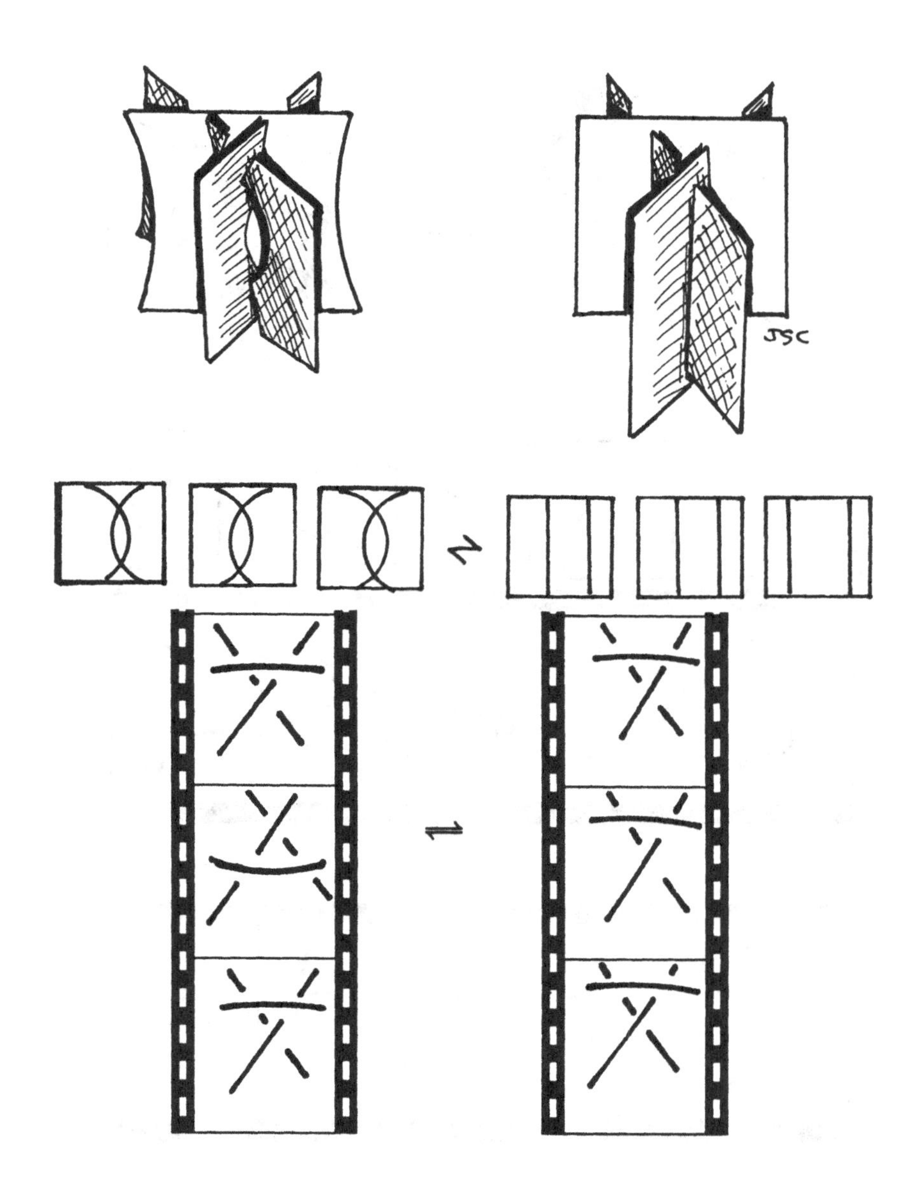

Figure 6.6: Canceling or adding a pair of triple points

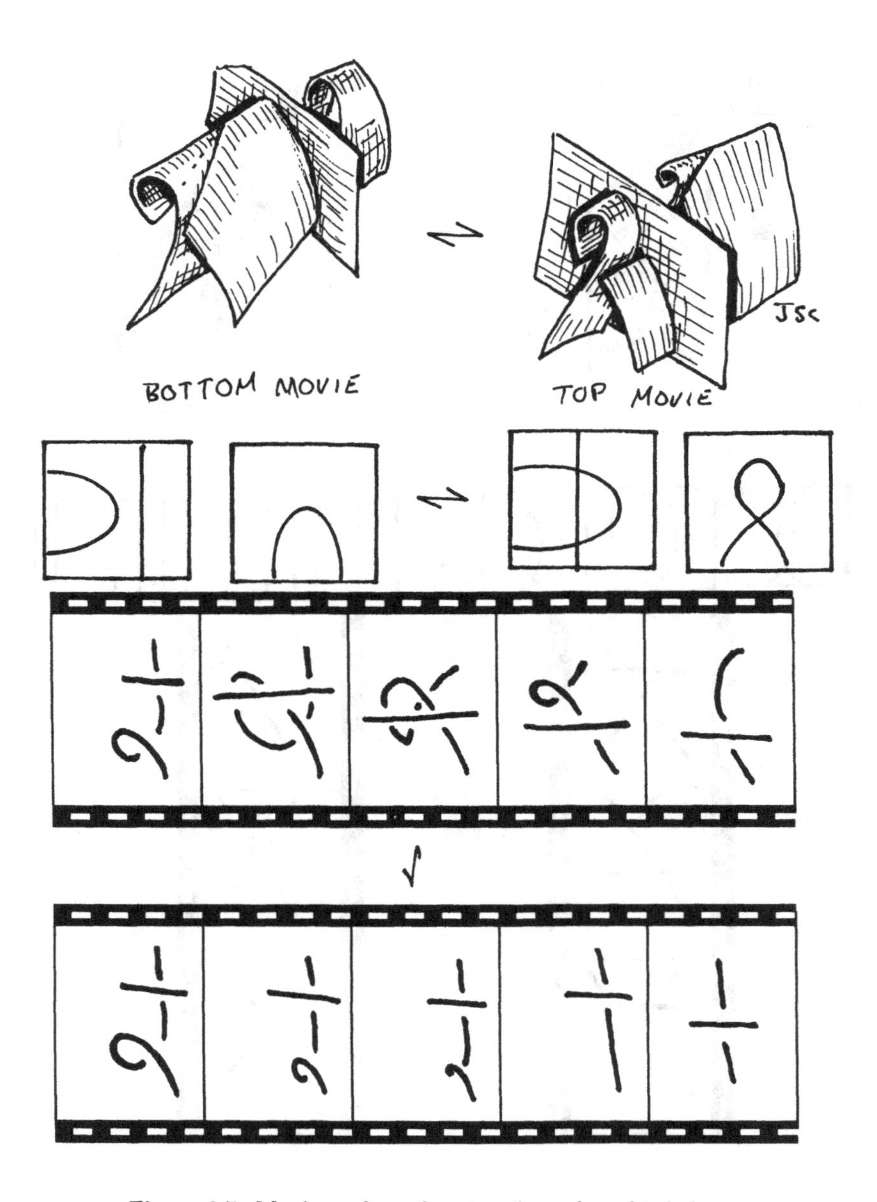

Figure 6.7: Moving a branch point through a third sheet

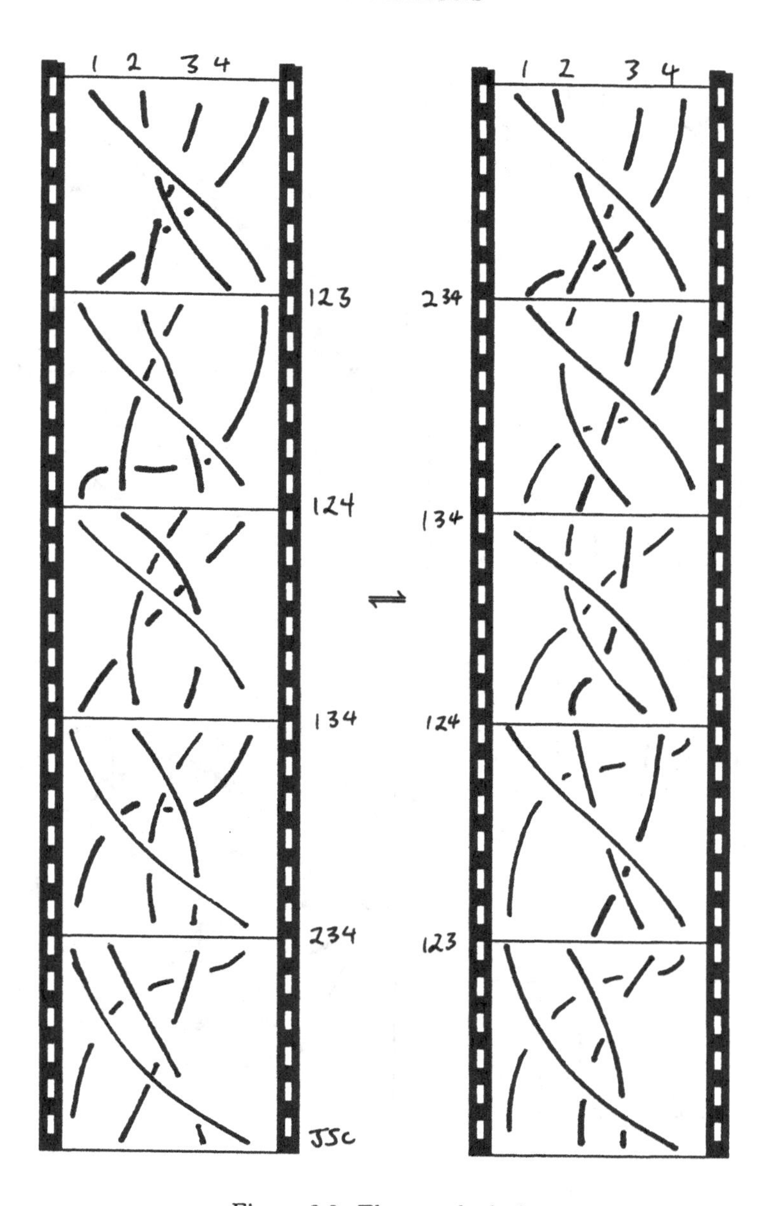

Figure 6.8: The tetrahedral move

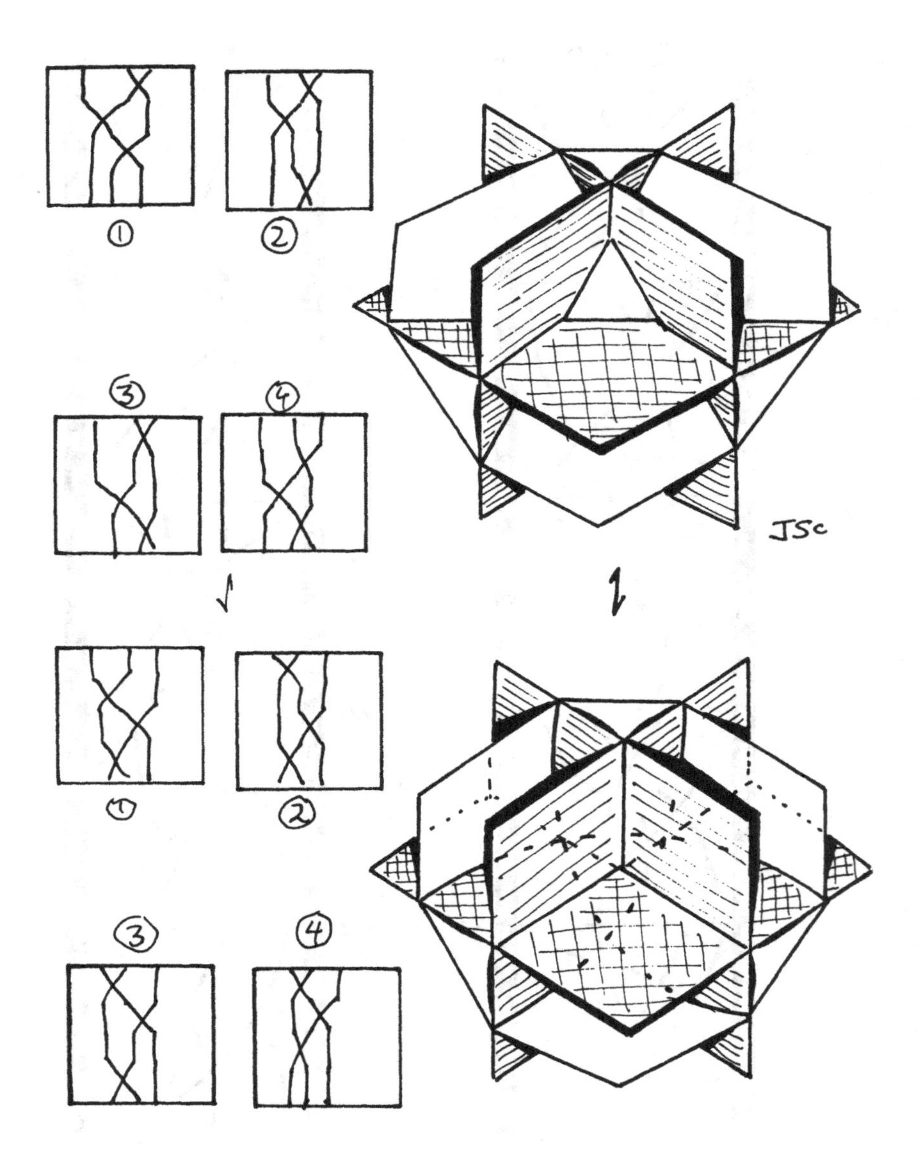

Figure 6.8b: The movie of the tetrahedral move

The Roseman/Homma-Nagase moves can be used to define invariants of surfaces that are knotted in 4-dimensions. Knotting phenomena tend to occur when the larger space has two more degrees of freedom than the smaller space. As with knotted curves, to define an invariant you define it by means of a representative diagram, and then check that the definition doesn't change under any of the seven moves. This approach was developed in detail in the papers [26] and [29].

(I should mention that Homma-Nagase show that only six of the moves are needed. But keeping the seventh move in the catalogue of moves simplifies some computations.)

The Movie Moves

At a conference in Santa Fe, New Mexico on new mathematical applications of mathematics to the study of DNA, Lou Kauffman asked me what are the moves that are necessary to transform one movie diagram of a knotted surface into another. I thought about that question for a while, and found the best way to answer it: Namely, I asked Masahico Saito the same question. Masahico and I worked out a set of 15 moves to movies that are sufficient to take one movie diagram of a surface into another (See the papers [22] and [25] for details of our techniques). These movie moves are analogous to the five Reidemeister moves. In the illustrations, projected surfaces indicate the changes in relative heights.

In Figures 6.2 through 6.8, movie versions of the seven Roseman moves appear. Figures 6.9 through 6.16 contain the remaining eight moves. Finally, Figure 6.17 illustrates a theorem discovered by John Fischer, that shows that 2 of the movie moves can be derived from the remaining 13.

In the stages that are intermediate to some of the movie moves, a local maximum, minimum, or saddle occurs simultaneously on the double point set and on the surface itself. Such intermediate stages were used in the depiction of Boy's surface in Figures 2.8 through 2.12. The simultaneity of critical points is achieved by mapping intersecting handles of lower dimension into higher dimensional handles.

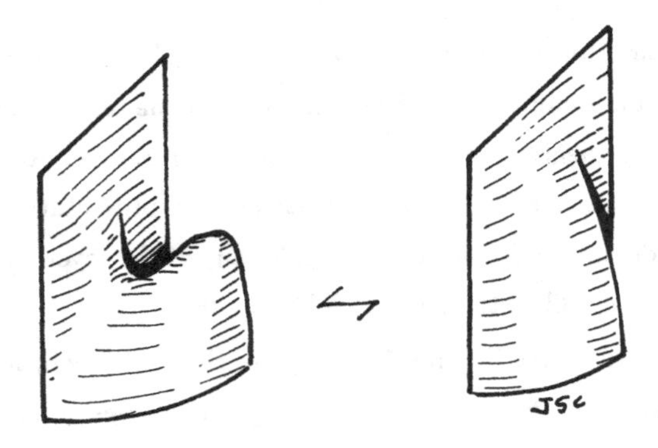

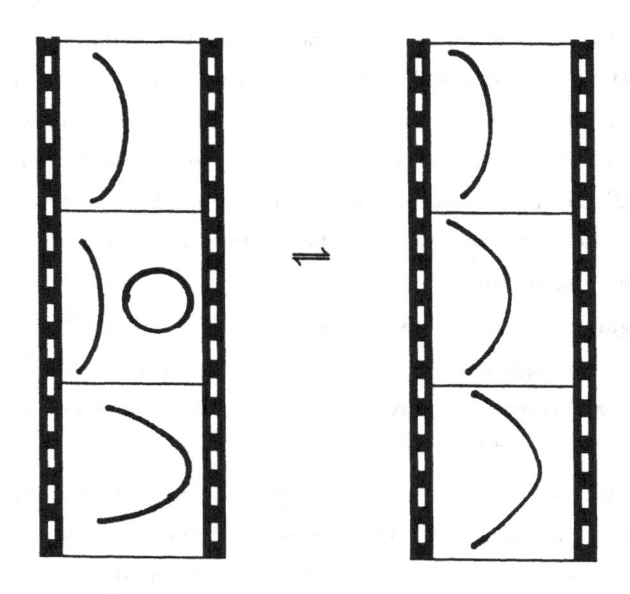

Figure 6.9: A mountain erodes

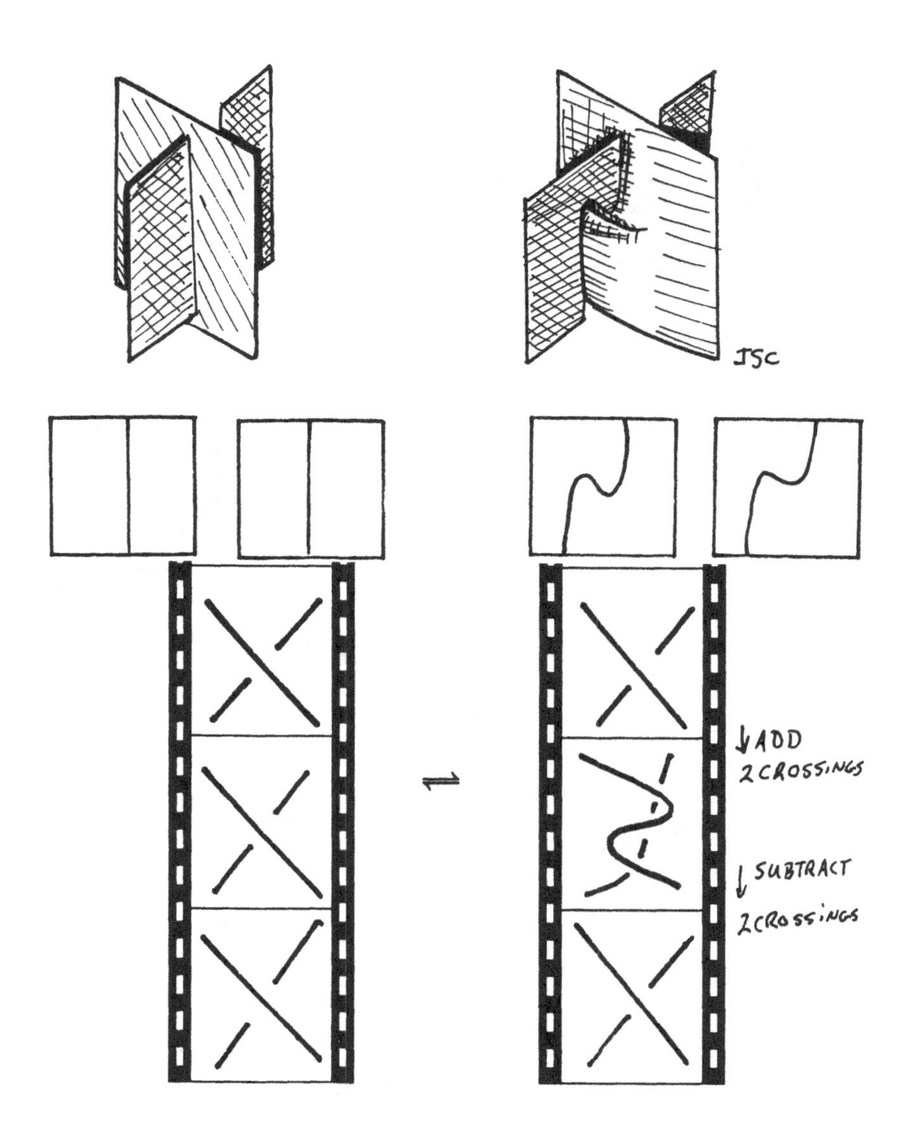

Figure 6.10: A type II birth cancels a type II death via a cusp

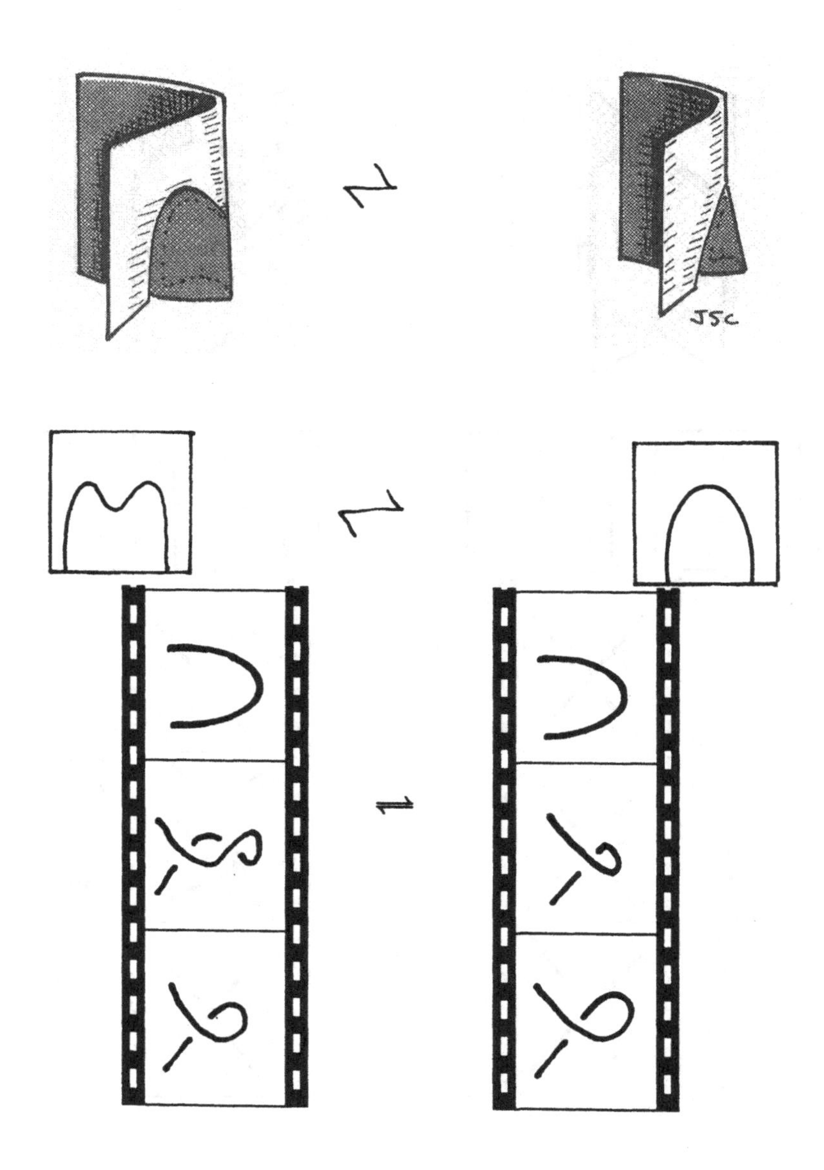

Figure 6.11: A type II birth cancels a type I death via a cusp

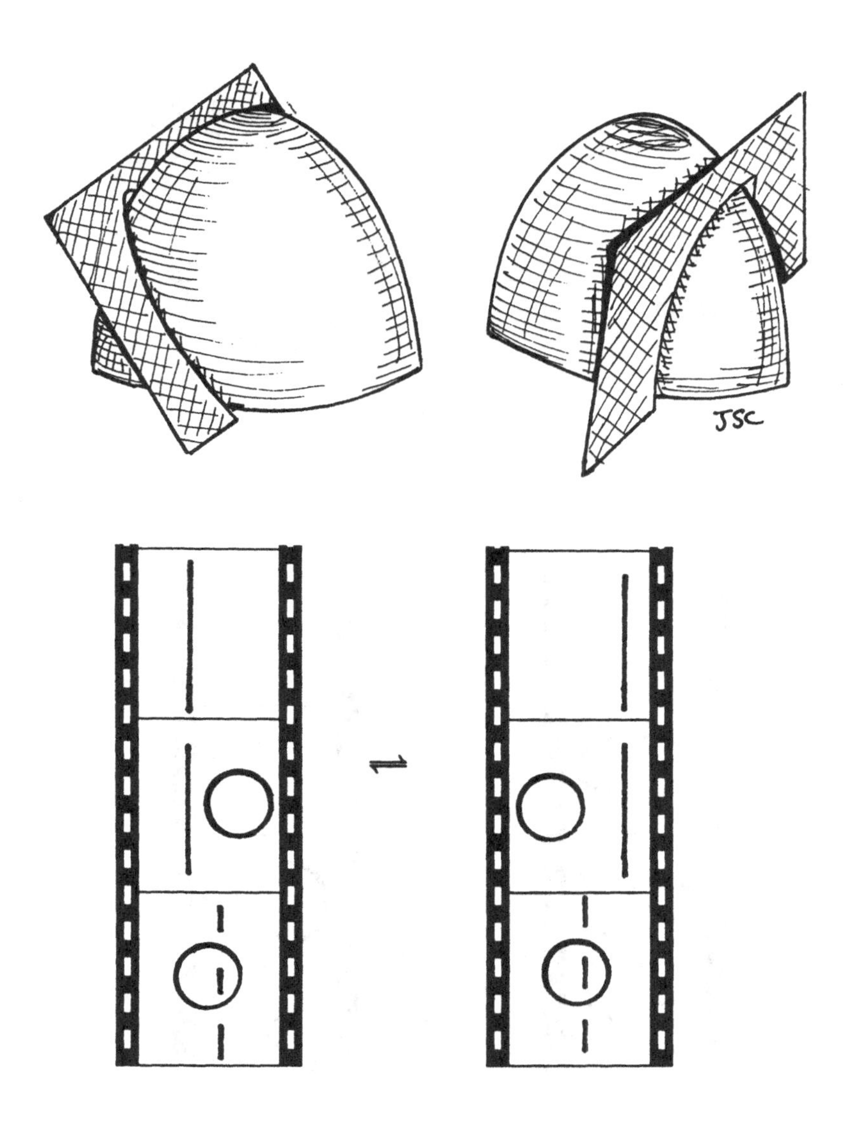

Figure 6.12: Moving a type II move to the other side of a mountain

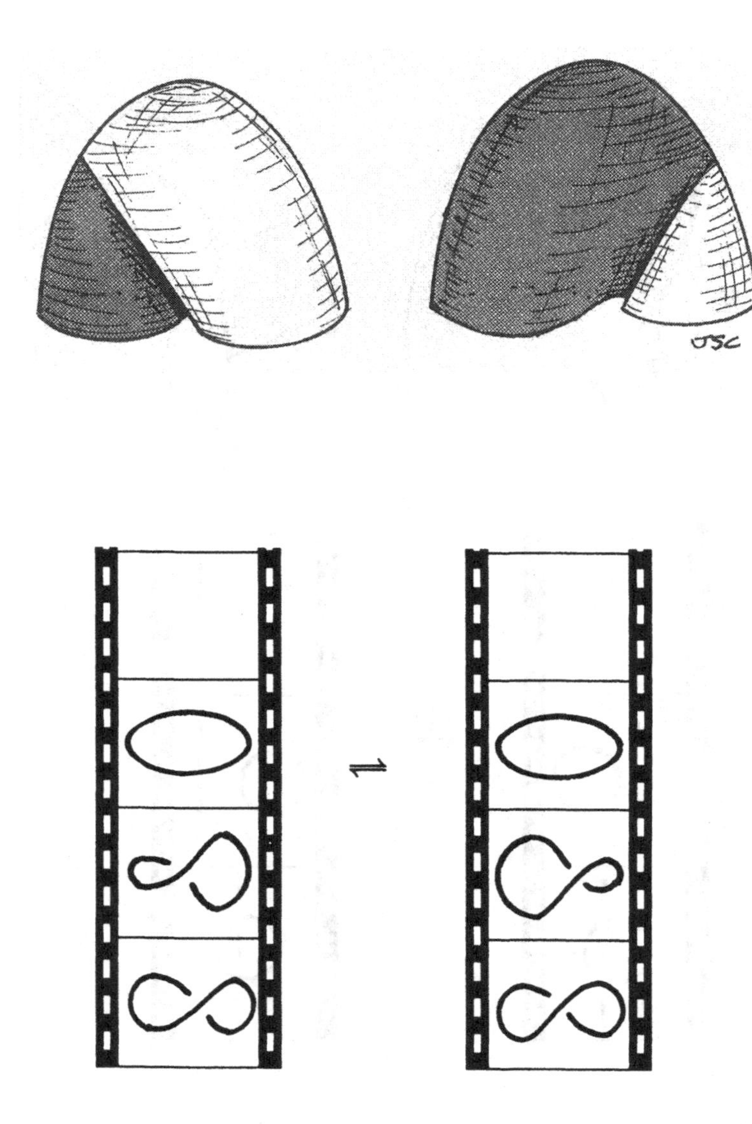

Figure 6.13: Moving a type I move to the other side of a mountain

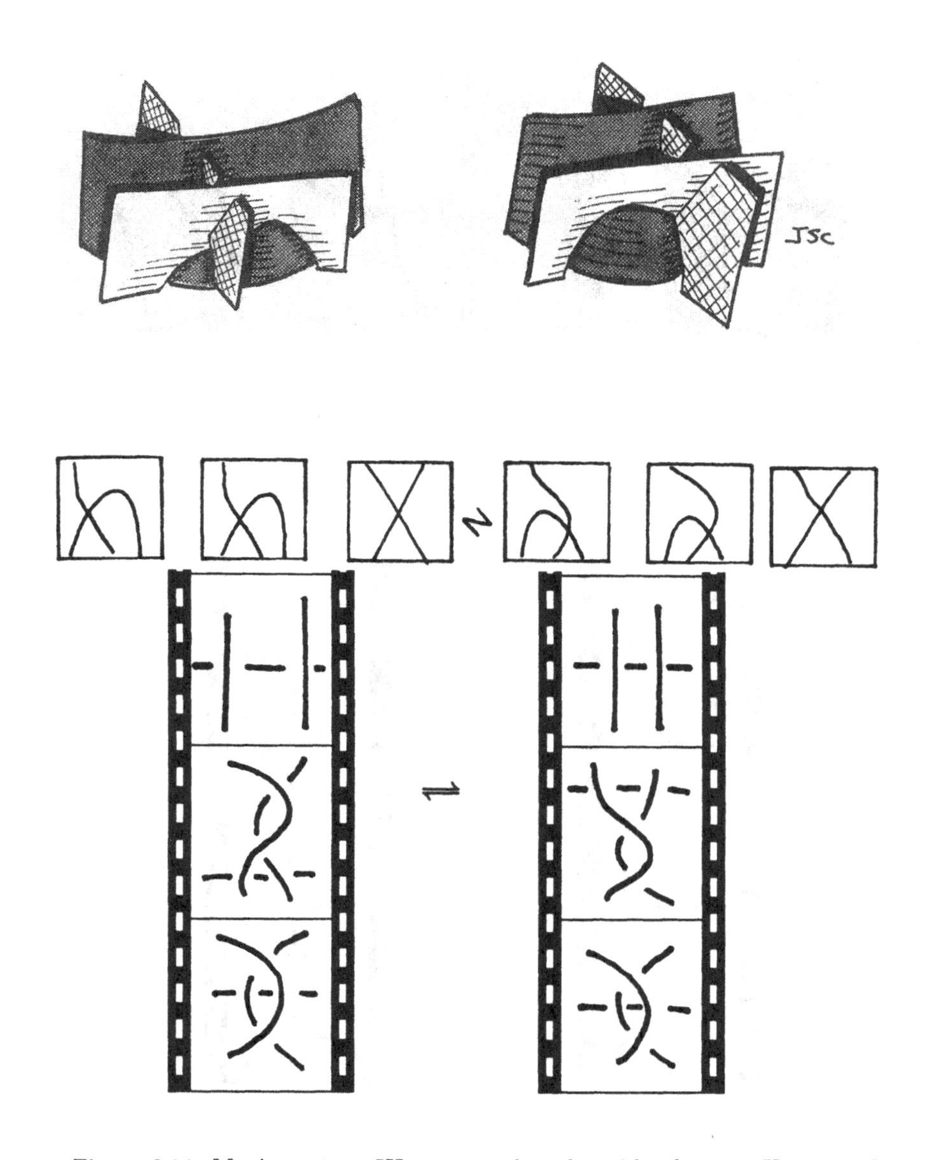

Figure 6.14: Moving a type III move to the other side of a type II mountain

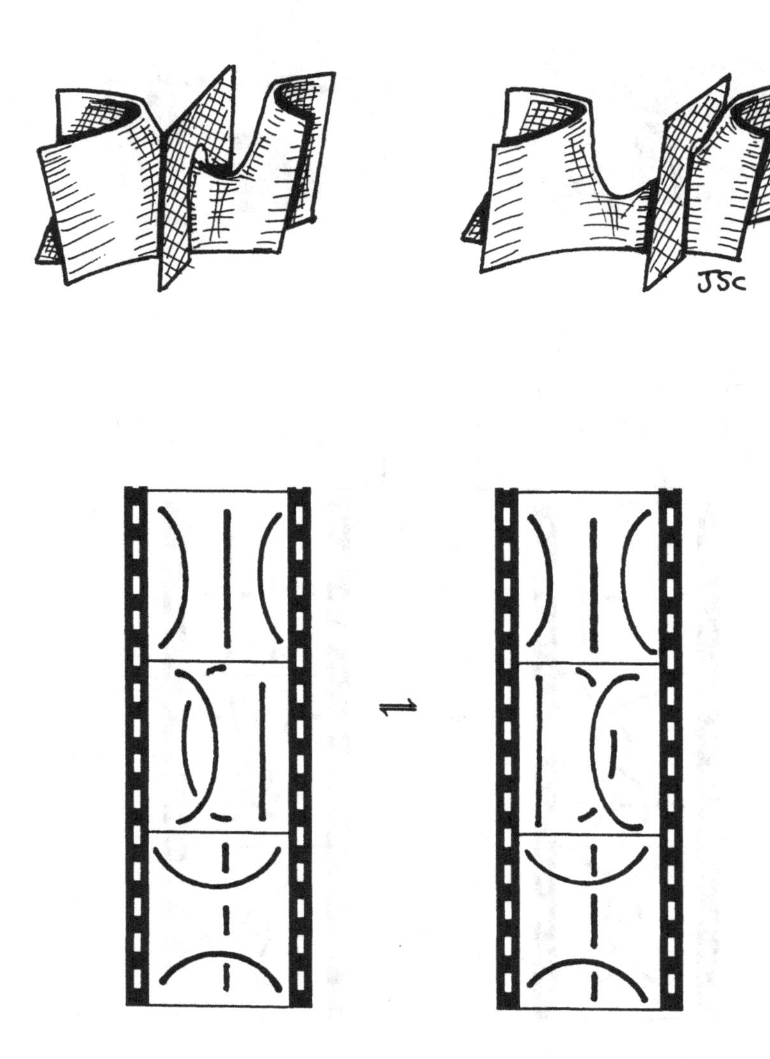

Figure 6.15: Moving a type II move to the other side of a saddle

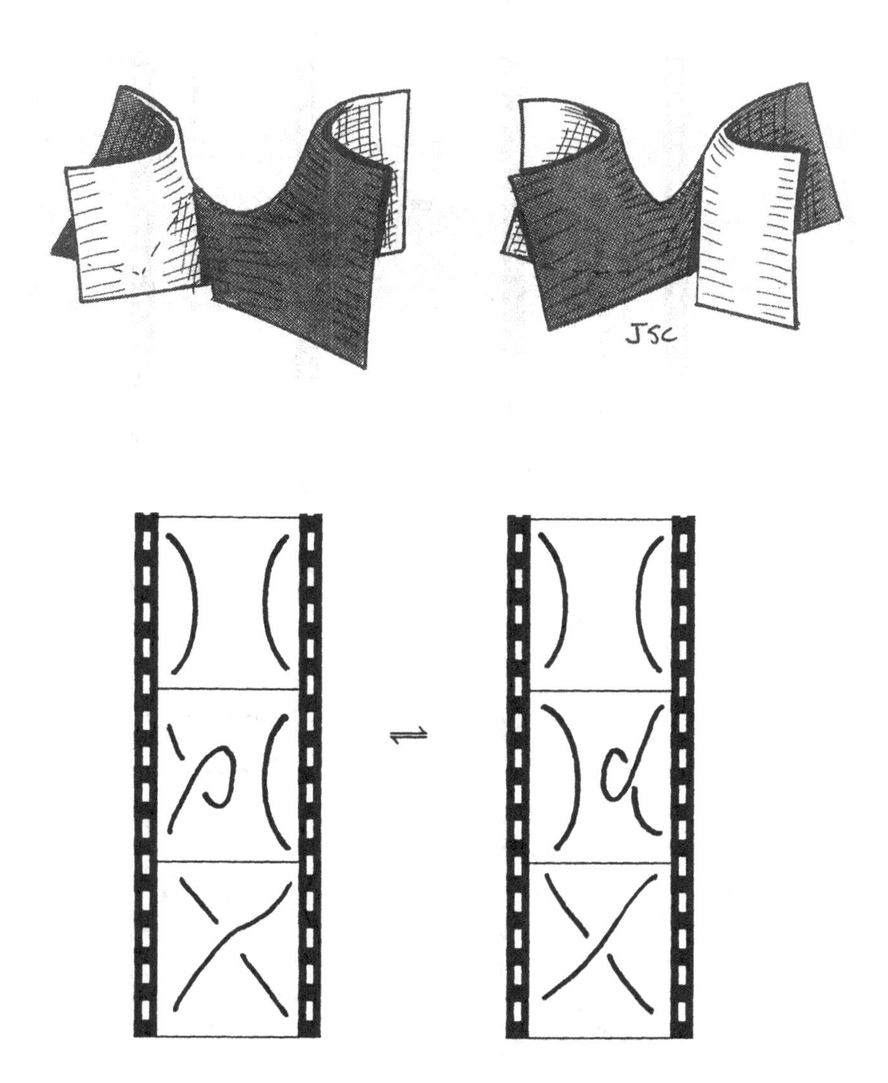

Figure 6.16: Moving a type I move to the other side of a saddle

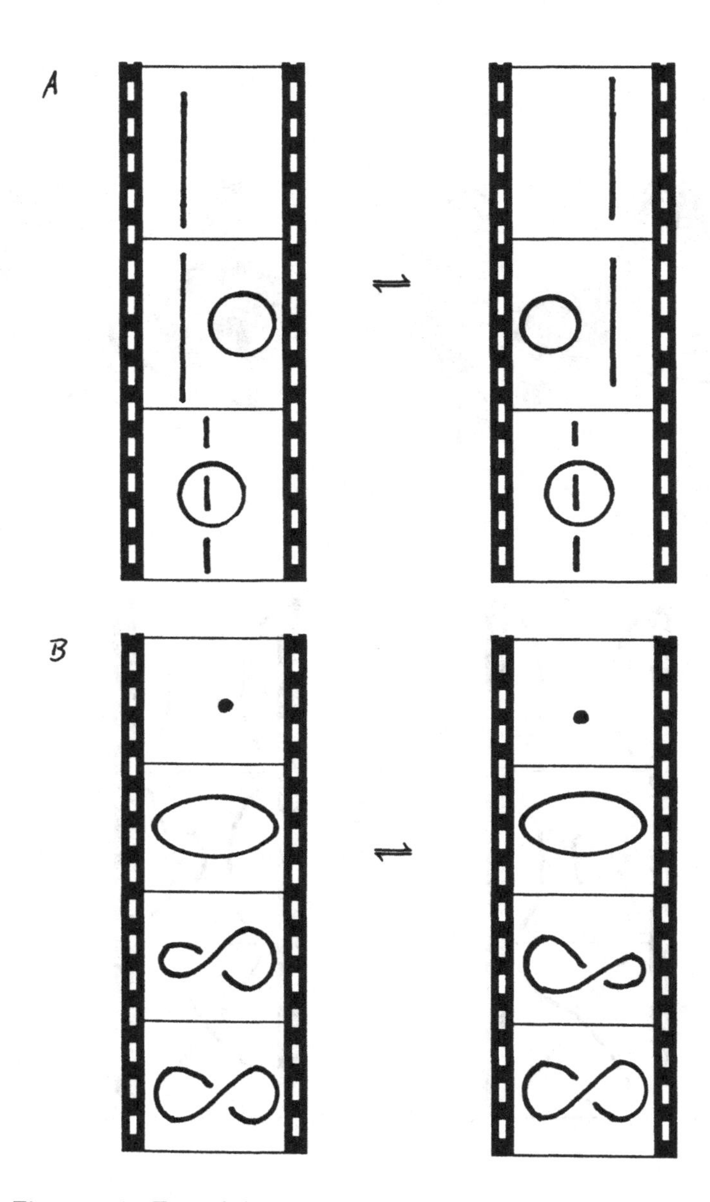

Figure 6.17: Two of the movie moves follow from the remaining ones

Proof of A.

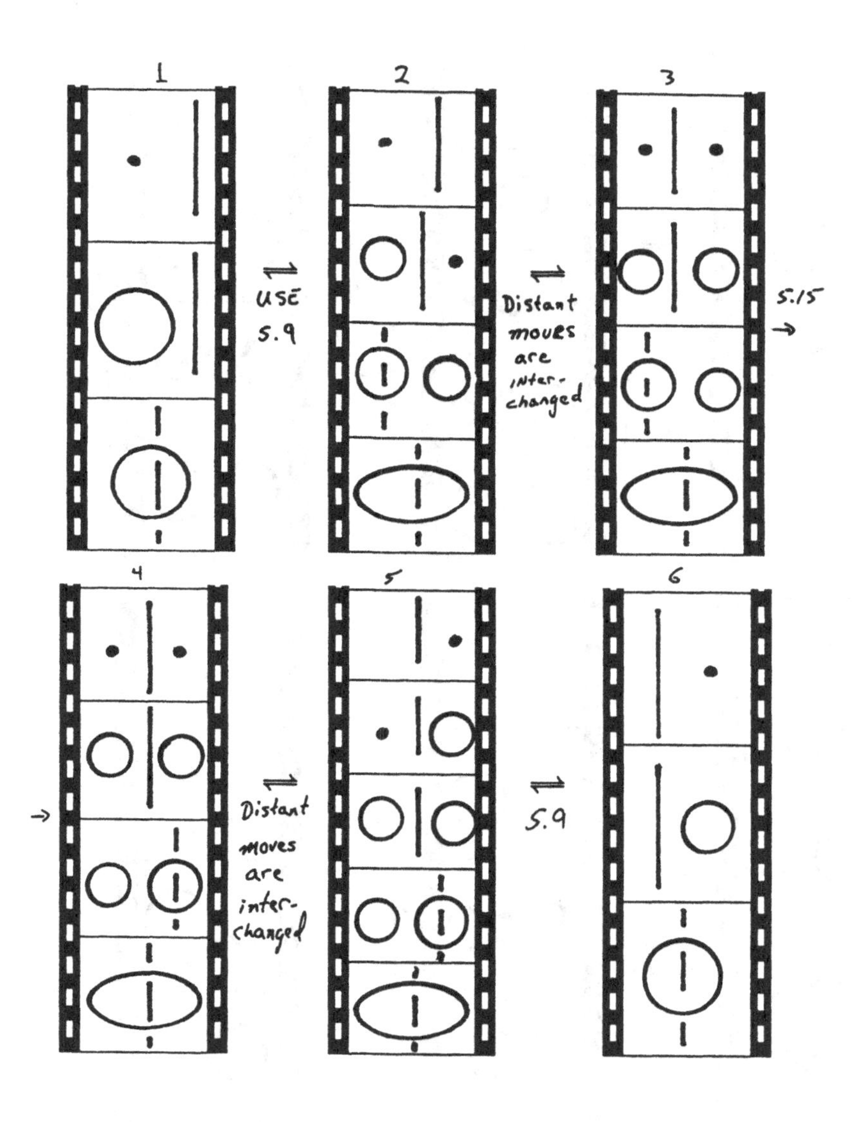
1
2
3
USE
5.9
Distant
moves
are
inter-
changed
5.15
4
5
6
Distant
moves
are
inter-
changed
5.9

Proof of B.

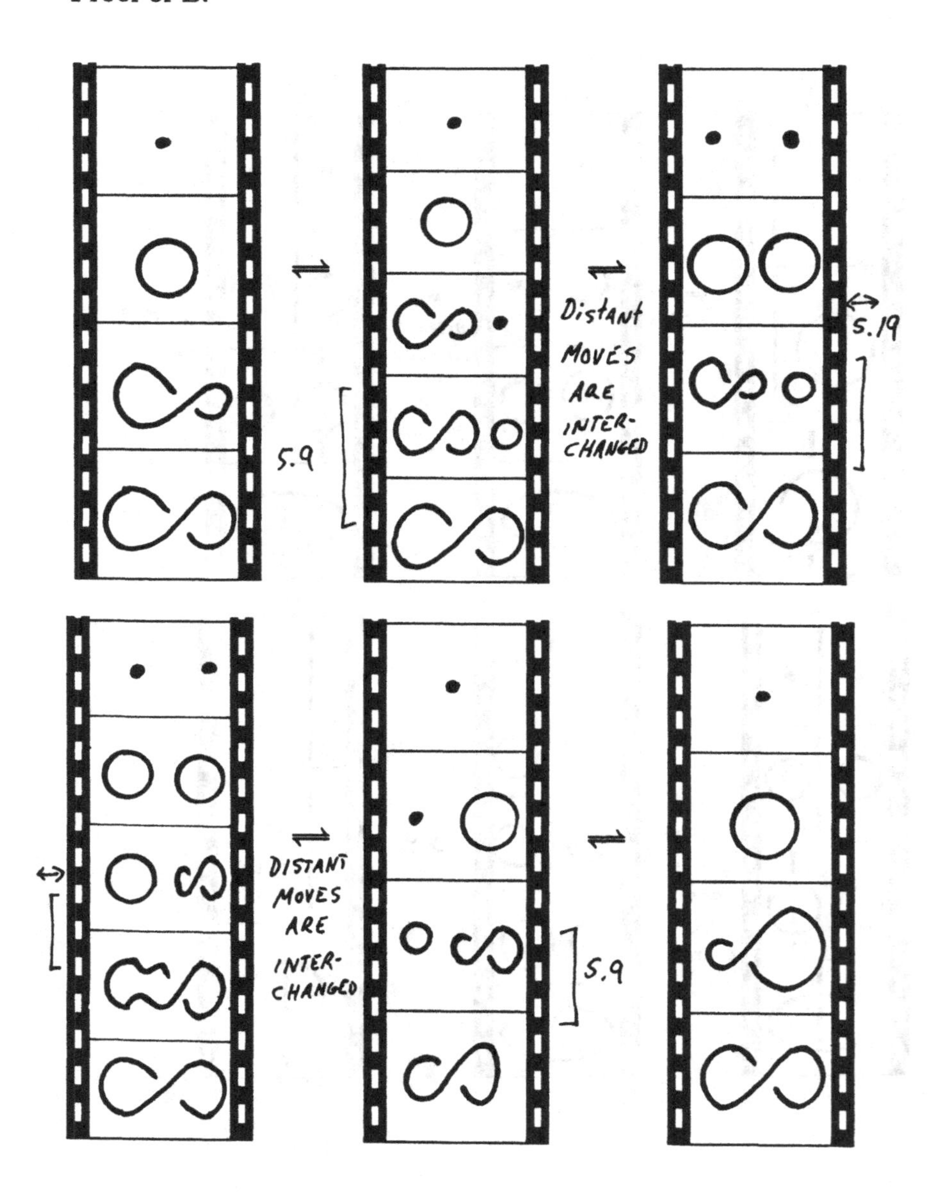

Distant
Moves
Are
Inter-
changed
5.9
5.19
DISTANT
MOVES
ARE
INTER-
CHANGED
5.9

There are other moves that are not depicted. These say that when two elementary string interactions occur on widely separated strings, then the order in which they occur is irrelevant.

A Remark about the Proofs

There is only one idea that is used in the proofs that the movie moves are sufficient when a height function is present and that the Roseman/Homma-Nagase moves are sufficient in the other case. The idea is general position. As I mentioned at the end of Chapter 1, the machinery of general position is complicated and delicate. But its usefulness is impeccable. First Roseman, and then Masahico and I, modified the general position machine to apply to the situation in which surfaces intersect themselves. There was a lot of technical work that we did to get these applications, but I think Dennis and Masahico will agree that the main idea of the proof was clear from the outset.

There are at least two types of proofs in mathematics. In the first type, you use established techniques to make slightly more progress on a given problem. In the second type of proof, a flash of insight into the problem appears. The proof springs, Athena-like, from your head — complete and fully grown. Ideas that spring forth, are the rewards of the painstaking day-to-day work. I am less familiar with the second type.

6.2 Klein Bottles

In this section, we use the movie moves to show that the Klein bottles depicted in Chapter 2 all represent the same surface in 4-space. The statements and proofs of this result are encoded entirely in pictures. In the pictures, the statement is given in terms of the projection of the surfaces into 3-space. Then a movie version of this projection is chosen and manipulated. The type of movie move that occurs and its location are indicated. When the movie move deformation is complete, the corresponding projection is shown. This figure is then rotated, to obtain a different movie parametrization, and the process continues.

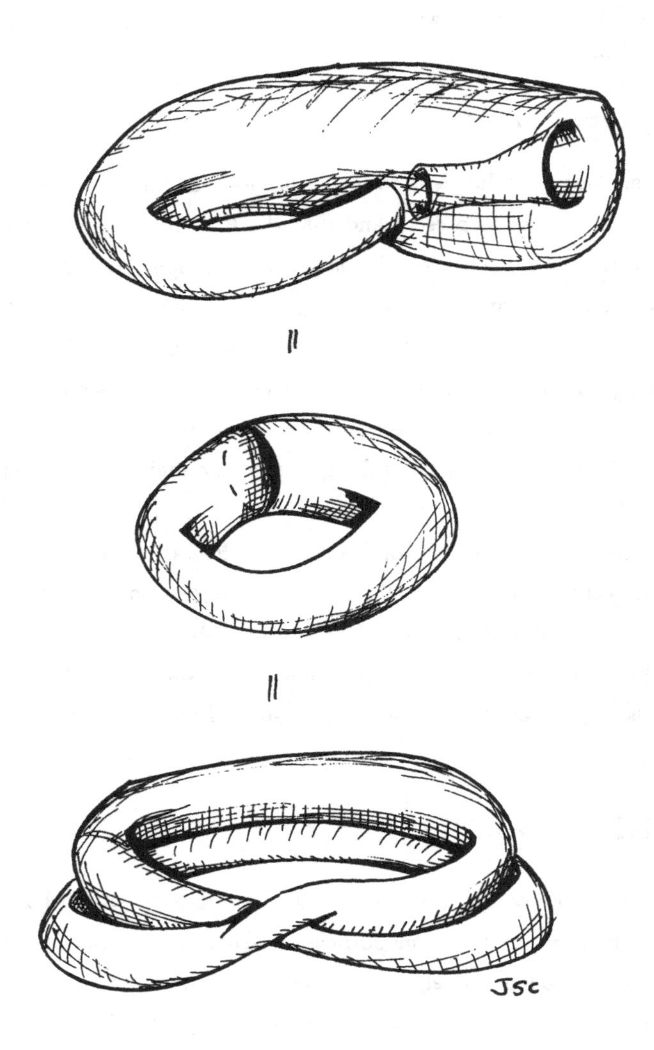

Figure 6.18: These Klein Bottles represent the same embeddings in 4-space

Proof.

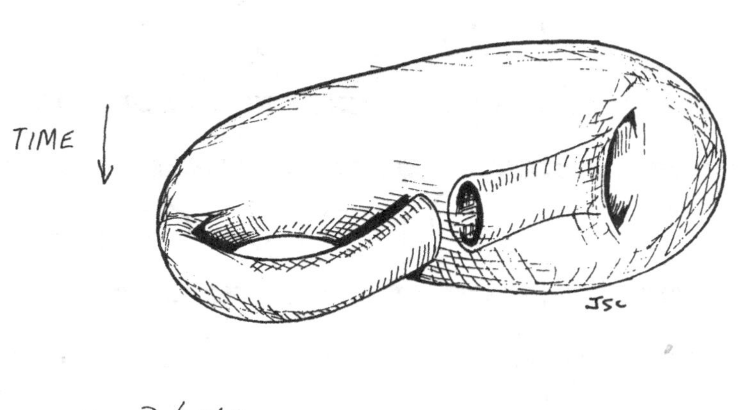

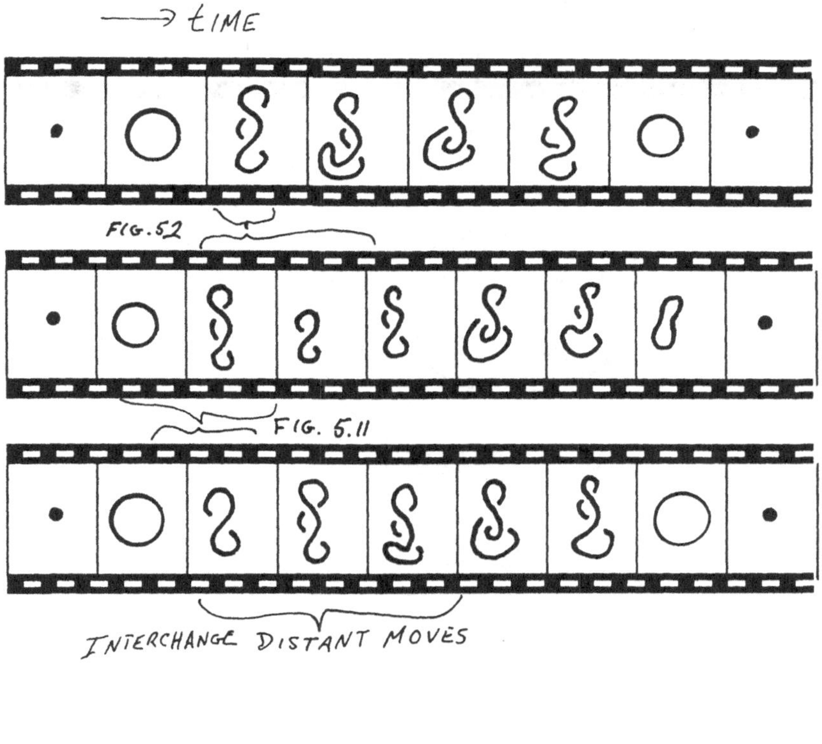

→ time

FIG.52

FIG. 5.11

INTERCHANGE DISTANT MOVES

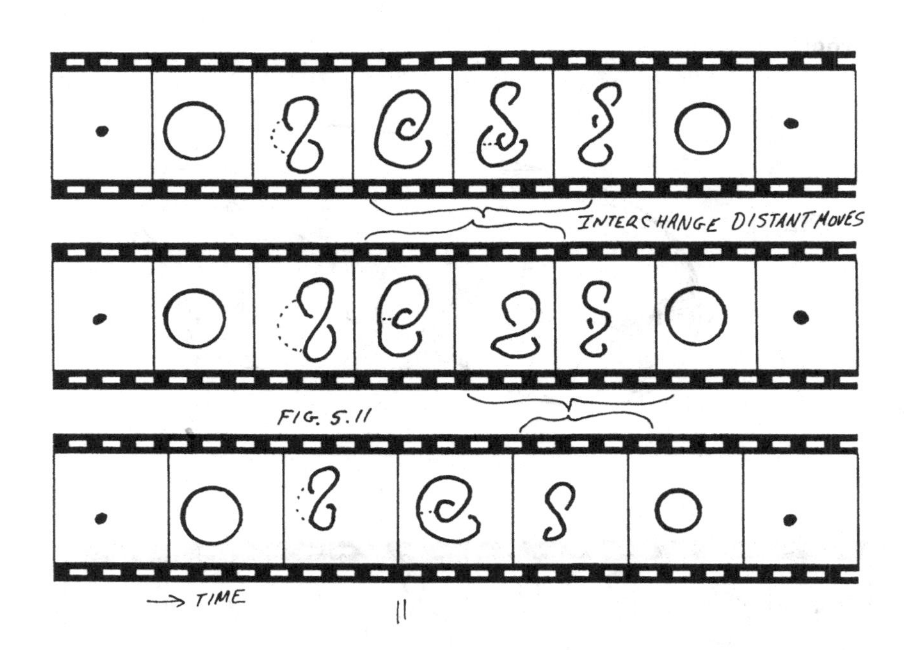
INTERCHANGE DISTANT MOVES
FIG. 5.11
→ TIME
||

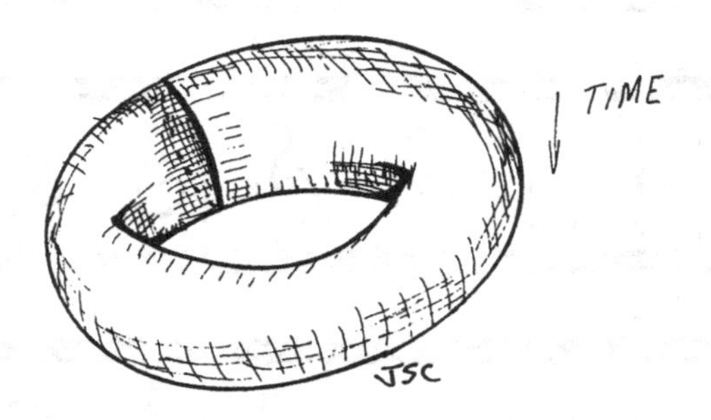
TIME
JSC

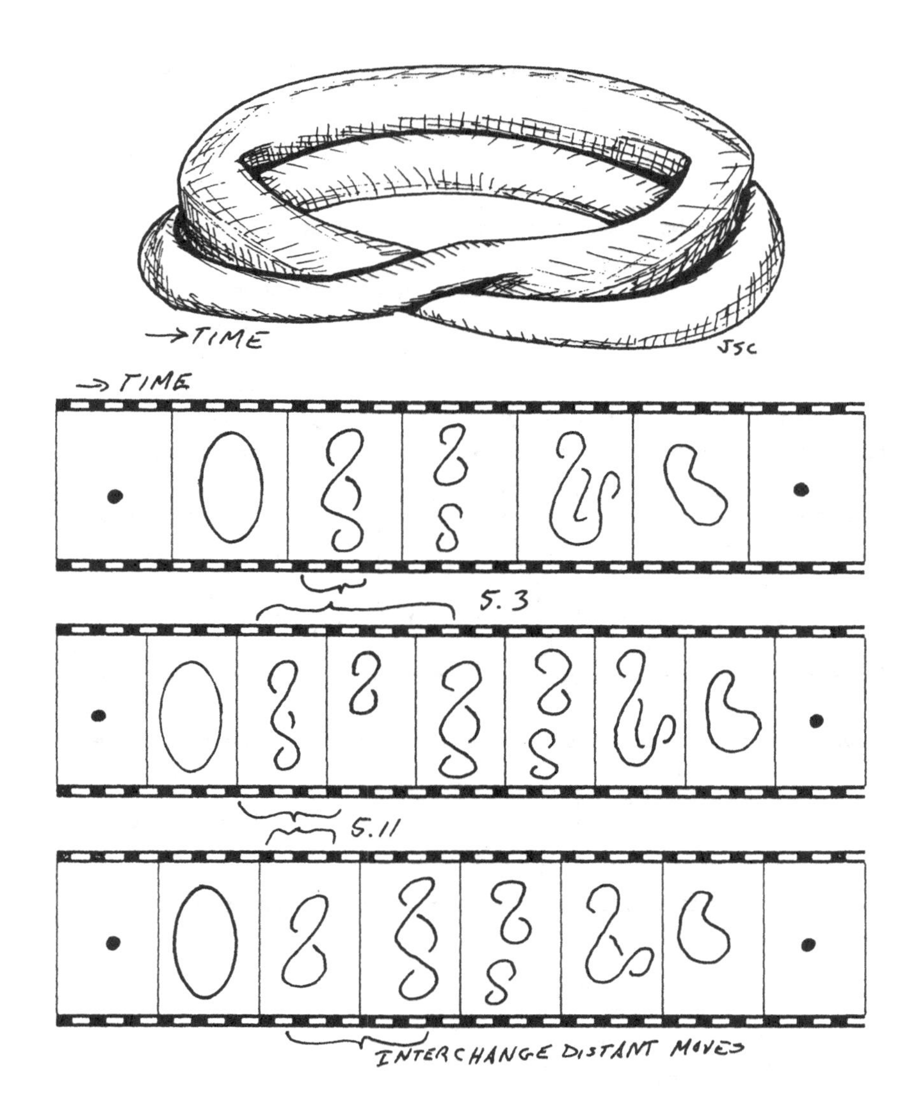
→TIME
JSC
→TIME
5.3
5.11
INTERCHANGE DISTANT MOVES

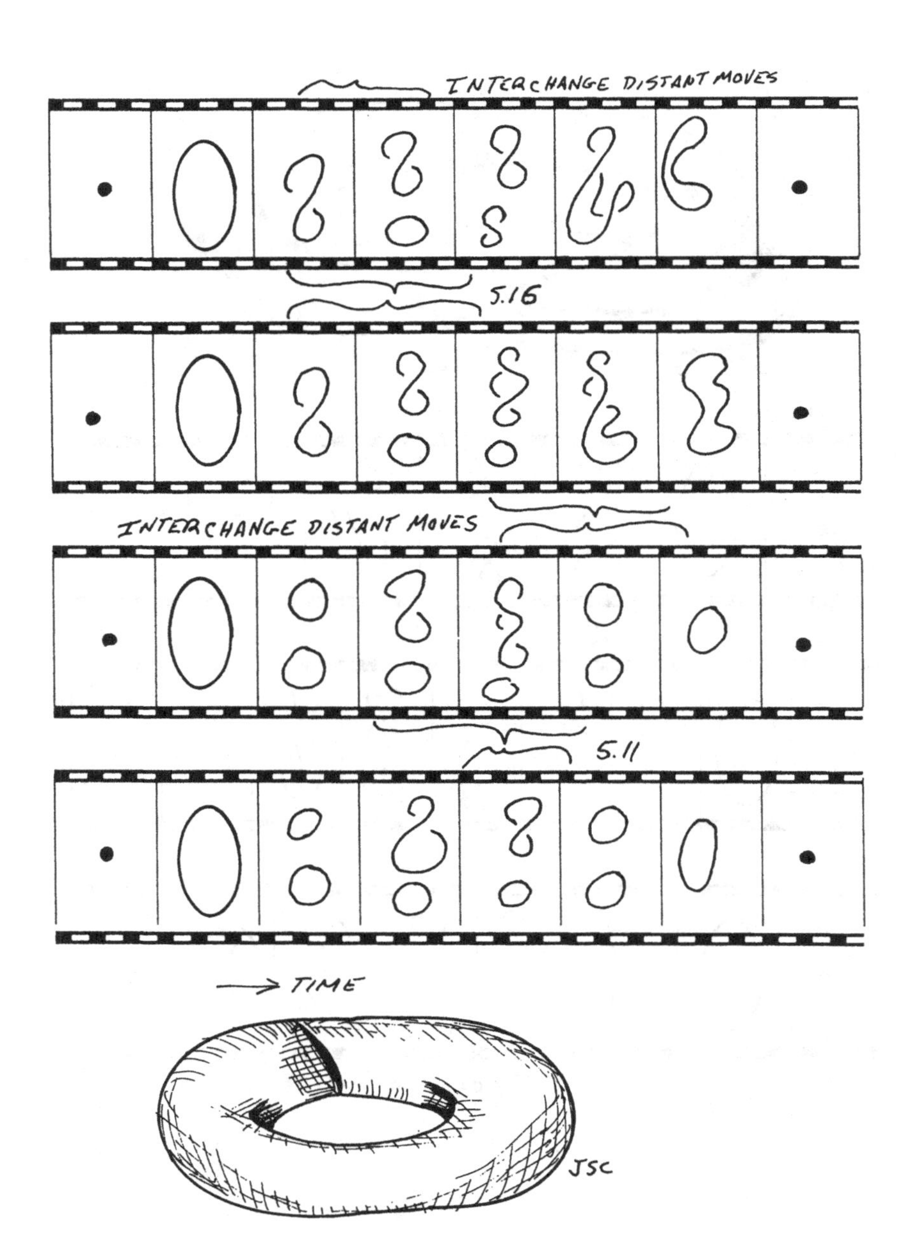
INTERCHANGE DISTANT MOVES
5.16
INTERCHANGE DISTANT MOVES
5.11
→ TIME
JSC

6.3 Projective Planes

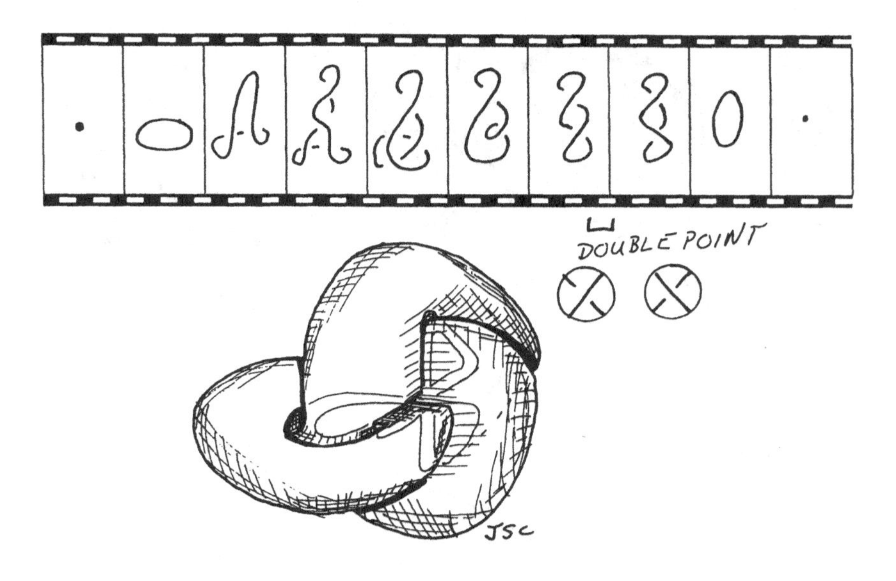

Figure 6.19: Boy's surface doesn't embed

Boy's surface is not the image of a projection from 4-dimensional space. In Figure 6.19, a movie of Boy's surface is depicted and this indicates a point at which crossings must be changed so that the surface can be pushed into 4-space. There is a deformation from Boy's surface to a cross cap that has a double point in 4-space. (This cross cap uses the silly double point of the disk bounded by the numeral 8). This deformation is depicted using the Roseman moves in Figure 6.20. Then, Boy's surface is deformed into a self intersecting Roman surface in Figure 6.21.

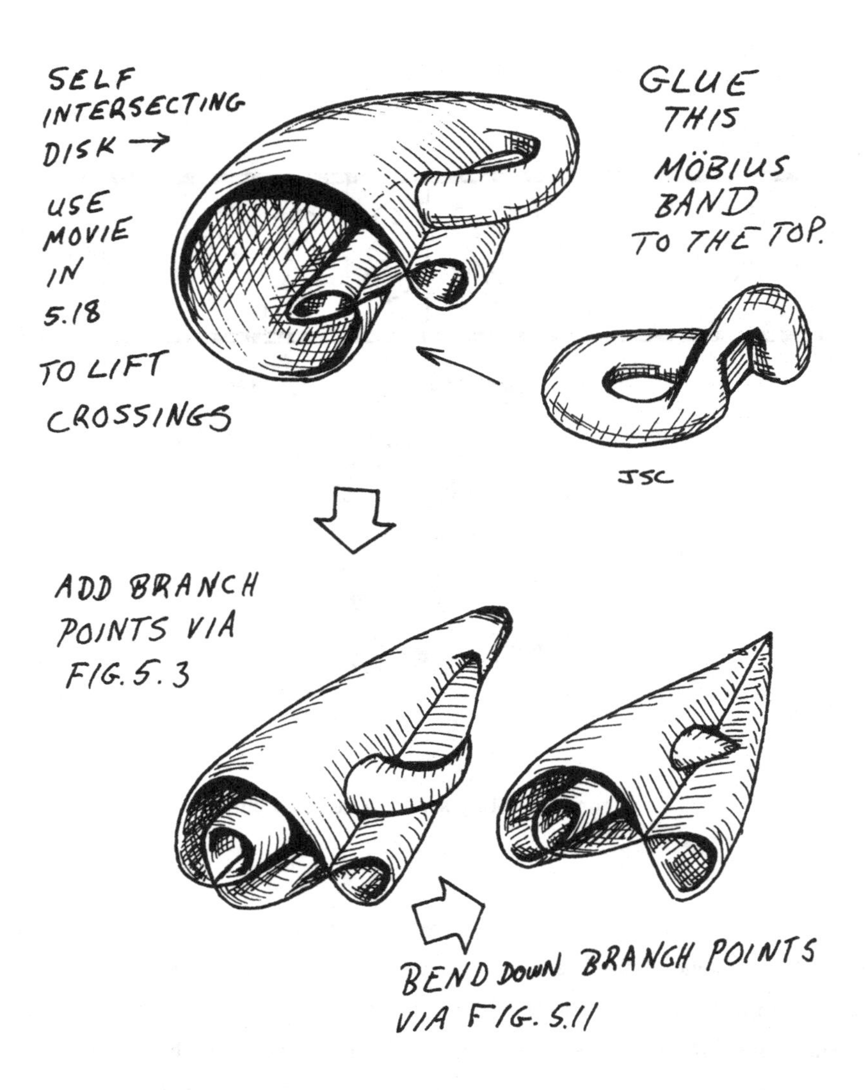

Figure 6.20: Deformation from Boy's surface to the self intersecting cross cap

FIG. 5.7
IS USED
TO REMOVE TRIPLE
POINT

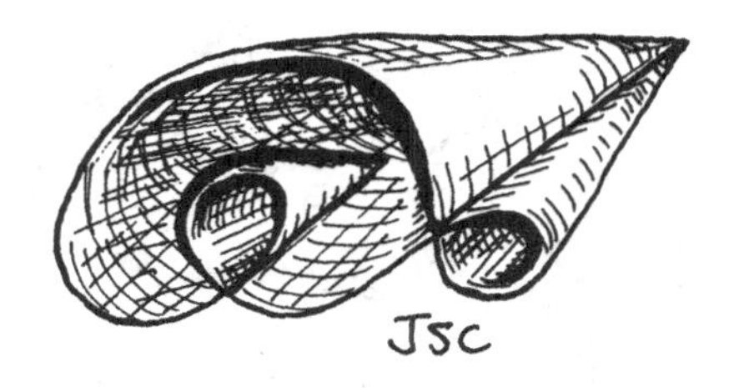

CROSSING
INFORMATION
IS SHOWN
FOR DISK

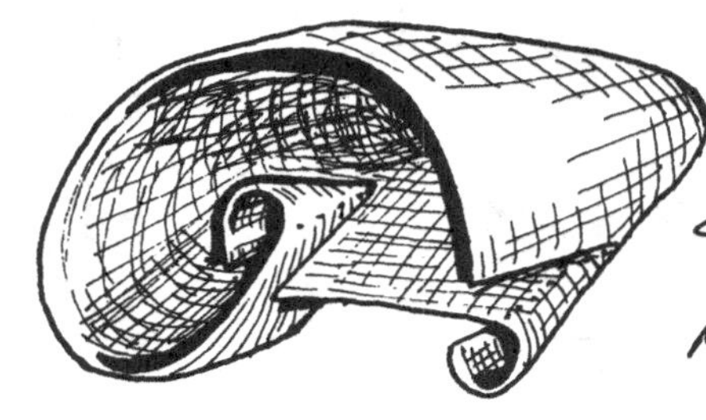

AND FOR
MÖBIUS BAND

NOTE DOUBLE
POINT TO THE
RIGHT

DISK + MOBIUS
BAND GIVE
CROSS CAP
MODEL
WITH
A DIFFERENT
HEIGHT FUNCTION.

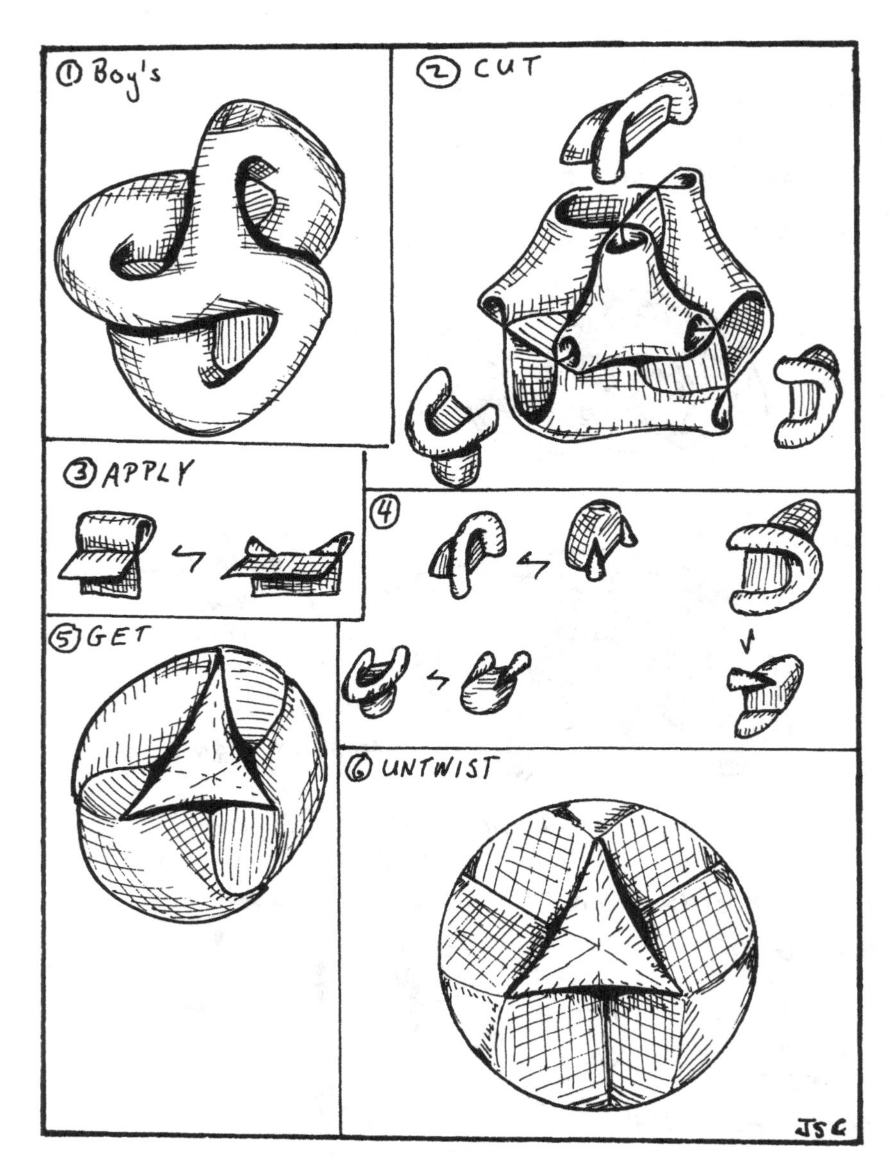

Figure 6.21: Deformation of Boy's surface into a Roman Surface

6.4 Everting the 2-sphere

The deformations among these non-orientable surfaces involved the introduction and canceling of branch points. Some surfaces can be deformed in ways that don't involve branch points. A startling result, due to Steve Smale [71], is that the 2-sphere can be turned inside out without any branch points. The word **eversion** is used to denote the process of turning a 2-sphere inside out.

Smale's original proof was not really constructive. That is, Smale showed that there is a way to turn the sphere inside out, but he didn't show how it could be accomplished. The best known construction of everting the sphere is due to Froissart and Morin [61]. Models of the Froissart-Morin eversion were constructed by Charles Pugh [63] when he was a graduate student. These models hung in a hallway in Berkeley until they were stolen. Nelson Max made some precise measurements of the models and made a computer animation of the eversion [58]. Since then John Hughes has taken the Max film and allowed a computer user to move the animation around in real time. There is a new film produced by the Geometry Center at the University of Minnesota called "Outside In," and Apery has some other animations of sphere eversions based on some new parameterizations of Morin.

Here I am returning to the low-tech realm of ink and paper, and I am illustrating the eversion in terms of the movie moves. I believe that these illustrations improve upon those given in Anthony Phillips's Scientific American [62] article because each change in the movie is understood in terms of a basic change. The draw back is that many more movies are needed to describe the process.

In the illustrations, each movie film represents a map of the 2-sphere as it is immersed. The figures at the top of the pages show the self intersection set of the sphere, and the regions where the changes will occur are marked. Since the process is given in terms of movies, the relative heights of the triple points and the maxima on the double point sets have to change often. When it won't cause confusion, some of the height changing movie moves will occur at the same time. The numbers on the double decker arcs indicate the Gauss codes of the corresponding frames in the movie. When arcs on the left interact with arcs on the right the arcs move around the back of the sphere. This corresponds to changing the starting point that is used to compute the Gauss code.

The north and south poles of the sphere are indicated on the preimage surface and within the movies, so that at the end of the movie we see that North and south have been interchanged.

I have made some efforts to make the movies topographically consistent with the Figures in the Max cinema of the eversion, but sometimes that is not possible if one is to see clearly the movie move that applies. The movie moves that are used do not contain crossing information. The entire process is occuring in 3-dimensions.

Near the end of the process, instead of unraveling the rest of the surface by means of movie moves, I draw the image, rotate in space, and then illustrate the rotation via a movie. This rotation simplifies the process a great deal since rotations are hard to achieve via movie moves. Similar rotations were used in deforming the Klein bottles. The rotation used here could have been performed shortly after the quadruple point, but the correspondences among the double decker sets would be more difficult to see. I am greatful to John Hughes for suggesting this move.

Hughes also has worked out a sequence of movies for the eversion, but his slices are with a different set of planes. The rotational symmetry of each still in the eversion is seen from top to bottom in the movies that I present. Hughes maintains symmetry in each still.

The way to read these figures is to play the game from the comic strip "Hocus Focus." Namely, examine a pair of successive figures and see how they differ. Similarly, examine how the successive stills from a movie differ. Most will differ by an elementary string interaction, but some will only differ by a topologically trivial move in which the topography has changed. When no topological change occurs, an equal sign appears.

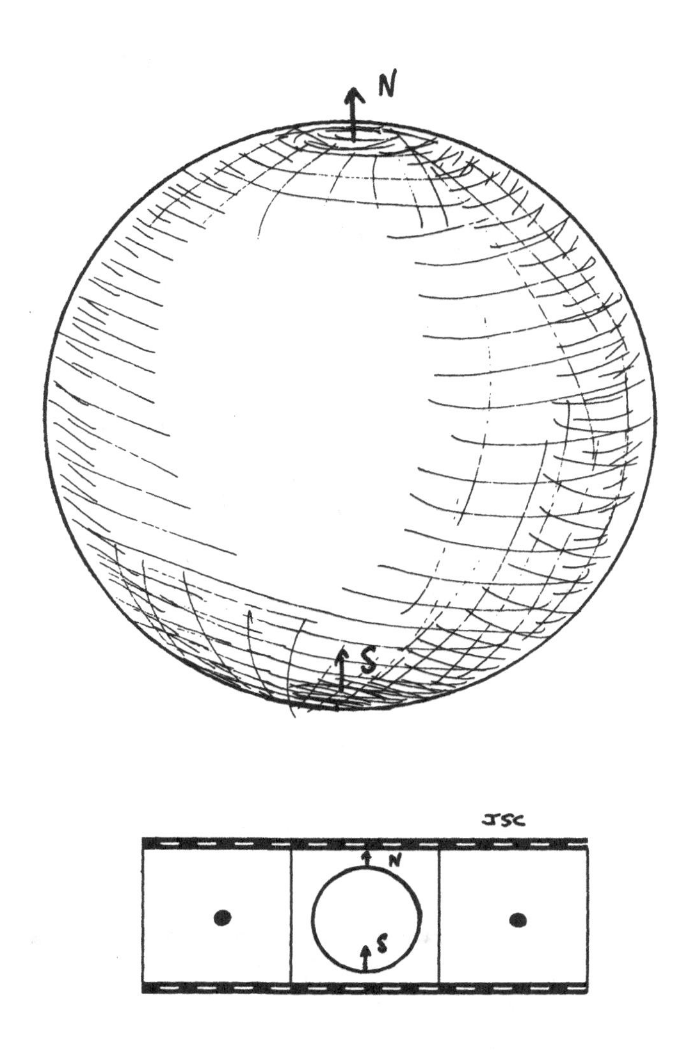

Figure 6.22: Everting the 2-sphere

... ADD CIRCLE OF DOUBLE POINTS via Bubble Move 5.4.

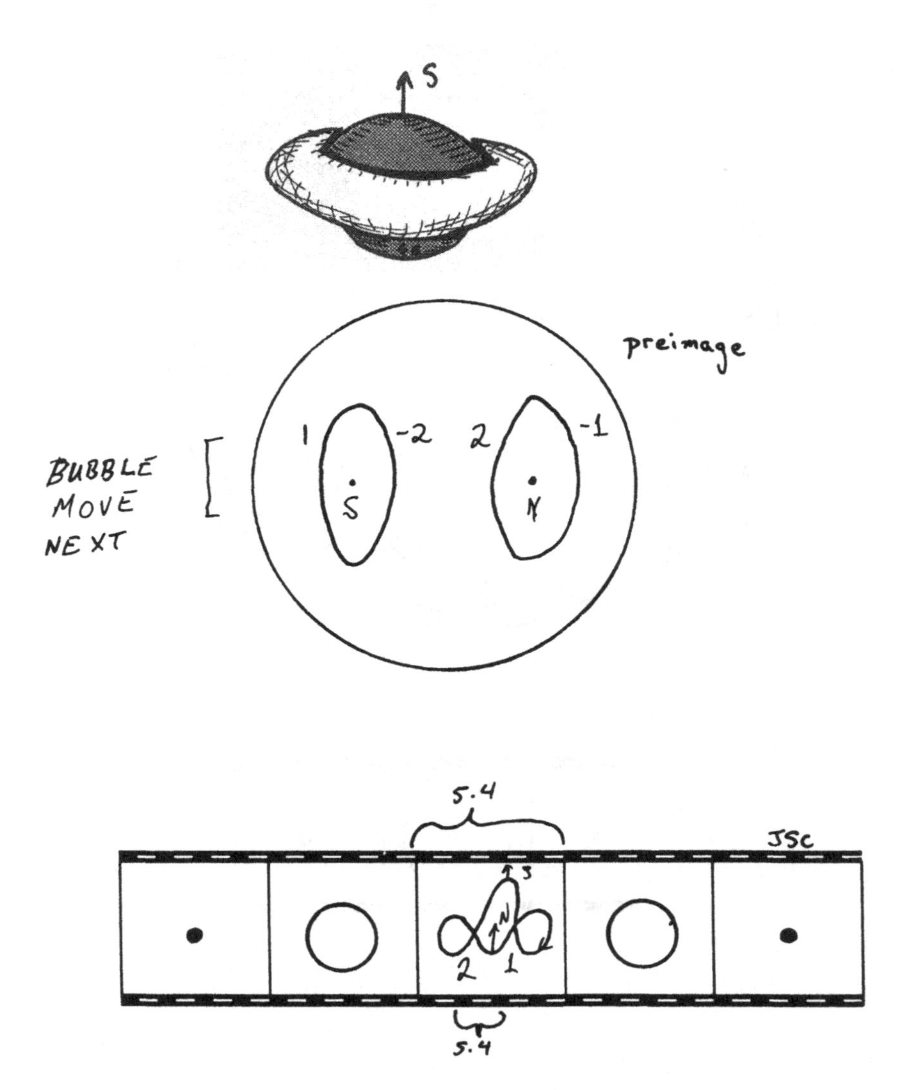

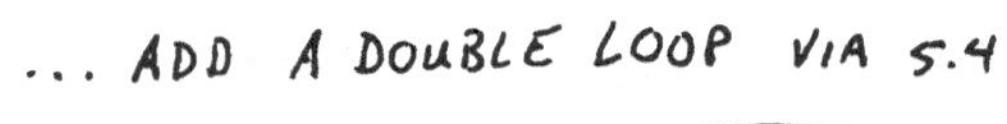
... ADD A DOUBLE LOOP VIA 5.4

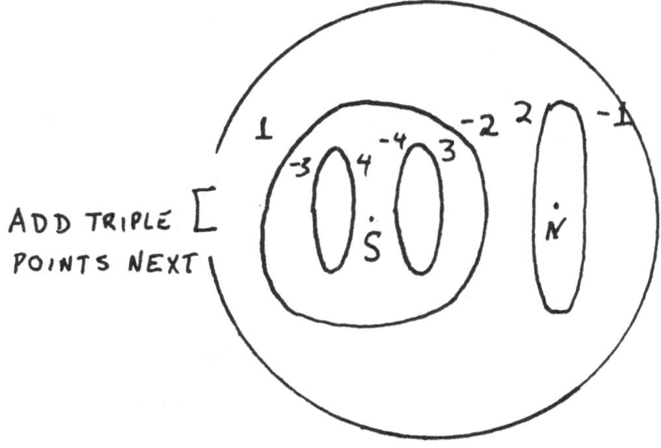
ADD TRIPLE POINTS NEXT
1
-3
4
-4
3
-2
2
-1
S
N

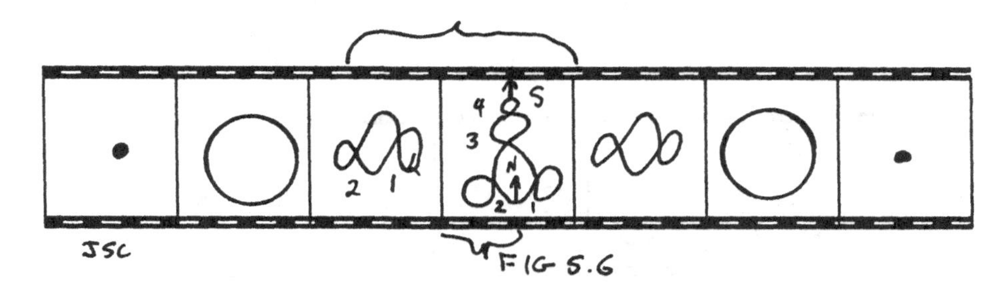
BUBBLE MOVE FIG 5.4
2
1
4
S
3
N
2
1
JSC
FIG 5.6

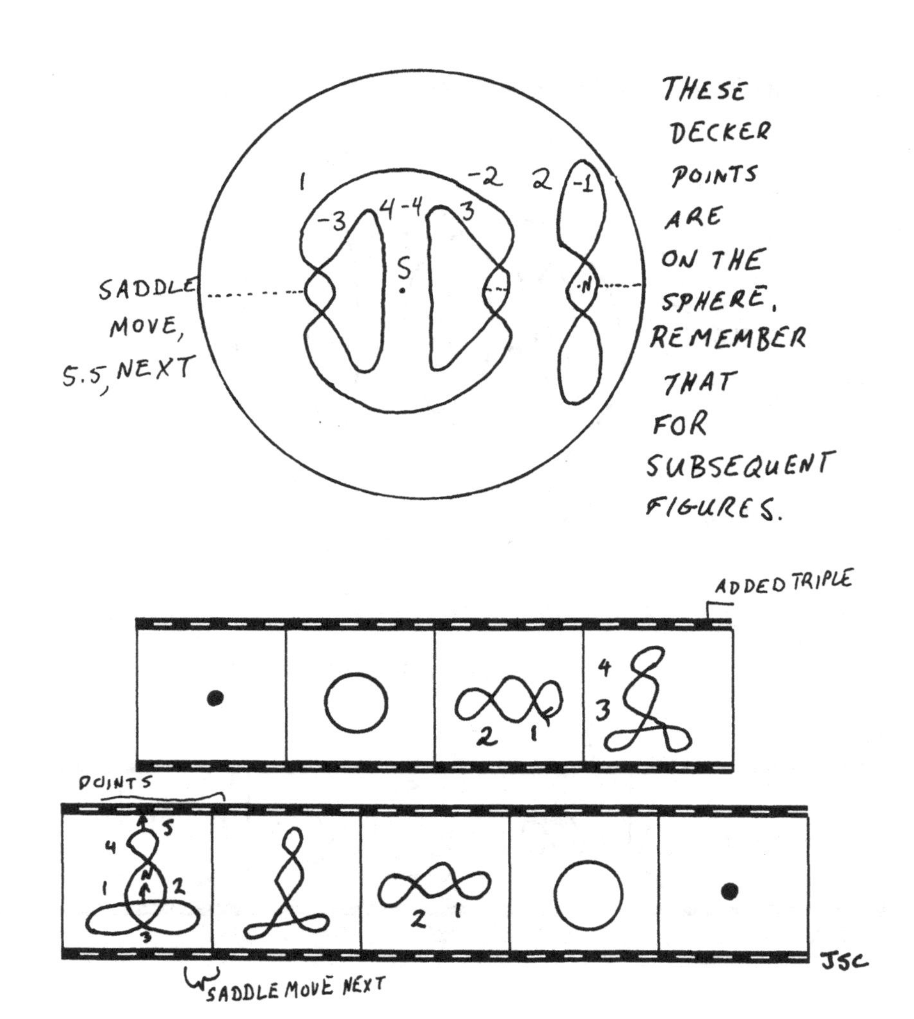
THESE DECKER POINTS ARE ON THE SPHERE. REMEMBER THAT FOR SUBSEQUENT FIGURES.
SADDLE MOVE, 5.5, NEXT
1
-2
2
-1
-3
4 -4
3
S
N
ADDED TRIPLE
4
3
2
1
POINTS
5
N
JSC
SADDLE MOVE NEXT

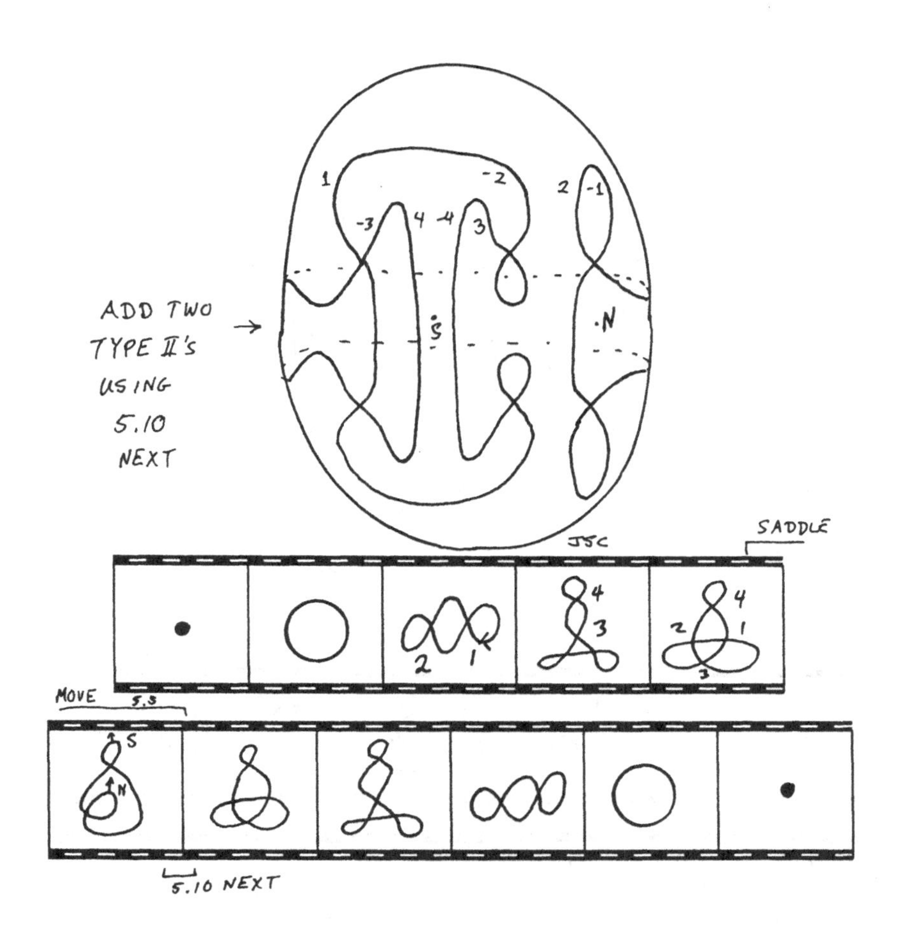
ADD TWO
TYPE II's
USING
5.10
NEXT
S
N
1
-2
2
-1
-3
4
-4
3
JSC
SADDLE
2
1
4
3
4
2
1
3
MOVE
5.3
S
N
5.10 NEXT

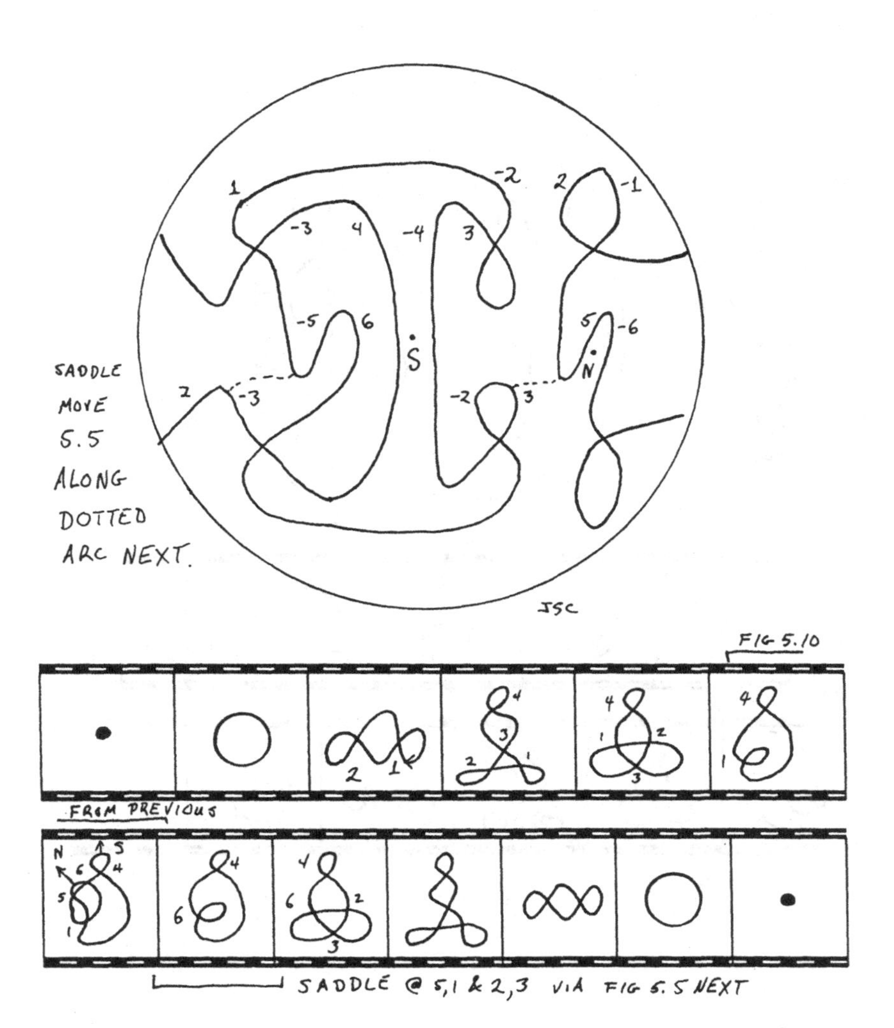
SADDLE
MOVE
5.5
ALONG
DOTTED
ARC NEXT.
S
N
JSC
FIG 5.10
FROM PREVIOUS
SADDLE @ 5,1 & 2,3 VIA FIG 5.5 NEXT

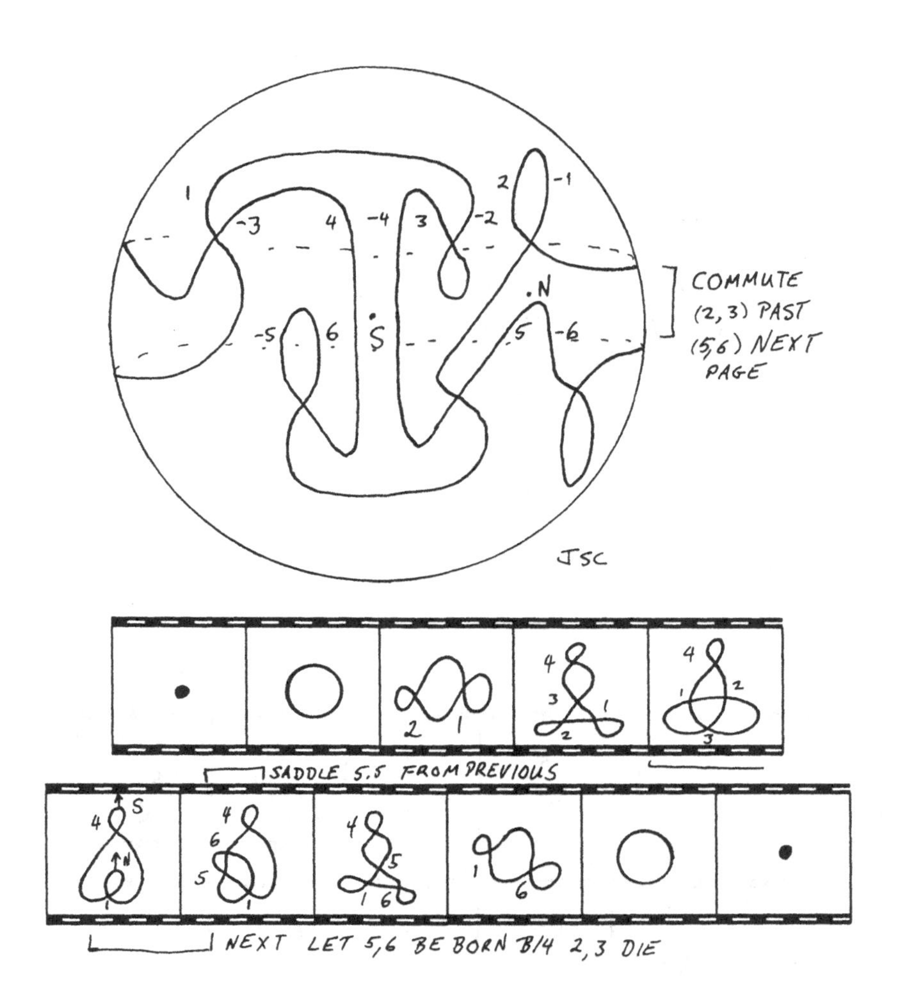

1
-3
4
-4
3
-2
2
-1
.N
-5
6
S
5
-6
COMMUTE
(2,3) PAST
(5,6) NEXT
PAGE
JSC
4
3
1
2
SADDLE 5.5 FROM PREVIOUS
S
N
6
5
NEXT LET 5,6 BE BORN B/4 2,3 DIE

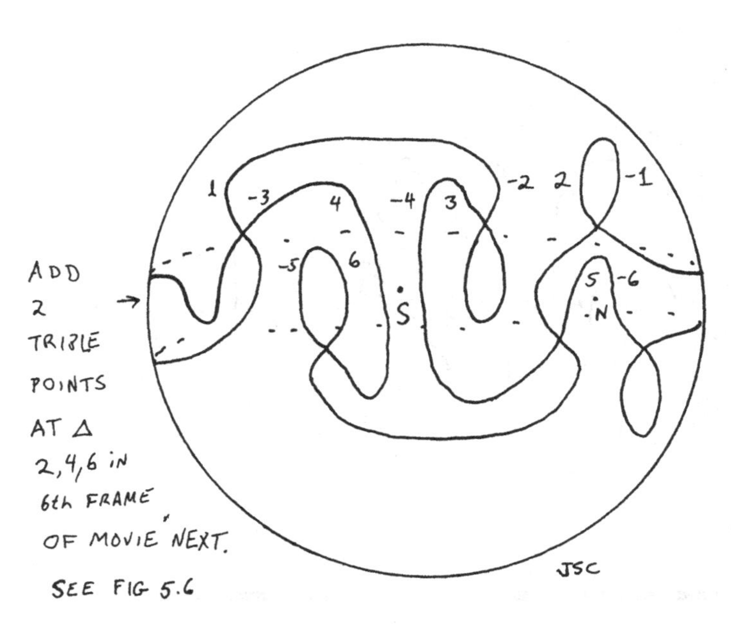

ADD
2
TRIPLE
POINTS
AT Δ
2,4,6 IN
6th FRAME
OF MOVIE NEXT.
SEE FIG 5.6
1
-3
4
-4
3
-2
2
-1
-5
6
S
5
-6
N
JSC

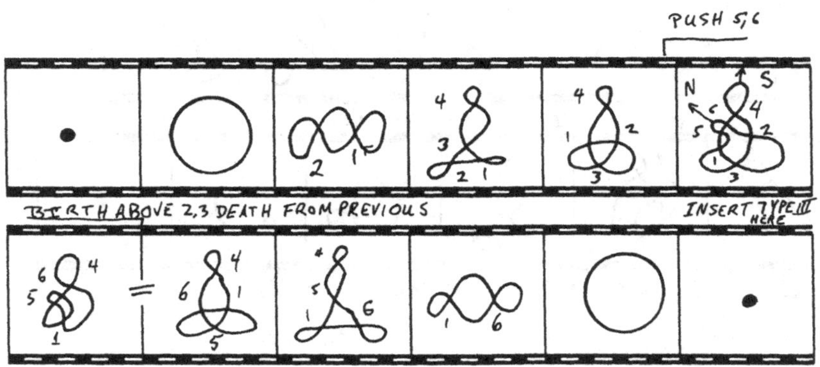

PUSH 5,6
BIRTH ABOVE 2,3 DEATH FROM PREVIOUS
INSERT TYPE III HERE
N
S

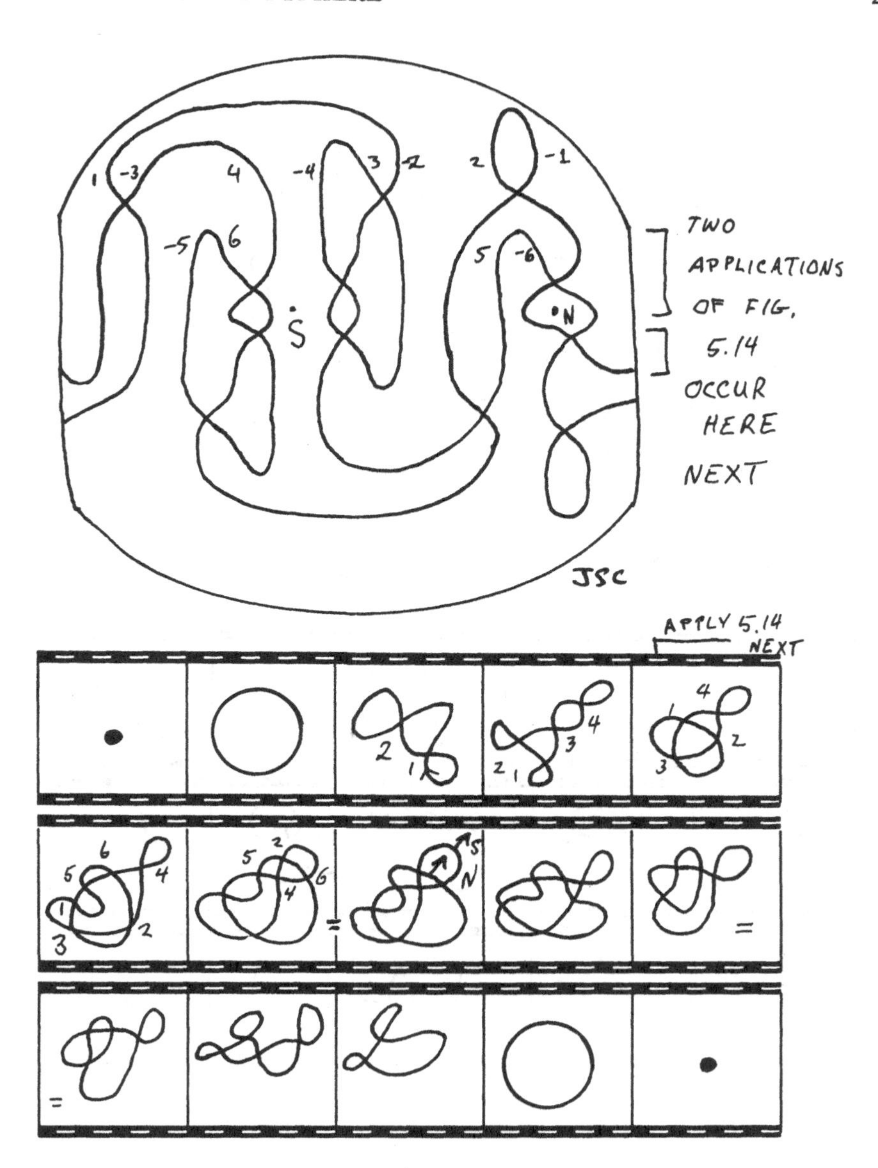
1
-3
4
-4
3
-2
2
-1
-5
6
5
-6
S
N
TWO APPLICATIONS OF FIG. 5.14 OCCUR HERE NEXT
JSC
APPLY 5.14 NEXT
4
1
2
3
4
2
1
2
1
3
6
5
4
1
2
3
5
2
6
4
=
S
N
=
=

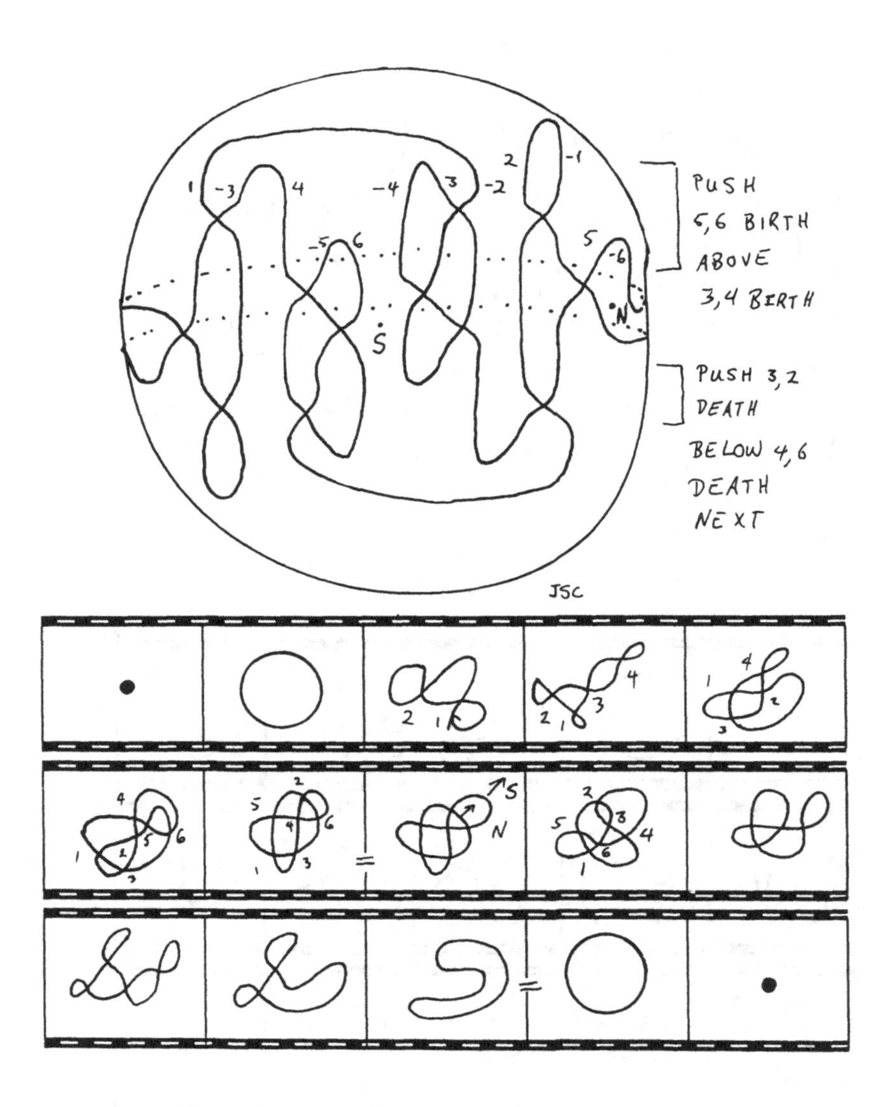
PUSH
5,6 BIRTH
ABOVE
3,4 BIRTH
PUSH 3,2
DEATH
BELOW 4,6
DEATH
NEXT
JSC

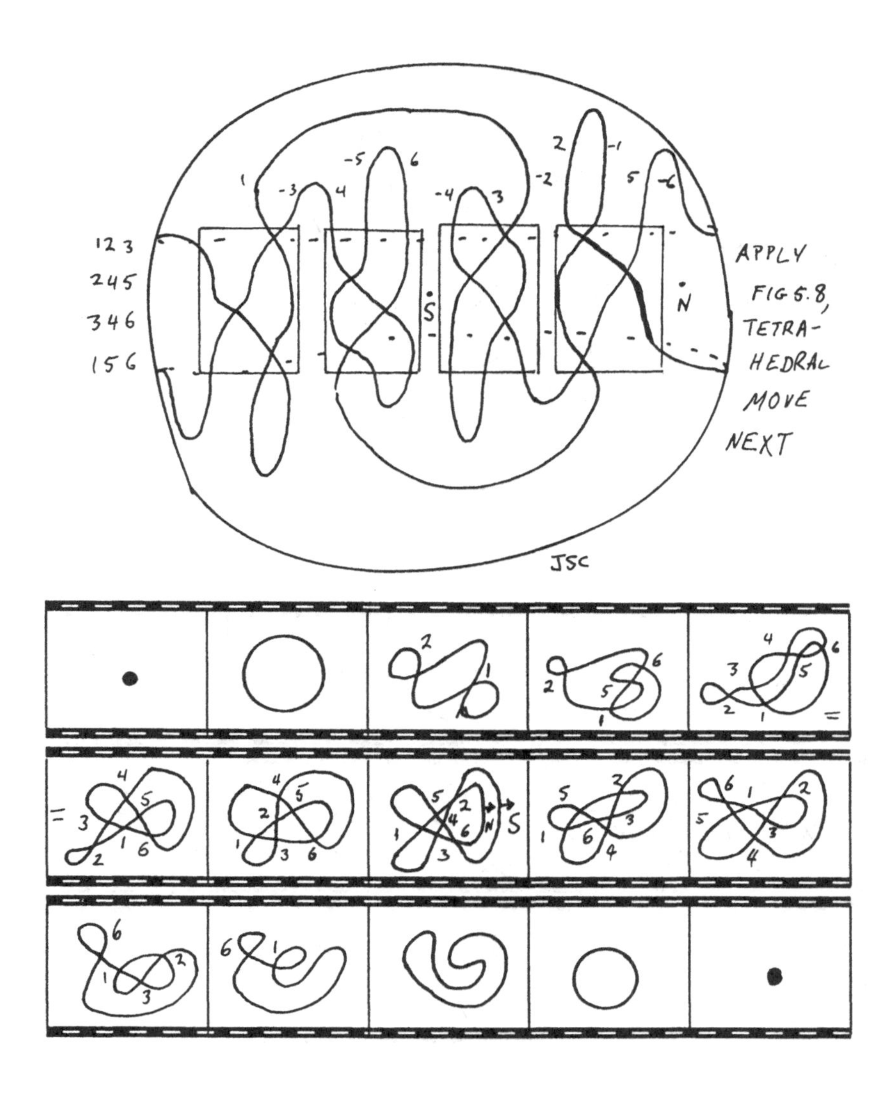
123
245
346
156
APPLY
FIG 5.8,
TETRA-
HEDRAL
MOVE
NEXT
JSC

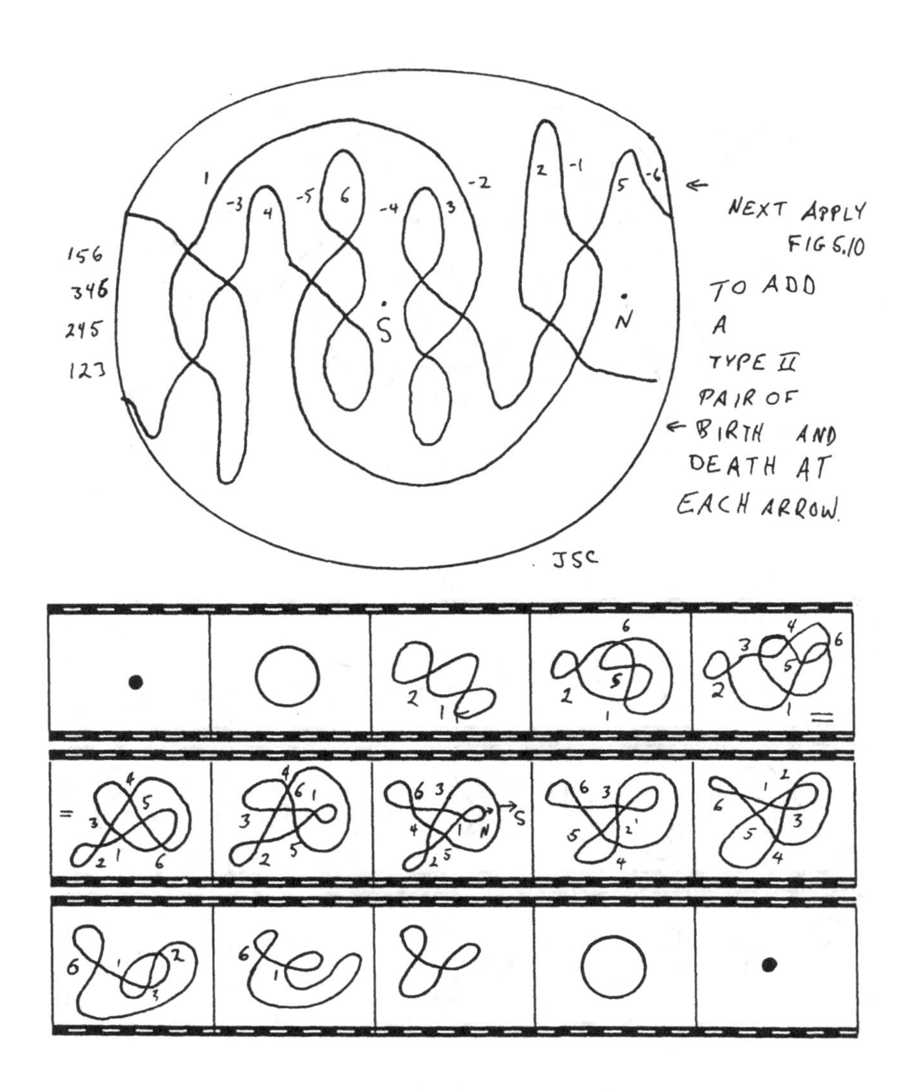
NEXT APPLY FIG 5.10
TO ADD A TYPE II PAIR OF BIRTH AND DEATH AT EACH ARROW.
JSC
156
346
245
123
S
N

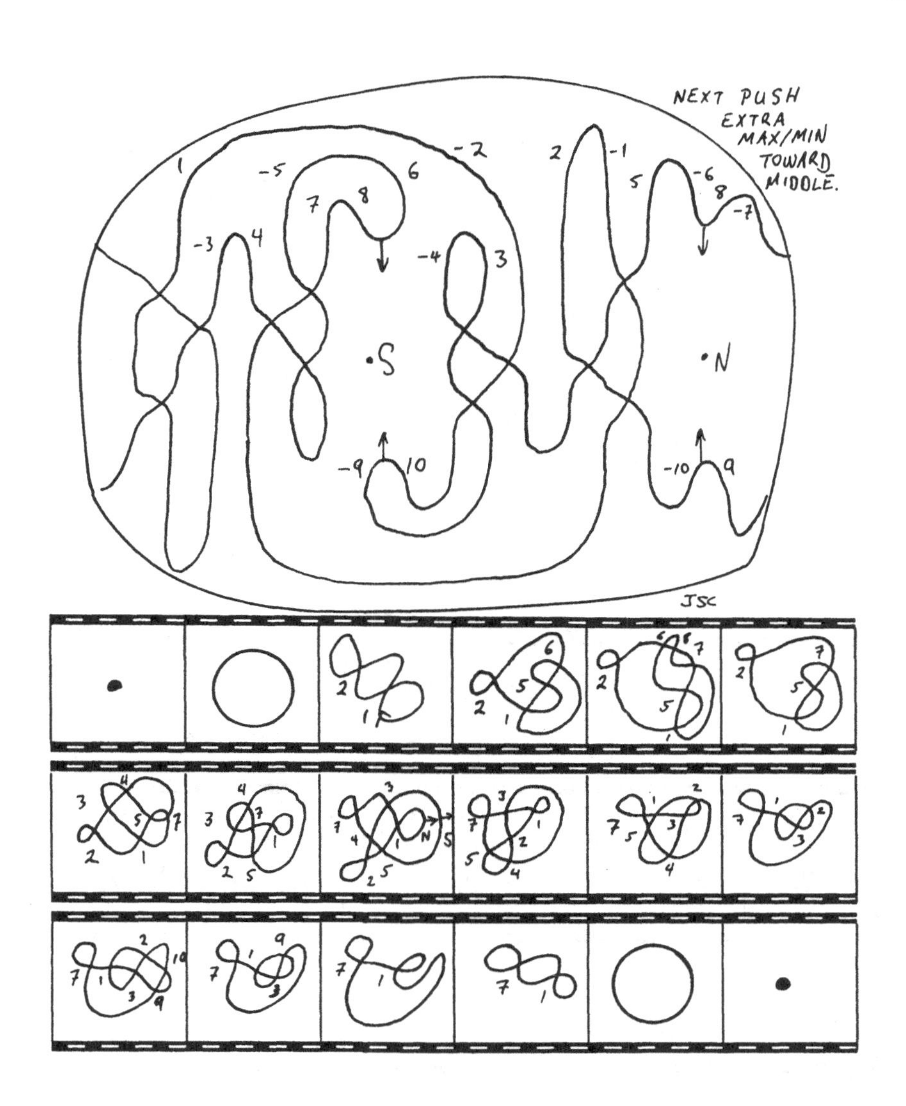
NEXT PUSH
EXTRA
MAX/MIN
TOWARD
MIDDLE.
S
N
JSC

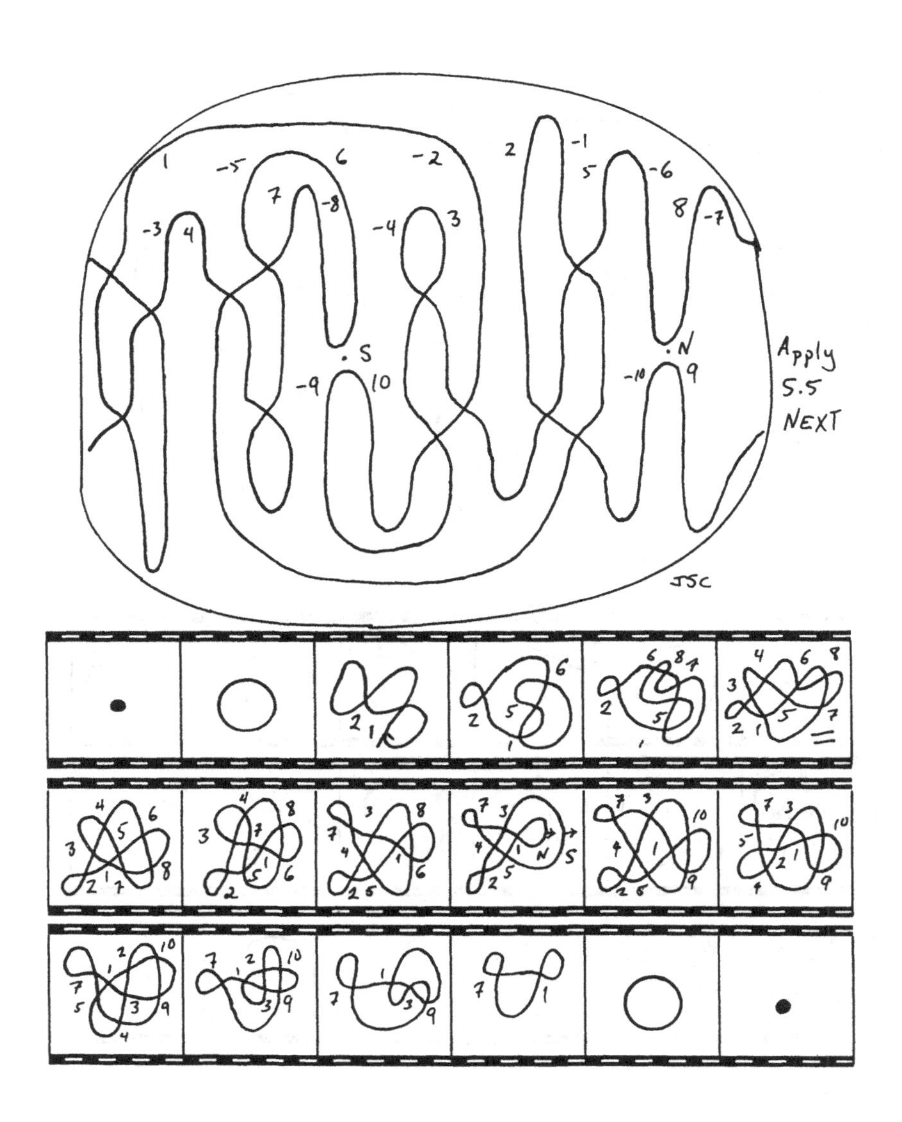
Apply
5.5
NEXT
JSC

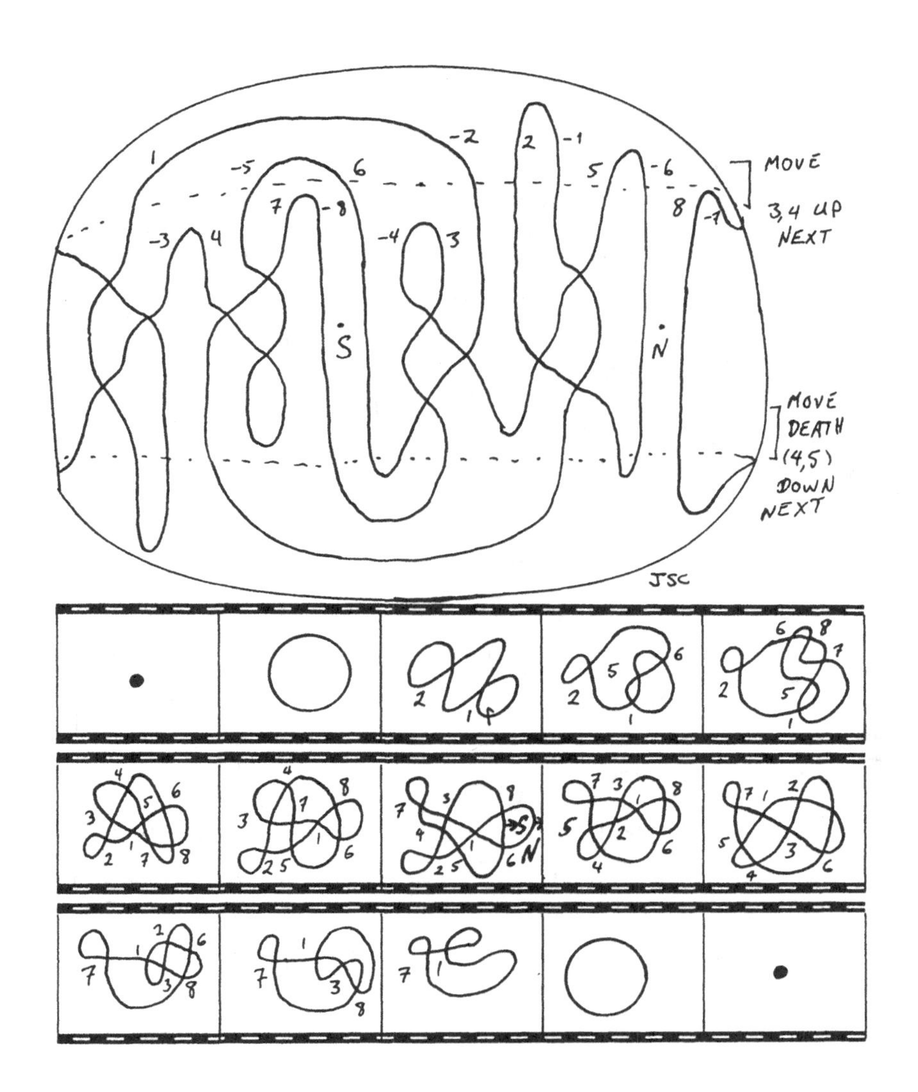
MOVE
3,4 UP
NEXT
MOVE
DEATH
(4,5)
DOWN
NEXT
S
N
JSC

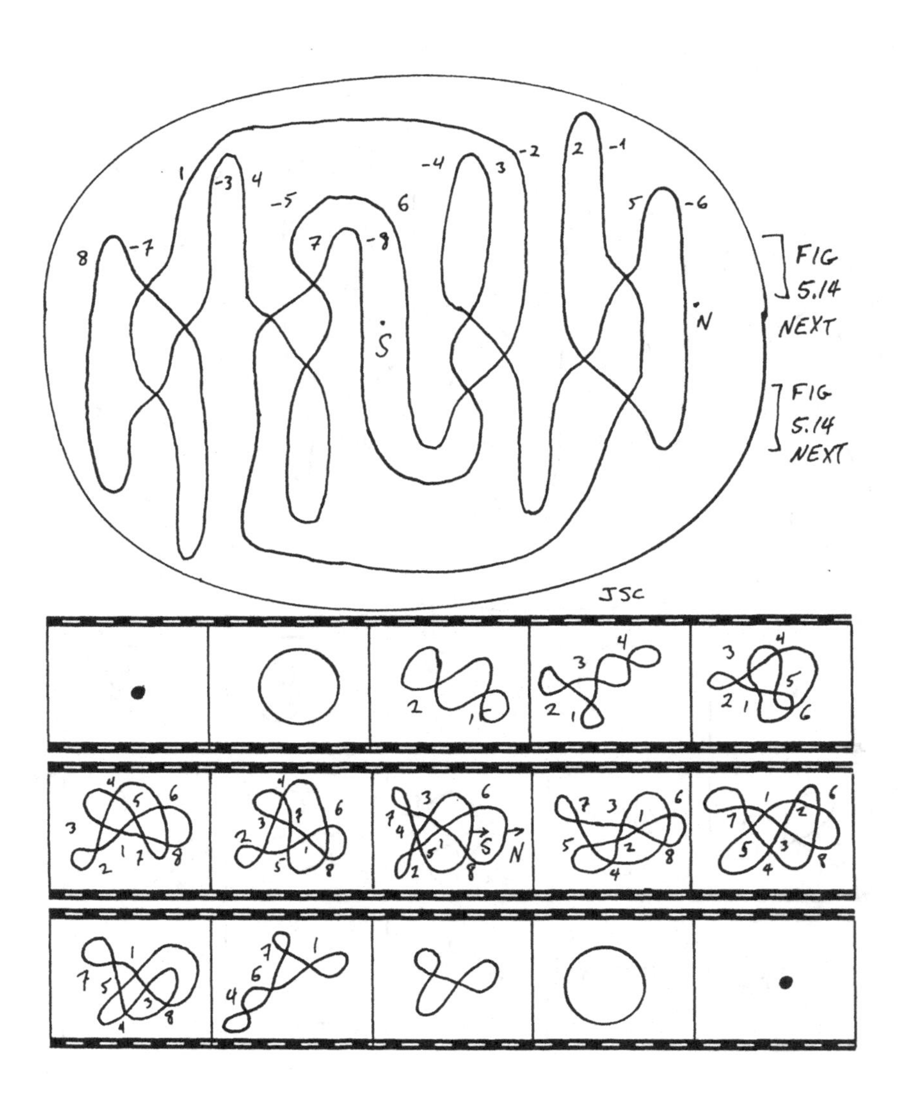
FIG 5.14 NEXT
FIG 5.14 NEXT
S
N
JSC

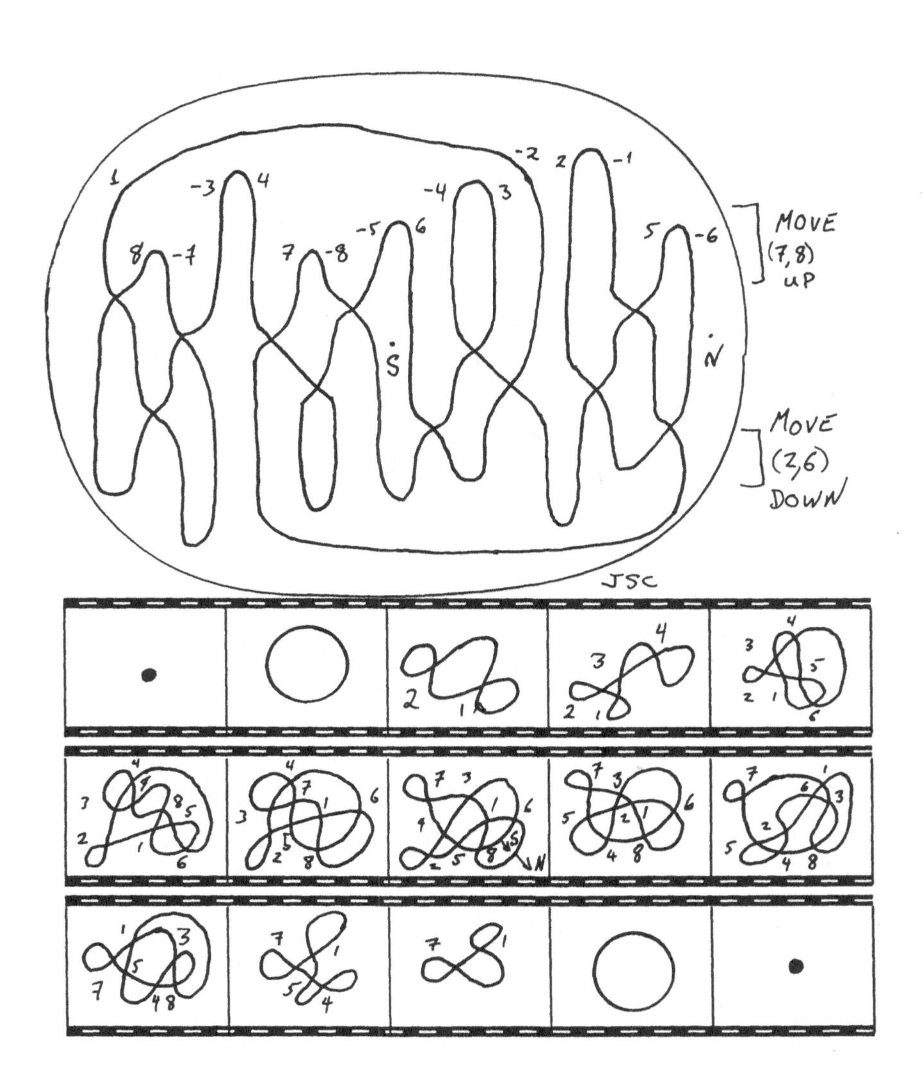
MOVE
(7,8)
UP
MOVE
(2,6)
DOWN
S
N
JSC

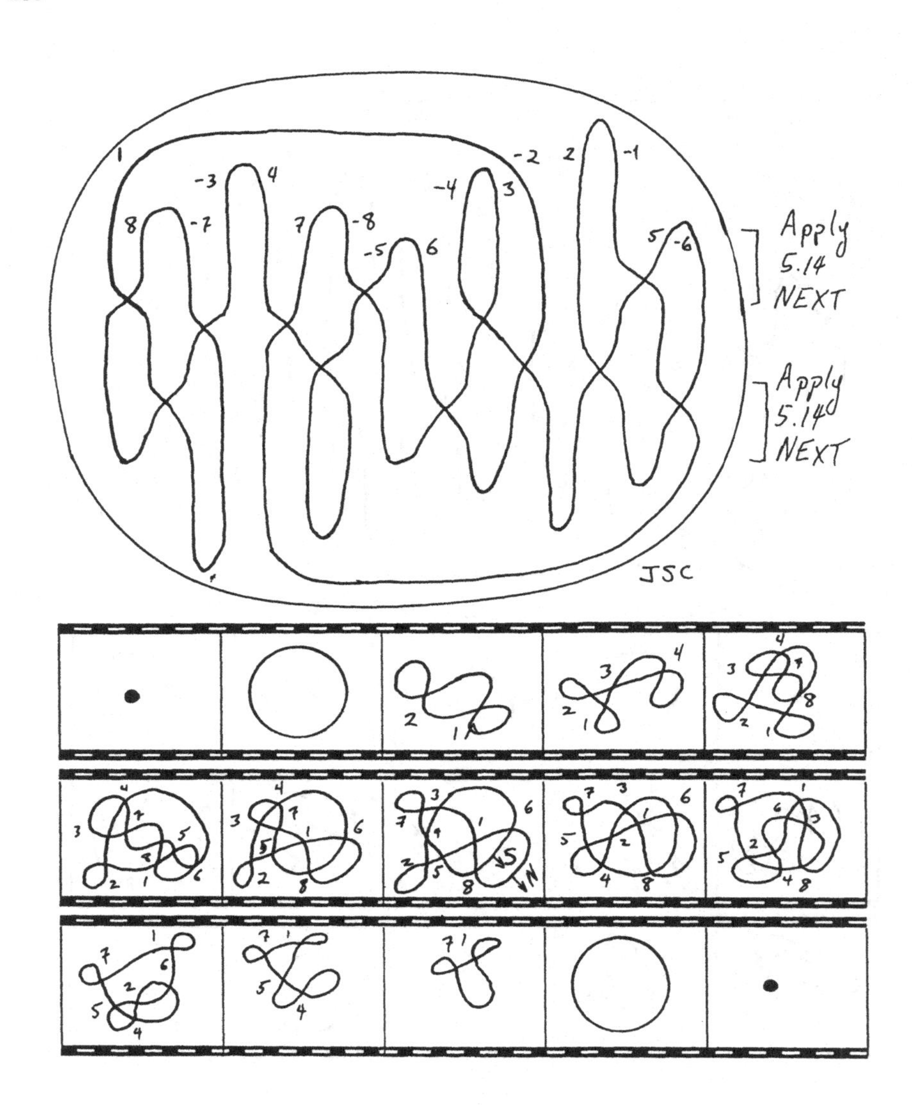
Apply
5.14
NEXT
Apply
5.14
NEXT
JSC

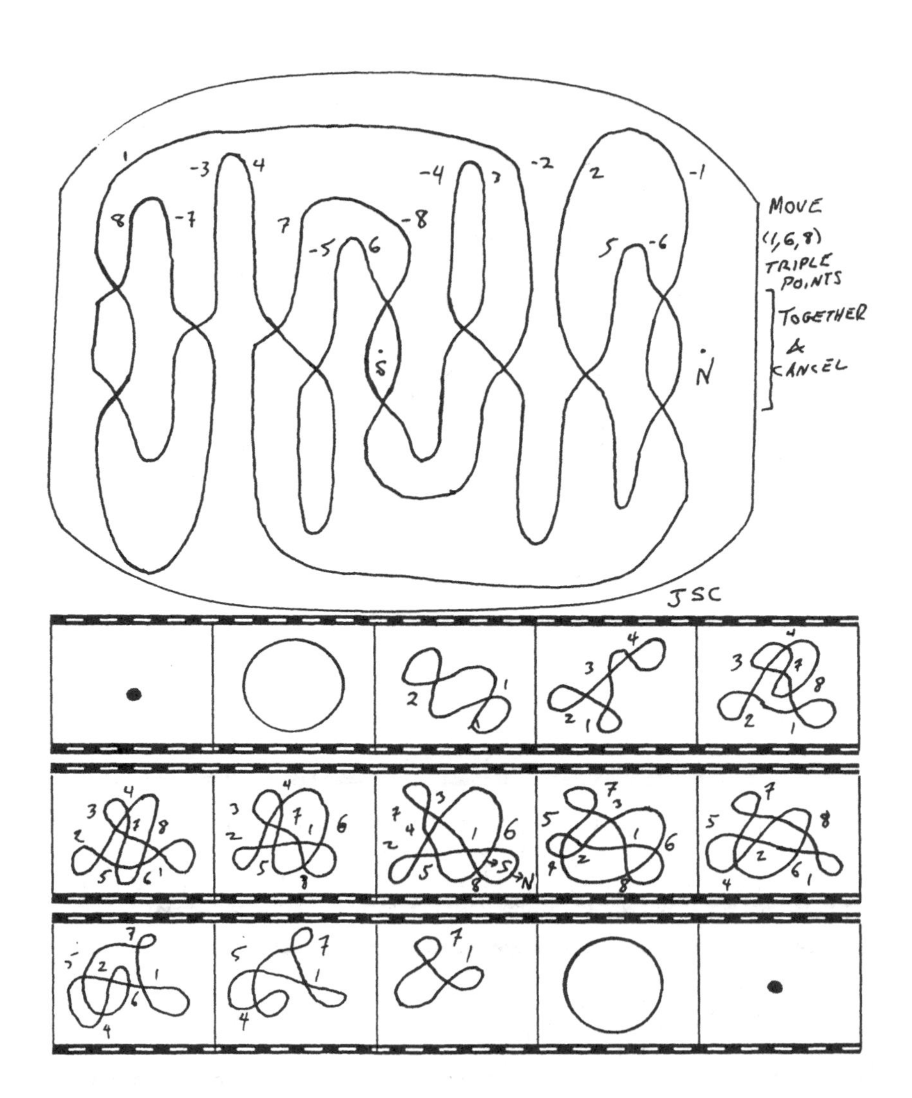
MOVE
(1,6,8)
TRIPLE
POINTS
TOGETHER
&
CANCEL
S
N
JSC

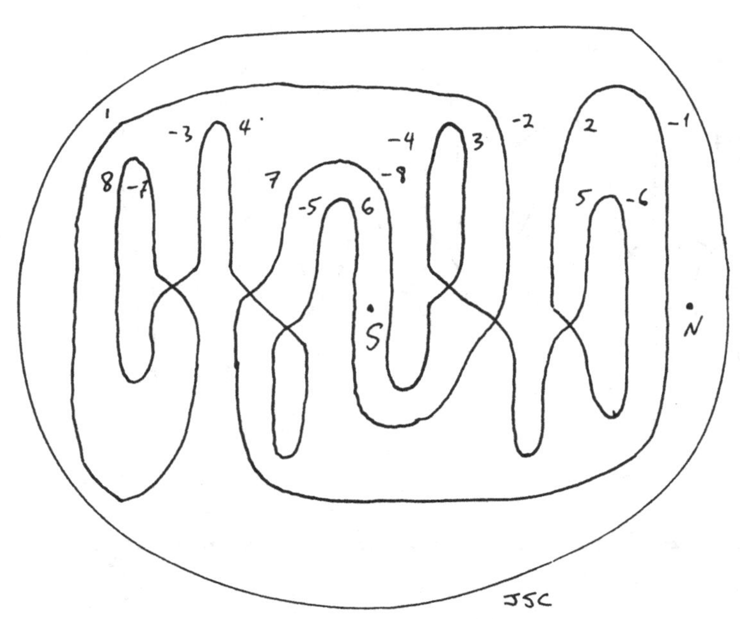

NEXT THE WHOLE SURFACE IS ROTATED IN SPACE, SO THE MOVIE CHANGES.

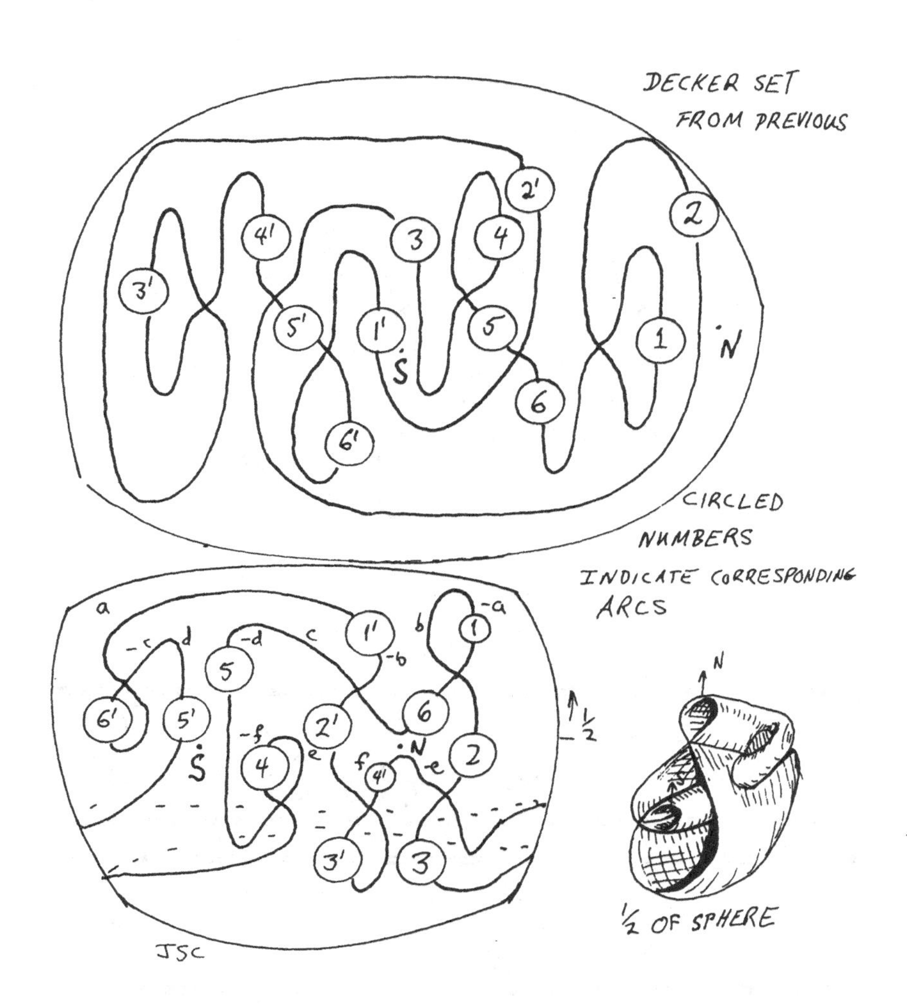
DECKER SET
FROM PREVIOUS
CIRCLED
NUMBERS
INDICATE CORRESPONDING
ARCS
1/2 OF SPHERE
JSC

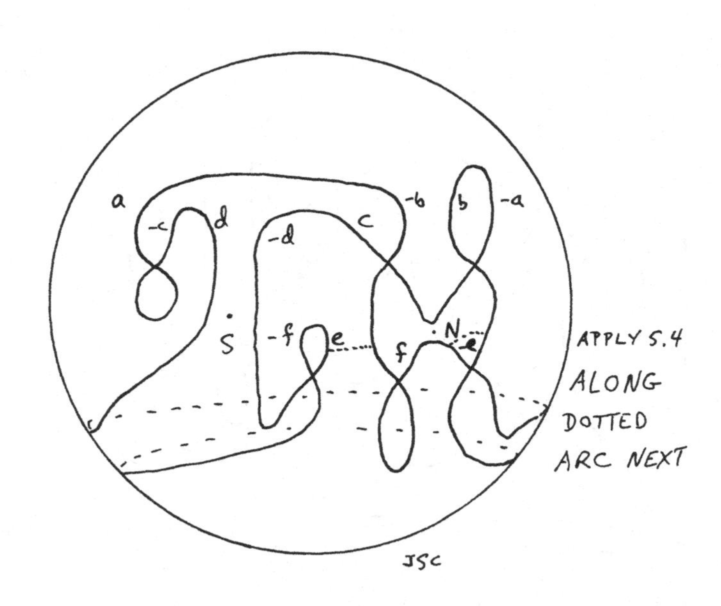
APPLY 5.4
ALONG
DOTTED
ARC NEXT
JSC

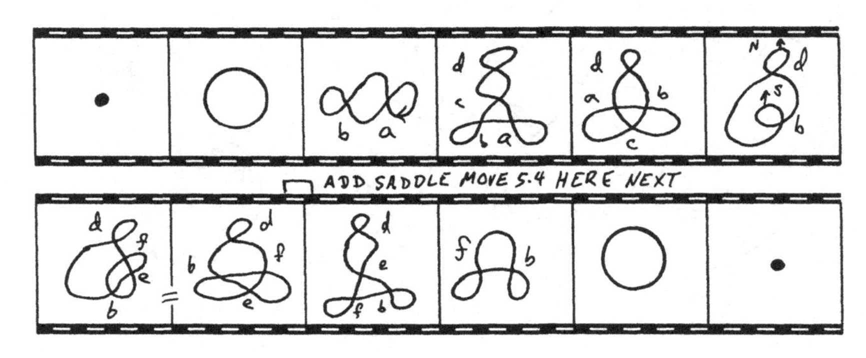
ADD SADDLE MOVE 5.4 HERE NEXT

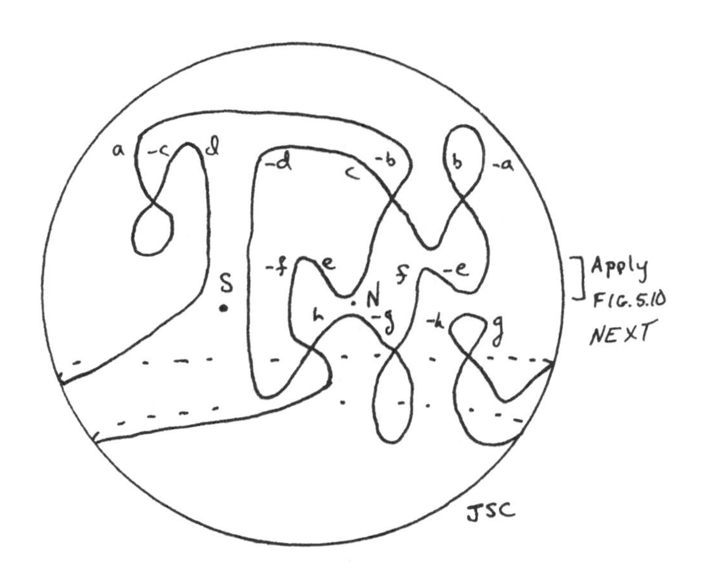
a
-c
d
-d
-b
c
b
-a
-f
e
f
-e
S
N
h
-g
-h
g
Apply
FIG. 5.10
NEXT
JSC

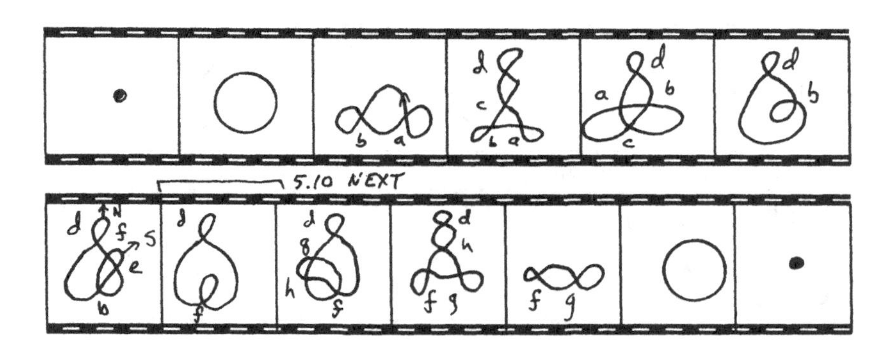
d
c
b
a
d
a
b
c
d
b
5.10 NEXT
d
N
f
S
e
b
d
f
d
g
h
f
d
h
f g
f g

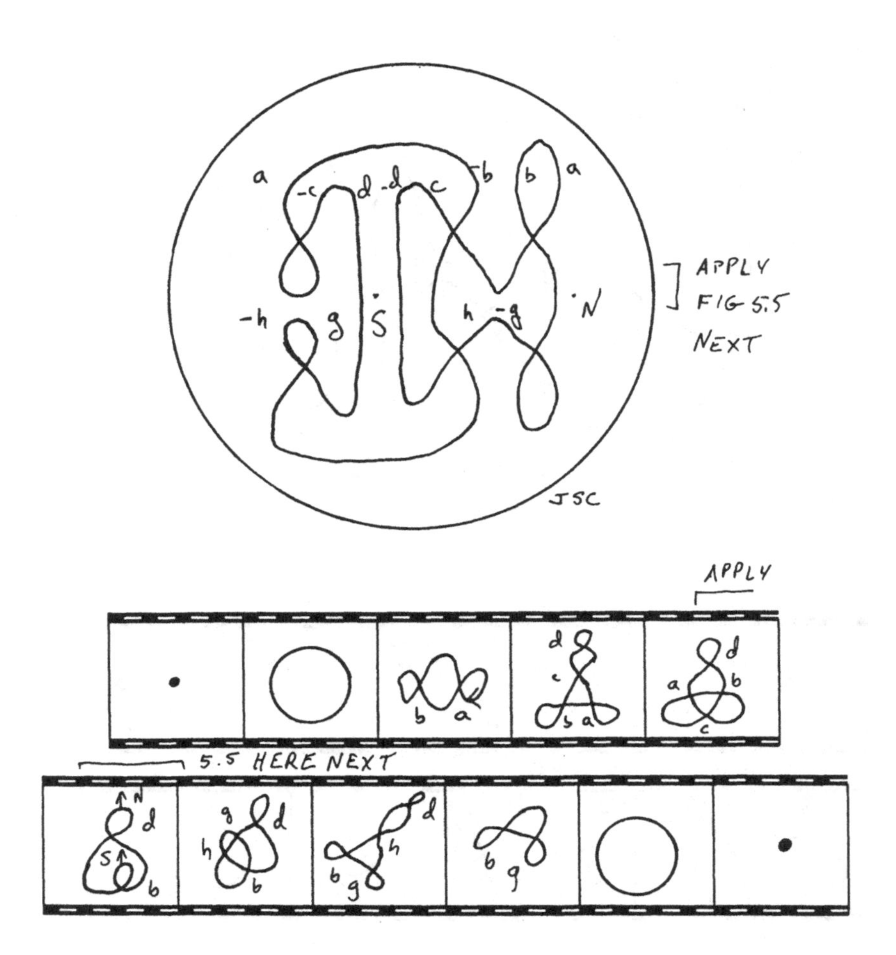

APPLY
FIG 5.5
NEXT
JSC
APPLY
5.5 HERE NEXT

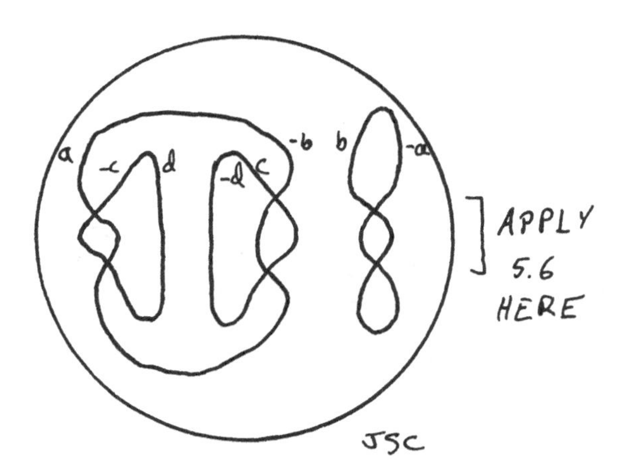
a
-c
d
-d
c
-b
b
-a
APPLY
5.6
HERE
JSC

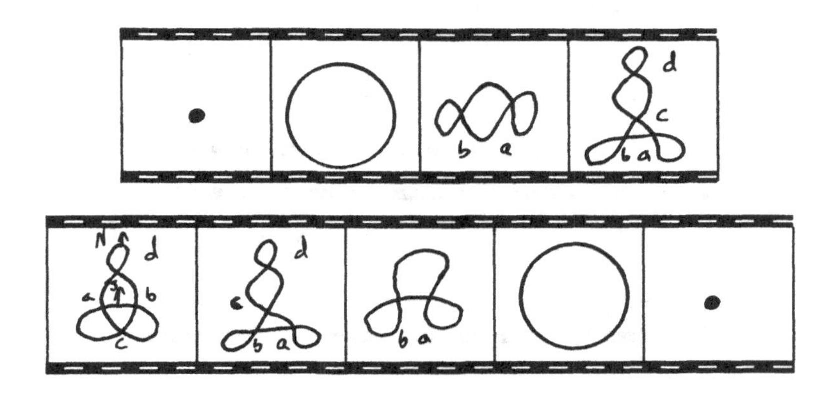
b
a
d
c
b a
N
d
a
b
c
d
c
b a
b a

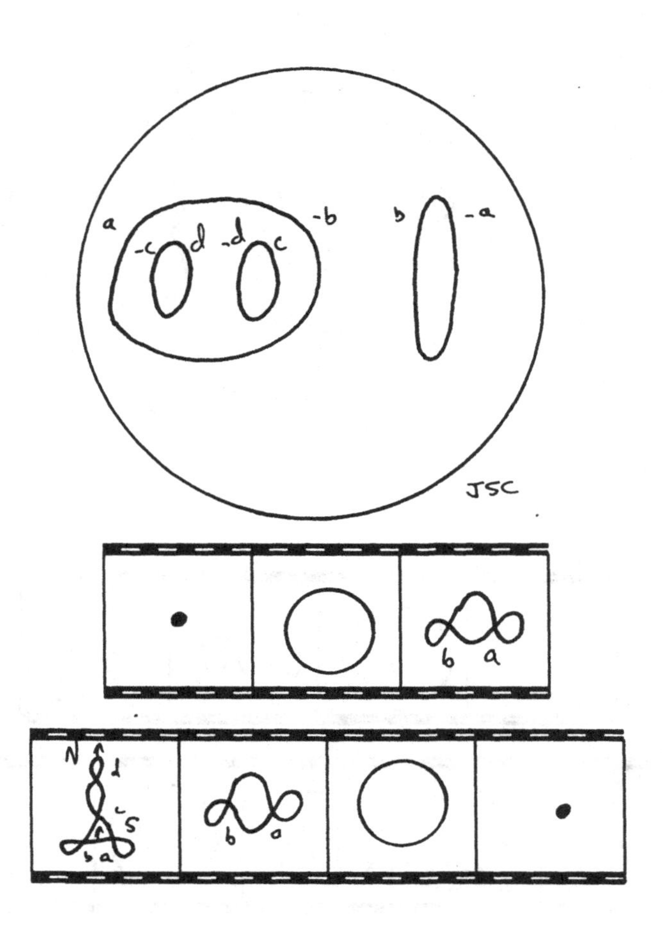
a
-b
b
-a
-c
d
-d
c
JSC
b
a
N
d
c
S
b
a
b
a

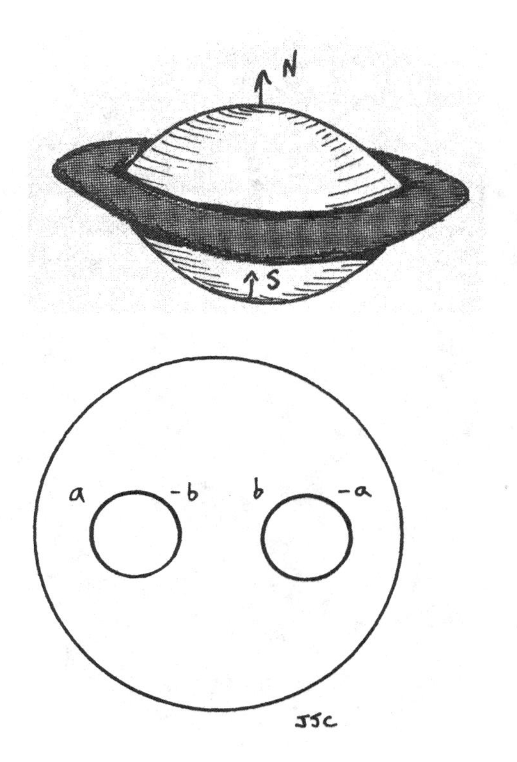
N
S
a
-b
b
-a
JSC

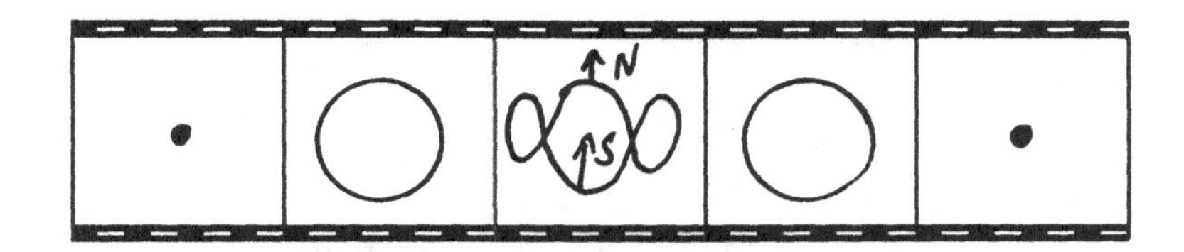
N
S

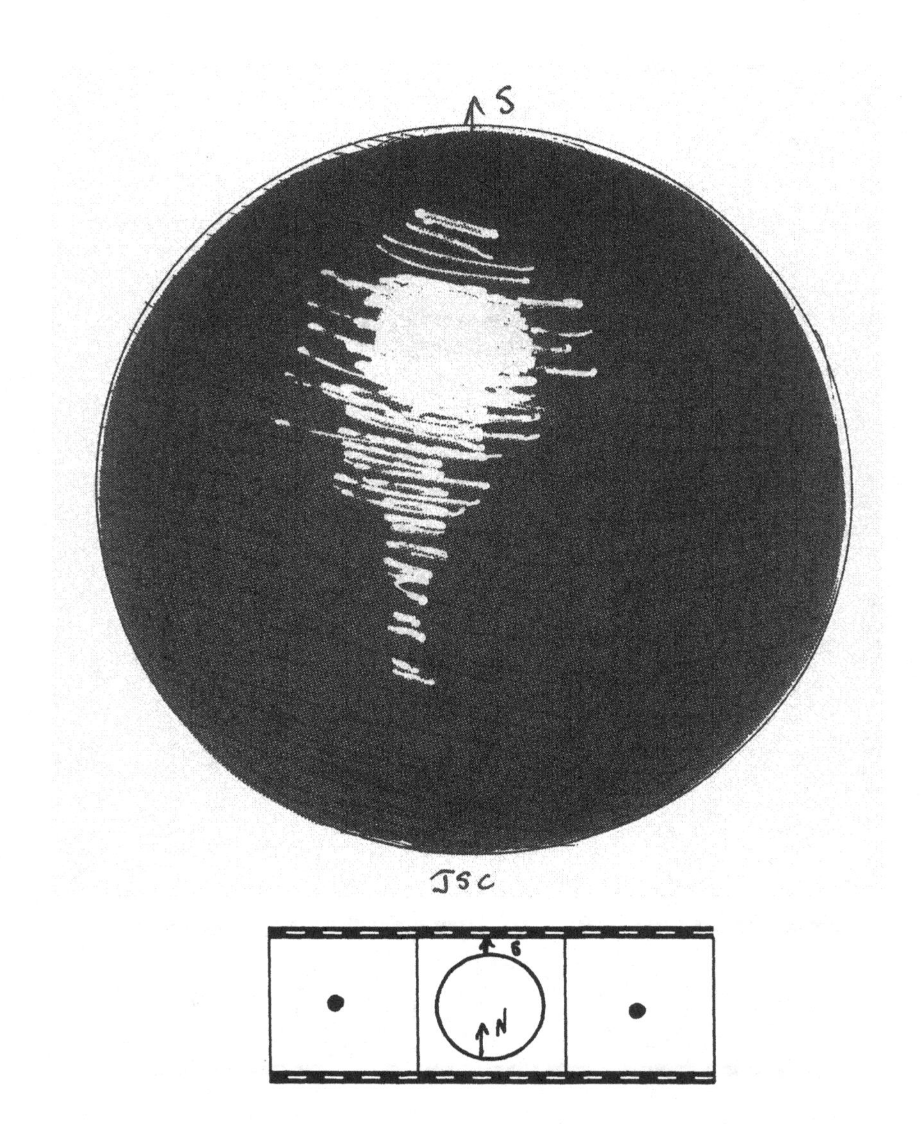
S
JSC
S
N

6.5 Notes

Surfaces in 4-space are important because they can be knotted just as curves in 3-space can be knotted. I have not covered knotting in any detail in this section because invariants of knotted surfaces are hard to find. The broken surface diagrams are an accurate depiction of the embedded surface within a small neighborhood of the knot. But these diagrams miss the larger structure of the space that is not near the knot.

To get a feeling for the amount of information that is lost in a movie diagram, try to imagine a movie of a knotted loop in space. The movie would consist of dots dancing in the plane.

For further reading, try Banchoff's book [6], Rucker's book [69], or Hilbert Cohn-Vassen [45].

Nearly all of this chapter is based on joint work with Masahico Saito. Our papers [24, 22, 25, 23, 26, 27, 28, 29] together contain a rather complete description of the quest for invariants.

Chapter 7

Higher Dimensional Spaces

This chapter is a brief excursion into the ideas of higher dimensions. In Chapter 4, n-dimensional space was defined as the set of n-fold sequences of real numbers. The distance from a point in this space to the origin is the square root of the sum of the squares of its coordinates; thus, the distance between any two points is measured via the Pythagorean Theorem. Other n-dimensional spaces are made by gluing together pieces of the standard n-dimensional space.

How do people think of these spaces? Well, most people draw lower dimensional pictures to gain intuition, and then apply mathematical induction to gain precise results. Throughout this chapter, n denotes an arbitrary positive whole number. You may think of n as being 2 or 3, but then you won't get any information beyond the earlier chapters. So you should think of n as being large. However when you get confused, think about what happens in the low dimensional cases. The text will guide you in the same way: The discussion will apply to a general whole number n, but the ideas will be exemplified in the visible dimensions.

The discussion of this chapter involves some notation. There is no way around it: If you want to talk about your friends, you should use their names, or else people will think that you are talking about someone else. I have done my best to keep the notation simple, self descriptive, and standard. Forgive me if these three goals are mutually exclusive.

Higher dimensional spaces called **manifolds** have the property that each point has a neighborhood that looks like ordinary n-dimensional space. To formalize the meaning of "looking like," we say there is a homeomorphism (continuous invertible function with continuous inverse) from the neighborhood to ordinary space. This function is called a **chart.** A collection of charts (*a priori* at least one for each point) is called an **atlas.**

(Further conditions specify the cases of a differentiable manifold, a piecewise linear manifold, or an analytic manifold. These conditions won't concern us here, but for the record the discussions of this and the previous chapters are in the differentiable category. Briefly, a differentiable manifold is one on which the notions of calculus make sense. In particular, a given function can be approximated by its derivative. When we need the calculus, I 'll tell you how it's used.)

The geographic terms "atlas" and "chart" cause me to operate under a misconception. When I open an atlas of the world, lines of longitude and latitude appear. These are curves that intersect at right angles in the charts. Also, when I think of ordinary space, I visualize the coordinate lines, planes, and hyperplanes therein. Consequently, when I think of a manifold, I tend to endow it with some extra structure. Namely, I tend to think of a manifold that contains certain subspaces — floors and walls, if you will. This book is filled with that prejudice.

There are reasons for thinking of space without coordinate frames of reference. First, there is no natural frame of reference in the physical world. Second, when you walk outside, you don't trip on the latitude lines drawn on the surface of the earth. Third, suppose that a frame of reference is chosen at a given point in a manifold. When this frame is transported around a loop in the manifold (by a Massey bug), the frame can become twisted with respect to the original frame. A most extreme case of this twisting occurs on a Möbius band: The transported frame is disoriented with respect to the original frame. Other kinds of twisting can occur, and such twisting is used to measure certain topological properties of the manifold.

On the other hand, *I* find that *my* frame of reference is natural for me. I haven't reached that pinnacle of perception from which I can see things from your point of view. So I put convoluted frames of reference inside the spaces that I study. No, change that: I study the convoluted frames of reference that can appear in a given space.

An example will help clarify this point of view.

7.1 The n-Sphere

The stereographic depiction of the 3-dimensional sphere is recalled in Figure 7.1. Use this picture to anchor the following discussion in reality.

The n-sphere is the set of points $(x_1, x_2, \ldots, x_{n+1})$ found within $(n+1)$-dimensional space that are a unit distance from the origin. The square of the distance is the sum of the squares of the coordinates, so $x_1^2 + x_2^2 + \cdots + x_{n+1}^2 = 1$. Inside this sphere, there are spheres of every lower dimension. From a coordinate point of view, the lower dimensional spheres that concern us are found by setting some of the coordinates in the sphere equal to 0.

For example, in the 3-sphere, we have the 2-dimensional spheres $\{(x, y, z, 0)\}$, $\{(x, y, 0, z)\}$, $\{(x, 0, z, w)\}$, and $\{(0, y, z, w)\}$. In any of these sets, the sum of the squares of the variables is 1. There are six coordinate circles, and four coordinate 0-spheres.

In the n-sphere, an $(n-1)$-sphere is determined by setting one of the coordinates equal to 0. These **equatorial** $(n-1)$-spheres are the boundaries of the coordinate disks in $(n+1)$-space.

An equatorial $(n-1)$-sphere has a coordinate that is constantly 0. When two such spheres intersect, two of the coordinates are constantly 0, and the intersection forms an $(n-2)$-sphere. Similarly, when several equatorial $(n-1)$-spheres intersect, the intersection is a lower dimensional sphere. The dimension of this intersection is the difference between n and the number of spheres that are intersecting. At most n equatorial spheres intersect in the n-sphere, and then the intersection is a 0-sphere.

Among the intersection sets, there are combinatorial patterns that are determined by the subsets of the set $\{1, 2, 3, \ldots, n+1\}$. Each k-element subset determines a

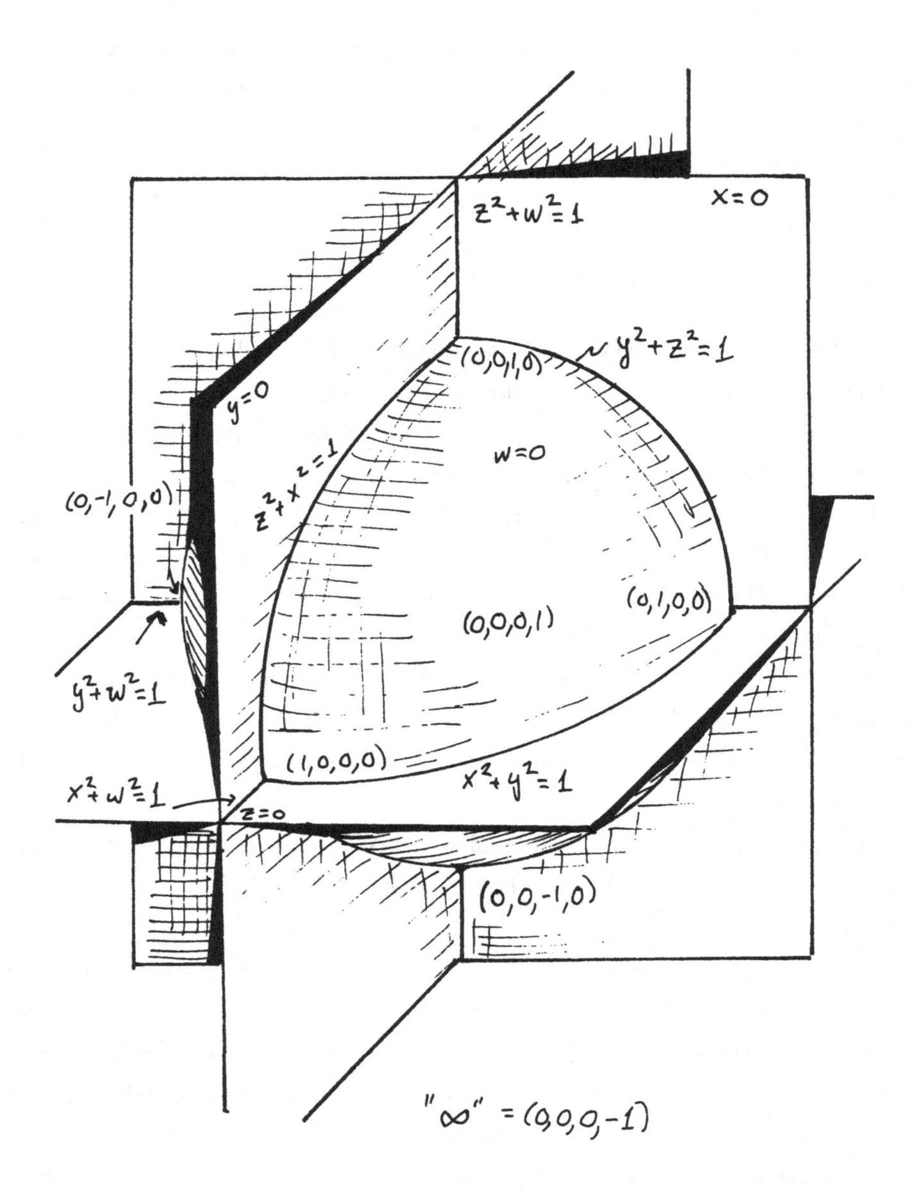

Figure 7.1: The equatorial $(n-1)$-spheres in the n-sphere

sphere of dimension $n-k$ that is the intersection of k equatorial $(n-1)$-spheres. The coordinates of this $(n-k)$-sphere that are always 0 are the indices which are in the chosen k element subset.

The n-sphere is an n-manifold. Consider a point in the n-sphere that is the intersection of a set of k equatorial $(n-1)$-spheres. This point has a neighborhood that looks like ordinary n-space. Moreover, when we consider a set of k coordinate disks in n-space, these intersecting disks map to a neighborhood of the intersection point under consideration. In particular, the intersection of n of the coordinate $(n-1)$-spheres consists of a pair of points, and in a neighborhood of each of these points, the intersecting spheres look like the collection of coordinate $(n-1)$-disks in the n-ball.

(Both the words **disk** and **ball** refer to the points in space that are less than a unit distance away from the origin. The word disk distinguishes the space of smaller dimension.)

The n-sphere is vast, empty, and desolate. When the coordinate $(n-1)$-spheres are inserted, the sphere takes shape. These form the walls of various chambers in the sphere. In the 2-sphere the chambers are triangles. In the 3-sphere the chambers are tetrahedra. What do the chambers look like in the n-sphere?

Other Types of Intersections

In the section on handles and duality, I will indicate why intersections among subsets are something that we should measure in a manifold. Now, I am going to discuss the dimensions among various types of intersections.

The coordinate arcs intersect in the plane at a point. In 3-space, two coordinate disks intersect along an arc while a coordinate disk and a transverse arc intersect in a point. On the other hand, a pair of straight lines will not intersect in 3-space unless they are in the same plane. Most pairs of lines in 3-space are not coplanar, so lines in space *tend* not to intersect.

How can we determine the dimension of a general position intersection between a pair disks of dimension k and ℓ in n-space? The rule is rather easy. Subtract n from the sum of k and ℓ. For example, disks are 2-dimensional, so the dimension of the intersection of a pair of disks in 3-space is $2+2-3=1$. Lines are 1-dimensional; since $1+1-3=-1<0$, lines don't usually intersect in 3-space.

***

Of course, lines can intersect, but arbitrarily close to a pair of intersecting lines in 3-space, there is a pair of lines in space that don't intersect. A completely different way expressing general position is as follows: A line in space is determined by a direction and a point. If two directions and two points are chosen at random, then the corresponding lines will probably not intersect. How high is that probability? 100%. When lines are coplanar, they are not a random sample in 3-space.

Strangely, an event that has 100% probability is not a certainty. A real number chosen at random has 100% probability of being irrational. However, most numbers with which humans are friendly are rational. There are rational numbers, but the amount of room that the rational numbers take up on the line is so small that we can't detect them in a probabilistic sense. Therefore, I can say that there is a 100% probability that a random number is irrational without being certain that it is. Similarly, a statement such as, "Random lines in 3-space don't intersect," is true even if you know of lines that do intersect. The statement is a probabilistic one.

Why is a random number probably irrational? Any number between 0 and 1 can be expressed as a decimal sequence. This is a sequence where each entry is taken from the set $\{0,1,2,3,4,5,6,7,8,9\}$. Take a pair of fair dice and roll them. If an eleven or twelve comes up, then roll again. Otherwise, record a digit from the numbers 0 through 9 as the value on the dice (0 will correspond to a roll of 10). A rational number is a sequence that eventually repeats. If your random sequence of rolls starts a repetitious pattern, then you would know that the dice weren't fair. Thus a random infinite sequence of numbers chosen from 0 through 9, represents an irrational number.

***

Let's consider a few more examples of generic intersections. A 3-dimensional ball and a 2-dimensional disk in 4-space will intersect along a $3+2-4=1$-dimensional set. A pair of 2-disks in 4-space intersects at a point since a point is 0-dimensional, and $2+2=4$. A pair of 3-dimensional balls in 4-space will generally intersect in a 2-dimensional disk.

The nice thing about the equatorial spheres, subspheres, and circles in the n-sphere is that their intersections inside the n-sphere represents the generic situation. Moreover, by fooling with coordinates we can easily count the dimensions of these intersections. For definiteness, consider the k-sphere that has coordinates in the n-sphere:

$$\{(x_1, x_2, \ldots, x_{k+1}, 0, \ldots 0) : x_1^2 + x_2^2 + \cdots + x_{k+1}^2 = 1, \}.$$

And consider the ℓ-sphere that has coordinates in the n-sphere:

$$\{(0, 0, \ldots, 0, x_{n-\ell}, \ldots, x_{n+1}) : x_{n-\ell}^2 + x_{n-\ell+1}^2 + \cdots + x_n^2 + x_{n+1}^2 = 1\}.$$

(In the case where $k = 1 = \ell$ in a 3-sphere these are the circles $x^2 + y^2 = 1$ and $z^2 + w^2 = 1$.) The intersection of these spheres consist of a sphere that has dimension $(k + \ell) - n$ provided that this whole number is bigger than or equal to 0. If the difference in dimensions is negative, then the spheres don't intersect; the circles that form the Hopf link don't intersect.

For another example, consider the 2-sphere $\{(x, y, z, 0)\}$ and the circle $\{(0, 0, z, w)\}$ as subsets of the 3-sphere. The intersection is the 0-sphere $\{(0, 0, z, 0) : z^2 = 1\}$, as we see in Figure 7.1.

The theory of general position allows us to approximate the intersections among certain subspaces according to the scheme above.

General Position in All Dimensions. *When manifolds of dimension k and ℓ intersect in an n-dimensional space. There is an approximation to their intersection that is of dimension $(k+\ell)-n$. In other words, we can shake the manifolds until the intersection is of this dimension.*

In the case of differentiable manifolds, we prove that there are general position approximations to any pair of intersecting manifolds by using the derivatives. The manifolds in question can be approximated by their tangent (hyper-)planes by means of derivatives of certain functions. The generic dimension of the intersection of a k-

plane and an ℓ-plane is $(k+\ell)-n$ because random planes of these dimensions intersect in that way.

(The case of the intersection is exemplified by random lines in the plane. It is hard to draw parallel lines because when the slope of a line is measured there is some experimental error. Consequently, when a line parallel to a given line is drawn, the measurement of its slope can differ by twice the measurement error. If there is an error, the lines drawn eventually intersect. Usually, this intersection is outside the plane of the picture.)

⋆⋆⋆

Mathematical induction has been playing a silent lead in the current section by means of the ellipsis (...). Another way to sing "99 Bottles of Beer on the Wall" is to sing, "99 bottles of beer on the wall, take one down and pass it around, DOT DOT DOT, no bottles of beer on the wall." The meaning of n-dimensional space has induction at its core.

Induction and intuition are not very distant relatives. We examine lower dimensional cases (*e.g.* the 0-sphere, the circle, the 2-sphere, and the 3-sphere) to gain intuition about the higher dimensional cases. Then we use induction to verify our intuition. If the induction doesn't work, we reexamine the lower dimensional cases to see point at which our intuition failed.

7.2 Handles and Duality

A **closed n-manifold** is a manifold that has no boundary, no pin pricks, and that can fit inside some finite volume region of a higher dimensional space. In this section, we sketch a construction of n-manifolds that mimics the constructions of closed surfaces and closed solids.

Balls, Cans, and Boundaries

The n-dimensional ball can be decomposed as the Cartesian product of lower dimensional disks. This decomposition was used in describing the disk as the inside of a square and in thinking of a 2-sphere as the boundary of a can.

We keep the notational convention of calling the n-dimensional ball, B, and a lower dimensional disk, D. We will be using disks of various dimension, so the k-disk will be denoted by D^k. Its boundary is a $(k-1)$-dimensional sphere, S^{k-1}. In general, superscripts indicate the dimension of the space, except when the dimension is clear.

The fundamental fact is that the n-ball is the Cartesian product of a k-disk and an $(n-k)$-disk. That is, $B^n = D^k \times D^{n-k}$, and the equality means topologically equivalent. The dimension k can range from 0 to n. When $k = 0$, D^k is a point and D^n is another name for the ball; the same thing happens when $k = n$.

The boundary of the n-ball is sphere, S^{n-1}, of dimension $(n-1)$, and the boundary of the product space is computed by the product rule. Since the boundary of an k-disk is an $(k-1)$-sphere, S^{k-1}, we have the polar/tropical decomposition

$$S^{n-1} = (S^{k-1} \times D^{n-k}) \cup (D^k \times S^{n-k-1}).$$

The union is taking place along the common boundary of these two sets which is the product of spheres, $S^{k-1} \times S^{n-k-1}$.

These decompositions are recalled for the visible dimensions in Figure 7.2 where a few other terms are indicated. First, the ball, when written as $D^k \times D^{n-k}$, is called a **k-handle**. Inside the product space, there are two special disks, $D^k \times \{0\}$ and $\{0\} \times D^{n-k}$. The former is called the **core disk of the k-handle**, the latter is called the **belt disk** or **co-core** . The boundary of the core disk is called the **attaching sphere**.

The range of handles includes 0-handles (where the core disk is a point at the center of the disk, and the belt disk is the entire disk) and n-handles (where the core disk is the entire disk and the belt disk is a point).

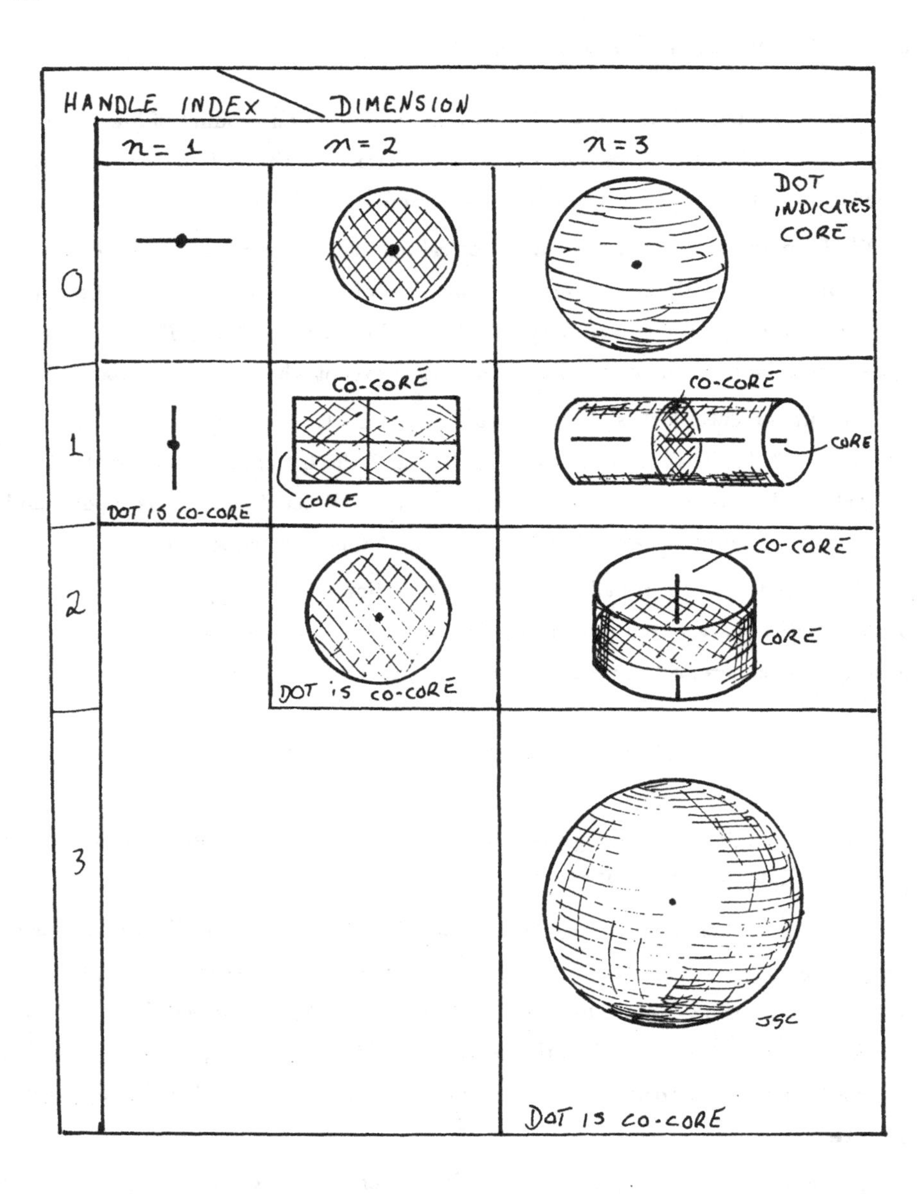

Figure 7.2: Handles in the visible dimensions

Building Closed Manifolds

A sensible way to build an n-manifold is to glue handles together along neighborhoods of attaching spheres. For example, we construct surfaces as a 0-handle, a collection of 1-handles, and a collection of 2-handles. The 0-handle is the disk whose boundary contains the pairs of mates. The 1-handles are attached to the disk in neighborhoods of the mates. The 2-handles filled the cookie-cutter holes on the outside; an attaching sphere of a 2-handle is the entire boundary circle of the handle. In the case of one boundary circle, there is only one 2-handle attached.

In the case of 3-dimensional spaces, the 0-handle is the inside ball. Each 1-handle is attached to the bounding sphere of this ball at a pair of disks. Each 2-handle is attached to the result along an annular neighborhood of a circle in the boundary; such a simple curve indicates where to map the attaching sphere of the 2-handles. The 3-handles are attached on the outside.

Every closed n-manifold has at least one decomposition into a finite set of handles. In fact, there are likely to be many such decompositions. Here is a very broad outline of the proof. The closed n-manifold is placed on a table in some higher dimensional space, and chocolate sauce is poured over the top. The way the sauce flows determines certain critical points. In general position, critical points are close enough to the critical points of quadratic functions, and the critical points of quadratic functions are well understood. The core disk of a handle comes from following the path of the chocolate sauce down from a critical point. The belt disk is the set of points from above that flow to the critical point.

A **critical point** is a notion from calculus. In calculus, we approximate a function by a linear function. The best linear approximation is given by the derivative. A critical point of a real-valued function (such as the height above the table of a point on an n-manifold) is a point at which the approximating tangent plane is horizontal or parallel to the table. Generic critical points are illustrated in the visible dimensions in Figure 7.3.

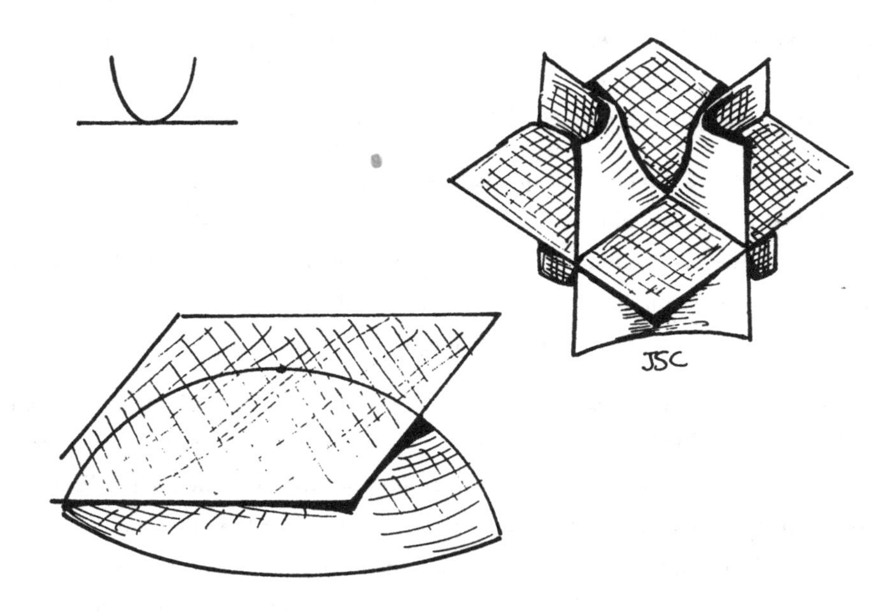

Figure 7.3: Critical points and tangents

Substance in n-Manifolds

Substantial sets in n-dimensional manifolds can often be constructed from the belt disks of the handles. However, not every belt disk is substantial. For example, we can give a handle decomposition of the 2-disk with a 0-handle, a 1-handle, and a 2-handle. This decomposition is depicted in Figure 7.4. The belt disk in the 1-handle can't be substantial in the disk because the disk has no substantial sets.

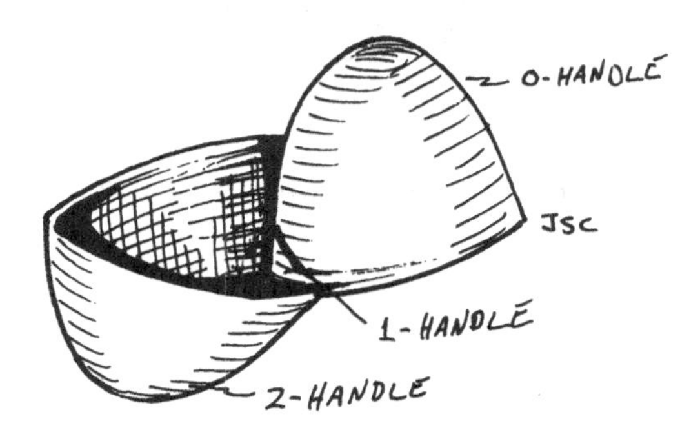

Figure 7.4: A handle decomposition of the disk

Let's think for a moment about an n-manifold that has a boundary and that has been constructed from a 0-handle and a collection of 1-handles. The belt disk of each 1-handle is substantial in the space. The proof is just the same as in the case of surfaces: When the manifold is cut along the belt disk, an n-ball results. The n-ball is the original 0-handle. The mates that are fused to form the belt disk are a pair of $(n-1)$-balls in the spherical boundary of the n-ball.

Suppose that after the 1-handles have been added, a 2-handle is attached. The attaching circle could cross the $(n-2)$-dimensional belt sphere of a 1-handle in a

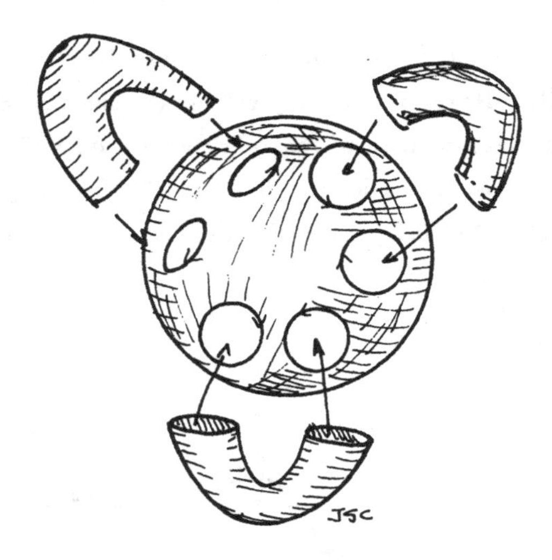

Figure 7.5: 1-handles attached to a 3-ball

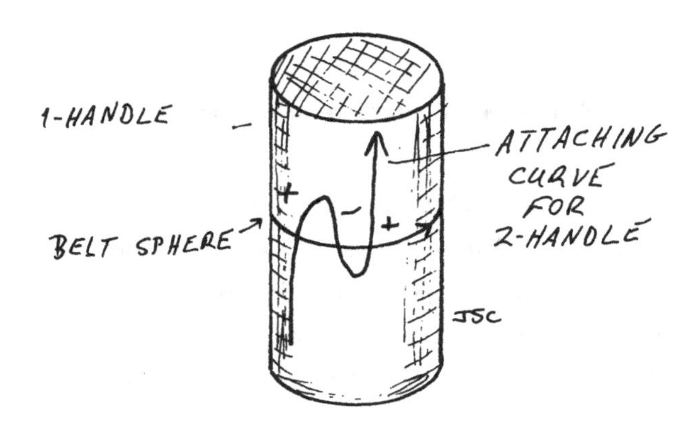

Figure 7.6: Counting intersections

discrete collection of points. These intersections can be counted to indicate how many times the attaching sphere travels over the handle.

In fact, we can do better than just counting the raw intersection points. There is a way to make sense of the sign ($\pm$) of such an intersection point in the case when the 1-handles have been attached in an orientation preserving manner. The net number of times a 2-handle crosses the belt sphere of a 1-handle is determined by looking at the sign of each intersection and adding the results. A net intersection needs to be computed as the diagram in Figure 7.6 indicates. In this way, we count only essential crossings.

Suppose that we know for each 2-handle how many essential times the 2-handle crosses each 1-handle. Then we can perform a similar computation when the 3-handles are added. For this computation, we count the intersections between the belt spheres of the 2-handles and the attaching spheres of the 3-handles. Furthermore, the process can be duplicated for all of the handles. Specifically, an intersection number can be determined that indicates the essential number of times the attaching sphere for a k-handle intersects the belt sphere of a $(k-1)$-handle.

A k-handle will remain substantial in the larger space provided that a $(k+1)$-handle does not pass over it exactly once.

In the general situation, we write down a matrix (a rectangular array of whole numbers) for each index k where k ranges from 0 to $n-1$. The columns of the array correspond to $(k+1)$-handles, and the rows correspond to k-handles. The entry in a given row/column tells how many times a $(k+1)$-handle crosses the belt sphere of the k-handle. By performing standard manipulations to the matrices, we can find a set of handles that represent the substantial sets in the larger space.

By the way, in the case that the 1-handles are attached in an orientation preserving manner, the attaching sphere for each 1-handle intersects the belt sphere of the 0-handle once with a $(+)$ sign and once with a $(-)$ sign (clockwise and counter-clockwise on the disk). In the case that the n-manifold is connected and orientable, we can arrange to have only one 0-handle, so the first matrix that we get is $[0, 0, \ldots, 0]$. And this row of zeros indicates that each 1-handle intersects the 0-handle with a plus and a minus sign.

The projective plane has exactly one handle of each index, 0,1,or 2. The intersection matrices are [2], [2]. Thus the attaching region of the 1-handle passes over the 0-handle twice. and the 2-handle wraps twice around the 1-handle.

The collection of the intersection matrices are used to measure the homology of the space. The **homology** encodes the substantial sets within the space.

The intersection matrices that are computed in each dimension are not invariants of the space in question. The intersections depend on the handle decomposition of the space, and different decompositions will give different intersection matrices. Different handle decompositions can be made by moving the manifold around on the table on which it rests. There are two standard techniques that are used to get from one set of intersection matrices to another. First, a canceling pair of handles can be introduced or removed as schematized in Figure 7.4. Second, handles can be slid over each other as in the case of surfaces, but in higher dimensions, handle sliding is achieved algebraically.

The homology of a space is defined in such a way that it is insensitive to the handle decomposition that is chosen to represent the space. Homology measures how various handles pass over each other, and it can tell which pairs of handles cancel. In short, homology is defined to suppress such technical issues.

Duality

Let's look once again at a k-handle. Such a handle is expressed in dimension n as the product $D^k \times D^{n-k}$. And the k-handle is attached along the polar regions $S^{k-1} \times D^{n-k}$. For example, a 1-handle is attached to a 0-handle at the arctic and anarctic disks on the bounding sphere. A 2-handle is attached to the boundary of the union of the 0-handle and the 1-handles along a neighborhood $S^1 \times D^{n-k}$ that we often describe as an annular or toroidal region. (It is annular or toriodal in the visible world.)

But in the case that a closed manifold is present, the tropics $D^n \times S^{n-k-1}$ are also glued to something. What are they glued to? Well, turn the handle description upside down, and consider each k-handle in the rightside-up world to be an $(n-k)$-handle in the upside-down world. Then the belt sphere of a rightside-up handle is the attaching sphere in the upside-down world.

Intersection matrices can be computed in the upside-down world. And if you set up the conventions for computing the upside-down matrices in just the right way, you'll find that Poincaré duality holds. (The spaces involved must be assumed to be orientable.) A rough statement of this duality is; *Each substantial set of dimension k intersects a substantial set of dimension $(n-k)$ in a single point.* In particular, in an even dimensional space (such as a surface) an even number of independent middle dimensional substantial sets appear.

7.3 Hypersurfaces in Hyperspaces

In this section, some of the intersection types that can occur are examined when branch points are not present. Each point in the manifold that is being mapped has a neighborhood which goes into the range in a flat one-to-one fashion. The tangent (hyper-)plane is not squashed in the process, and the map is said to be a **general position immersion.**

General position immersions represent some interesting topological phenomena. This section will begin with examples that point to a nice question. The question turns out to be equivalent to a major unsolved problem in algebraic topology. Equatorial spheres and intersecting handles inside higher dimensional handles suggest a method of solving the problem. The connection to algebraic topology is sketched.

Examples

Here is a sequence of examples that suggest something that is not true. The numeral 8 represents a continuous map of the circle into the plane that has a single double point. Boy's surface is a continuous map of a closed surface into 3-space that has a single triple point. The 2-sphere eversion can be interpreted as a map from a thick 2-sphere into 4-space that has a single quadruple point. On either side of the eversion, cap the bounding spheres with 3-balls. The result is a 3-sphere mapped into 4-space that has one quadruple point.

You might guess that there is a closed n-manifold mapped without branch points into $(n+1)$-space with exactly one $(n+1)$-tuple point, but there isn't.

There is no 4-manifold that is mapped into 5-space with exactly one quintuple point. In fact, the dimensions $(n+1)$ in which there is such a map are rather rare. There are known to be examples in dimension $(n+1) > 4$ only when $(n+1) = 6, 14, 30$, and 62, and the last two cases are not explicitly constructed. In higher dimensions, examples can only exist when n is 3 less than a power of 2. These results are consequences of Theorems of Koschorke [56], Eccles [33], Freeman [35], and Lannes [54].

So we know that for "most" dimensions, n, there is not a closed n-manifold mapped into $(n+1)$-space without branch points and with only one $(n+1)$-tuple point. And we know the possible dimensions for which there could be such a manifold. We don't know if there is a closed 125-dimensional manifold in 126-space with one 126-tuple points. And we don't have an inductive argument to determine that there is such a map of a $(2^j - 3)$-dimensional manifold.

Question

Is there is a closed $(2^j - 3)$-manifold in $(2^j - 2)$-space with exactly one $(2^j - 2)$-tuple point?

This question is equivalent to a 30 year old problem in algebraic topology. Before explaining the algebraic meaning of the question, I'll discuss some geometric techniques for attacking this problem and its relatives.

Intersecting Handles

A k-handle is a decomposition of the n-ball as a Cartesian product $D^k \times D^{n-k}$. The language is not completely descriptive because the dimension, n, of the handle does not appear in its name. Throughout this chapter we have been working in an arbitrary dimension. As n varies so does the k-handle: A 2-dimensional 1-handle is a rectangular strip of golden fleece. A 3-dimensional 1-handle is a cylindrical can and its contents. The boundary of a 4-dimensional 1-handle consists of a pair of 3-balls (the attaching region) and a thick 2-sphere (the belt region).

Inside an n-dimensional k-handle, I like to put the intersection of $(n-1)$-dimensional k-handles. The handle is a product, $D^k \times D^{n-k}$, and in the factor D^{n-k} we can consider the intersection of coordinate disks of dimension $(n-k-1)$. Then we take each coordinate disk and multiply it by D^k. The result in dimension 3 is the intersection of two strips on the inside of a can, or three coordinate disks in the ball.

But all of the coordinate planes don't have to be inside the handles. For example, in the handle decomposition of projective space, only one disk appeared inside the 0-handle, and only one strip appears inside the 1-handle. At the other extreme, the equatorial spheres in the n-sphere can be used to give a handle decomposition of the n-sphere in which each 0-handle has n-intersecting $(n-1)$-disks inside it, each 1-handle has $(n-1)$ intersecting 1-handles inside it, and so forth.

By putting intersecting handles within a handle, I am sometimes able to push the limits of perceptions beyond dimensions 3 and 4. The belt sphere of a k-handle is an $(n-k-1)$-dimensional sphere. So this dimension is already smaller than the dimension, n, of the bigger space. In the belt sphere, consider the equators that are the belt spheres of the lower dimensional handles intersecting within the given handle. These equators and their intersections are of smaller dimensions still. So when the intersections of the belt spheres are on the limits of perception, pieces of the bigger handle are being studied. There is a science in deciphering the intersections by means of the lower dimensional pictures. When changes occur in the regions of the intersections, that science becomes an art.

If a closed n-manifold is intersecting itself in $(n+1)$-space (as Boy's surface does in 3-space), then near a point in the intersection set the n-manifold looks like intersecting handles. For example, the triple point of Boy's surface is the intersection of coordinate disks in a 0-handle of 3-space. Each arc of double points that starts and ends at the triple point is the core of solid 1-handle, and a pair of 2-dimensional 1-handles intersect along these arcs. The rest of Boy's surface consists of four disks; these are the cores of 2-handles in ordinary 3-space. Figure 7.7 indicates this decomposition.

The question of the existence of a closed n-manifold in $(n + 1)$-space can be attacked by means of handles that are intersecting inside handles, but this method has only gone so far. I rely too heavily on my ability to see, and I haven't been able to formalize all of the geometric ideas into concise algebraic statements. Anyway, here is a description of how this goes.

The $(n + 1)$-equatorial $(n - 1)$-spheres in the n-sphere form the boundary of the coordinate planes as they intersect in $(n + 1)$-space. The various intersection sets are to be removed in a controlled way. Handles are attached, and these have their own equatorial immersions on the inside.

Figure 7.8 shows how to eliminate the self intersections of the equatorial 2-spheres in 3-space using a construction that involves Boy's surface in an essential way. In this figure, all of the triple points except 2 are removed by running handles between the triple points on the round sphere in the picture. A circle of intersection results and this is removed because it bounds a disk. After that removal two Boy's surfaces result that are separated by the 2-sphere. Handles are attached to gradually simplify the picture until an embedding results.

Three key points are to be made. First, the sequence of pictures describes a movie of a 3-dimensional object in 4-space. Second, the object has a unique quadruple point at the origin. Third, the handles that were used to connect the triple points of the two Boy's surfaces passed over each other in an equatorial fashion.

The problems of finding an n-dimensional manifold in $(n+1)$-space with given self intersection behavior are fascinating because they involve pushing perception beyond the visible dimensions.

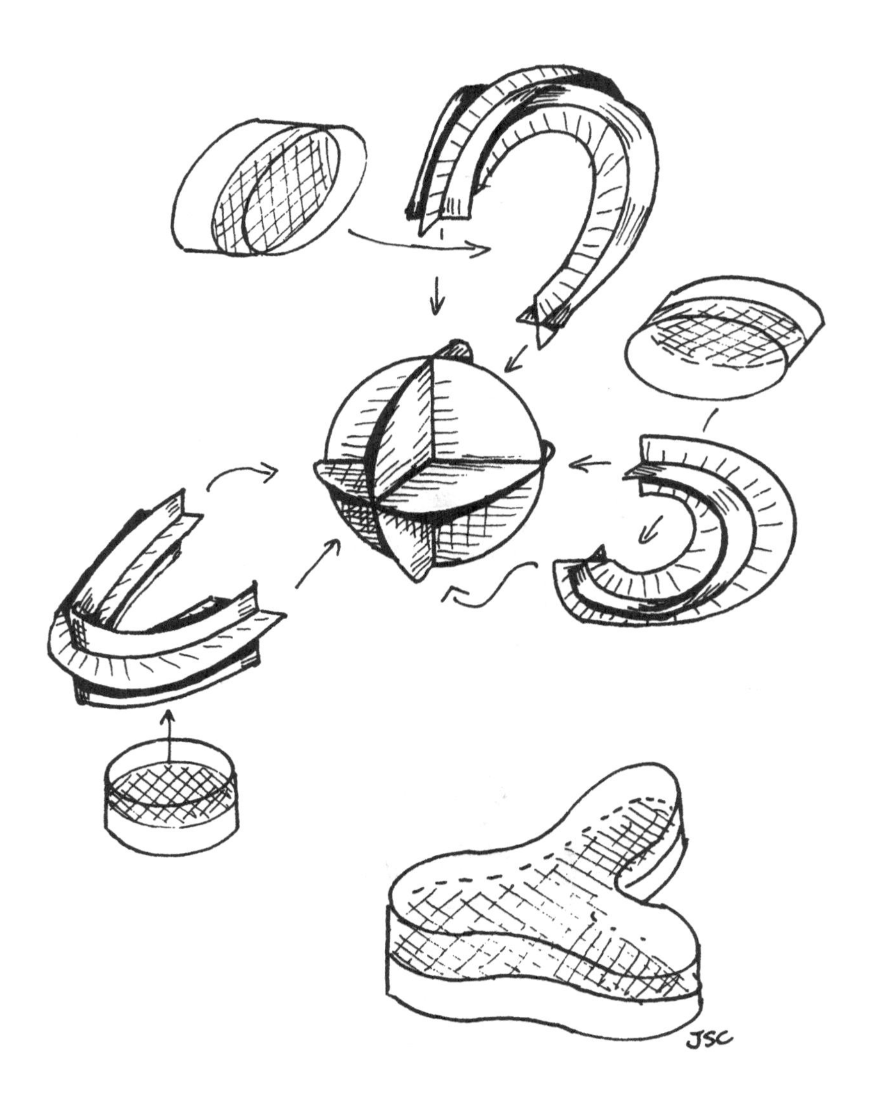

Figure 7.7: Handle decomposition for Boy's surface

Next we examine the structures that occur on the other self intersection sets.

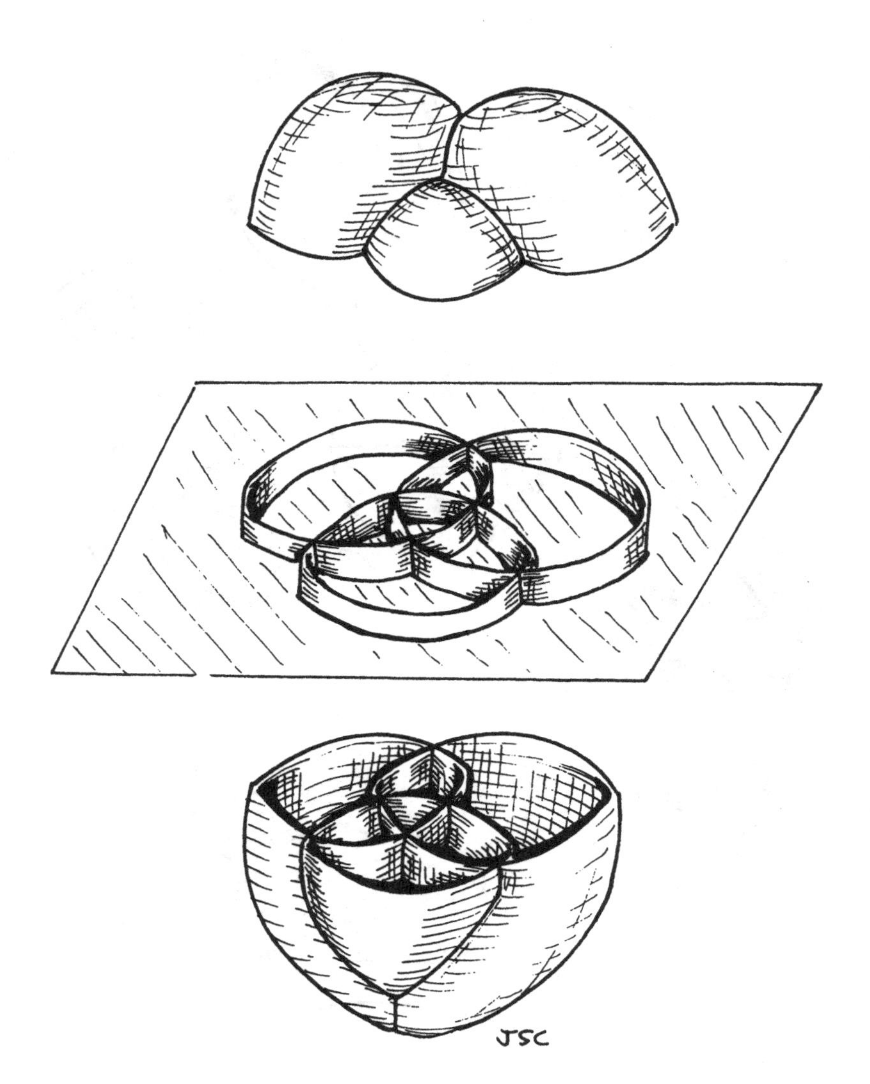

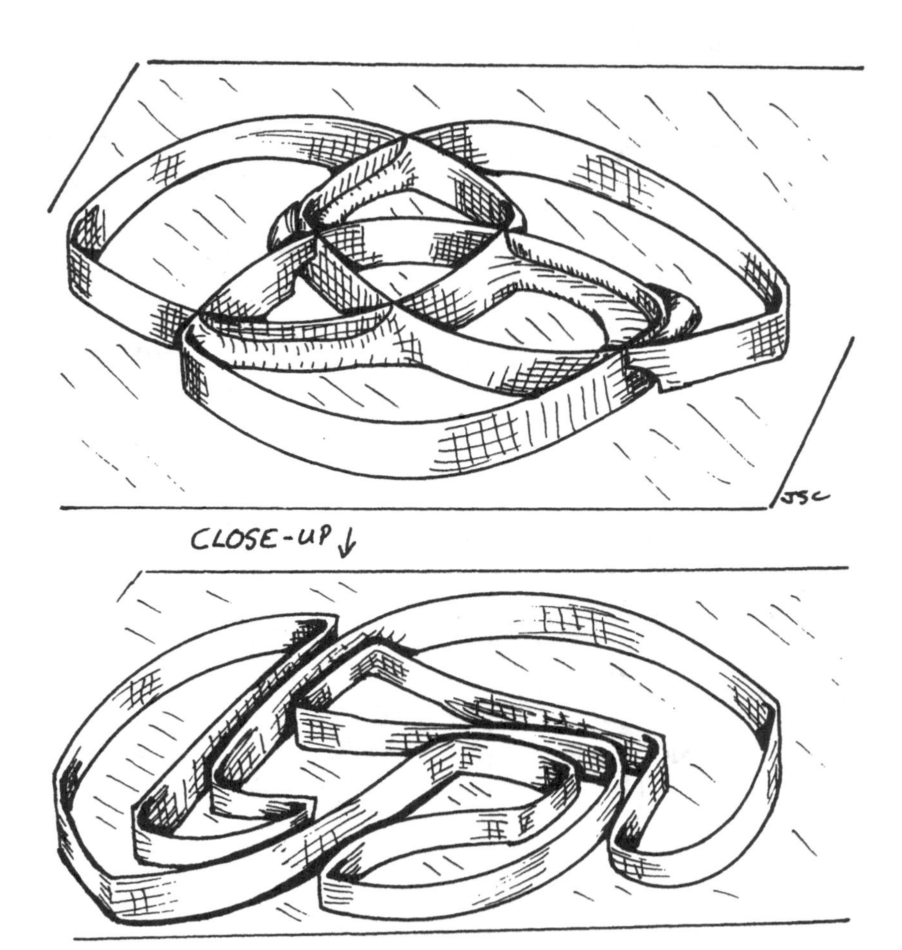
JSC
CLOSE-UP

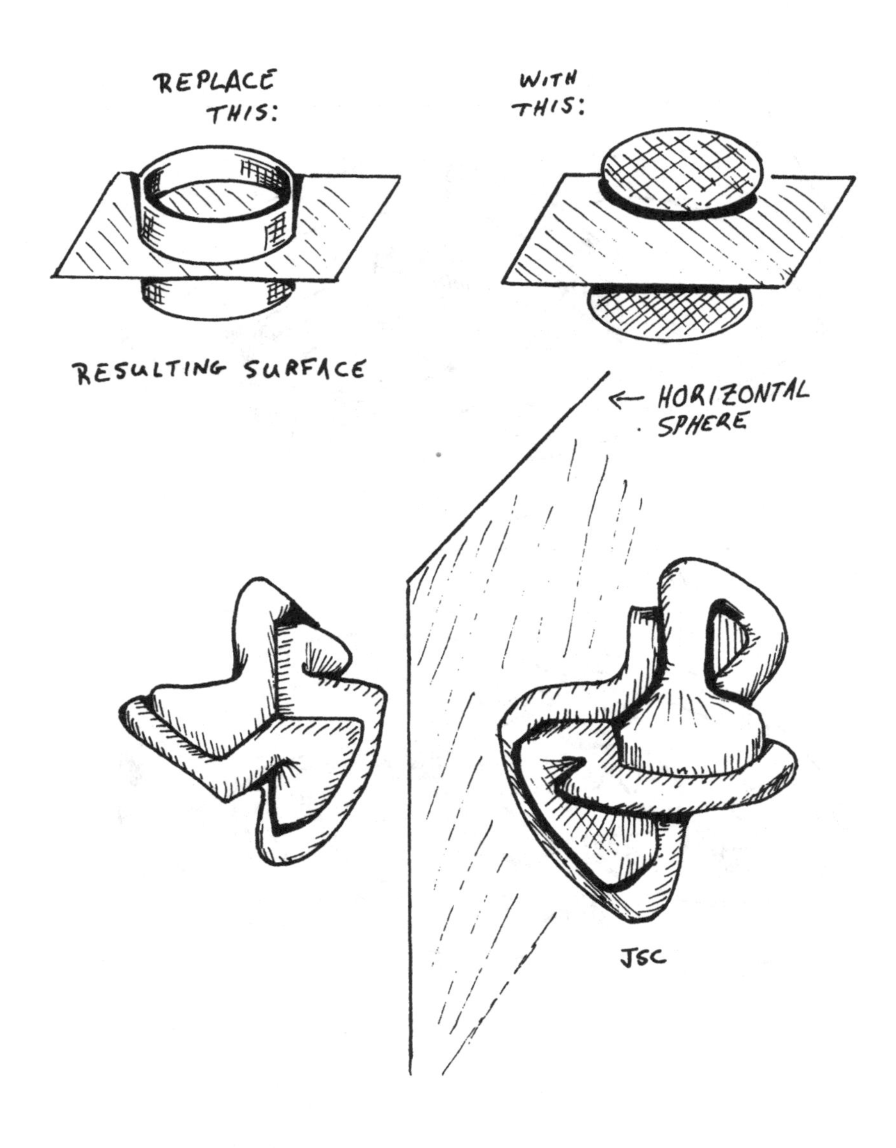
REPLACE
THIS:
WITH
THIS:
RESULTING SURFACE
← HORIZONTAL
SPHERE
JSC

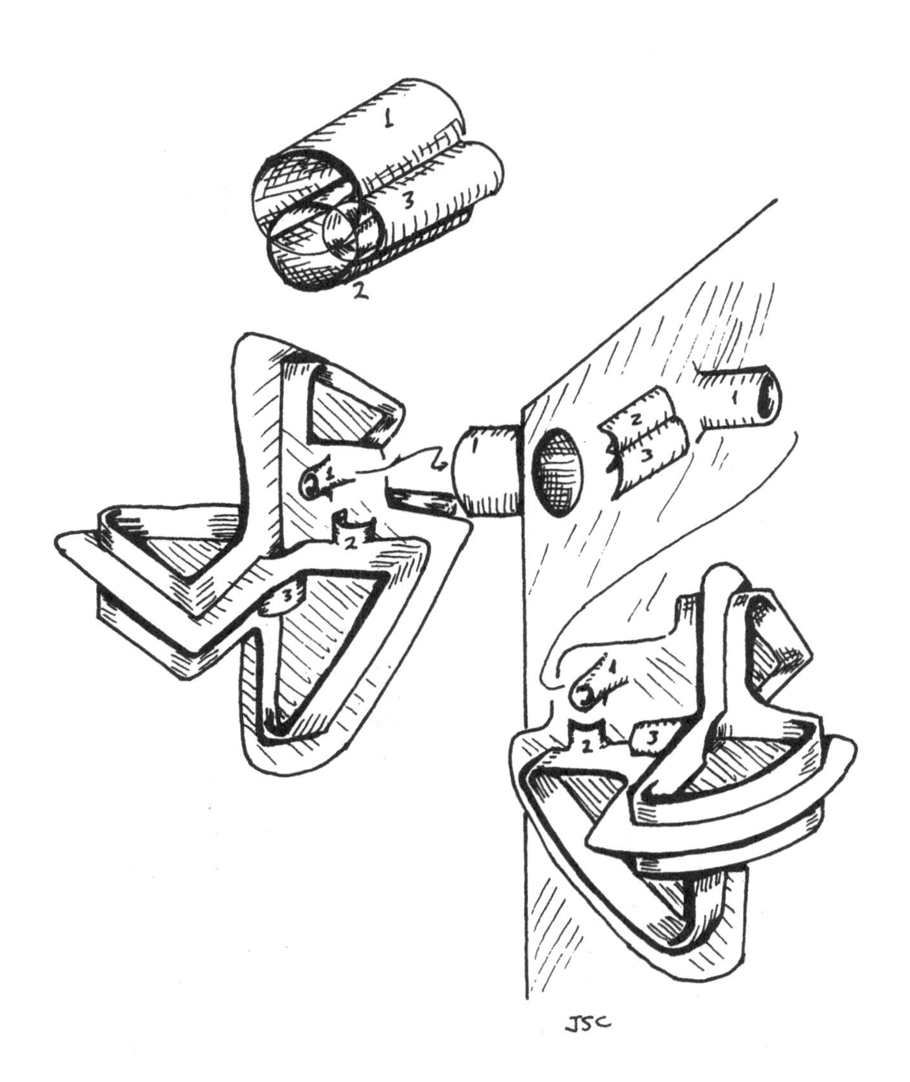
1
3
2
1
2
3
1
1
2
3
1
2
3
JSC

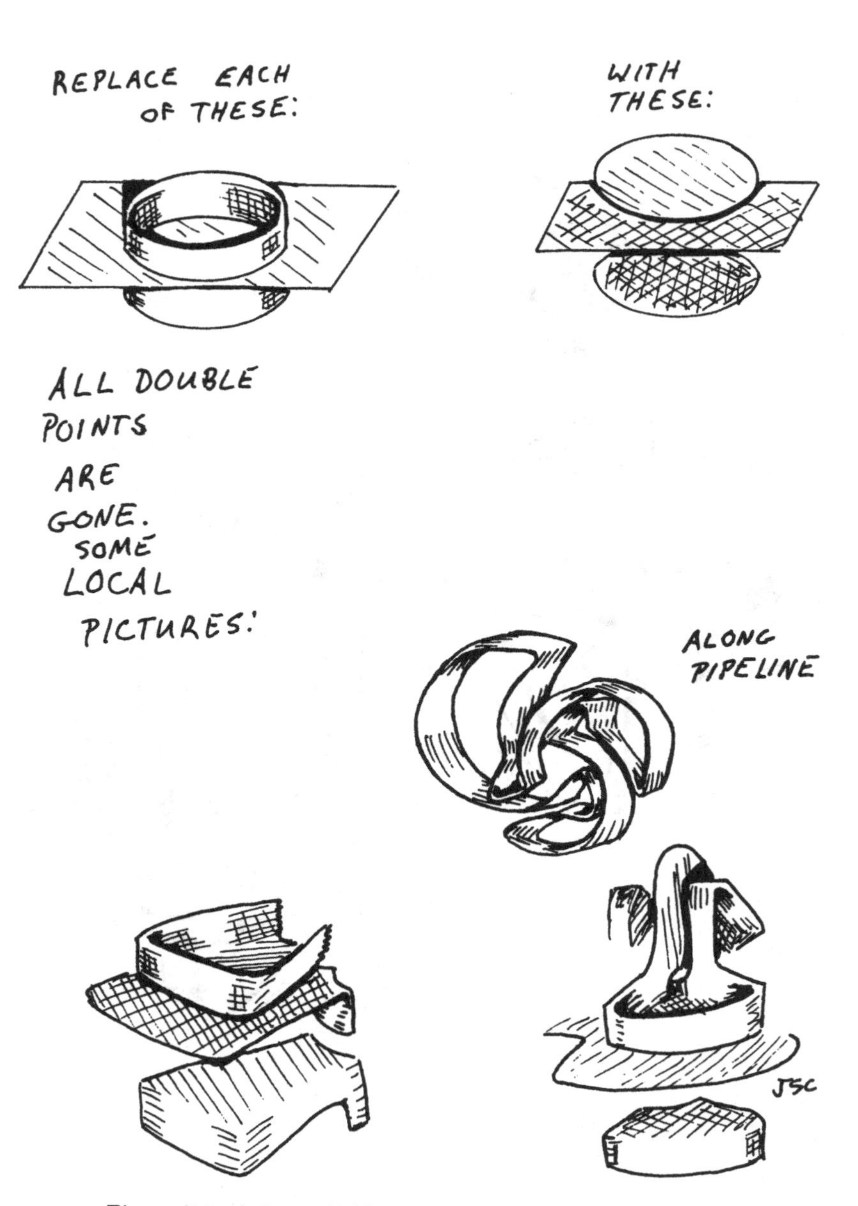

Figure 7.8: A 3-manifold in 4-space with one quadruple point

General Intersection Information

Boy's surface has double arcs and a triple point. The 2-sphere eversion has a quadruple point, arcs of triple points, and a surface of double points. A generically immersed n-manifold in $(n+1)$-space could conceivably have intersections of every possible dimension. The equators in the n-sphere illustrate the possibilities.

The intersection sets can be seen as the non-generic image of lower dimensional manifolds. The dimension of such a manifold is the difference in n and the multiplicity of the intersection. As the various planes in the n-manifold intersect, they provide coordinate frames of reference on the intersection manifold. Given a loop in the intersection manifold, the frame of reference twists along that loop, and so the intersecting sheets become permuted. The intersection set together with its twisting frames of reference provide a measurement of how complicated the original map can be.

For example, despite the complication of the intersection sets, the equatorial spheres bound the coordinate disks, and there is no twisting along the loops in the spheres.

An example of a surface in 3-space that has normal twisting is provided by the torus that is a figure 8 with one full twist. This immersion is depicted in Figure 7.9.

Immersions That Bound

If we have an n-manifold immersed in $(n+1)$-space, we can think of it as being in the $(n+1)$-sphere by means of stereographic projection. Under what conditions is an n-manifold the boundary of an $(n+1)$-manifold that is mapped into the $(n+2)$-ball?

First, let me say that there are n-manifolds that are not the boundaries of other manifolds. The projective plane is not the boundary of any 3-dimensional manifold. Second, in the case that there is a single $(n+1)$-tuple point, the immersed manifold is not the boundary of an immersed $(n+1)$-manifold. The figure 8 is the example to consider.

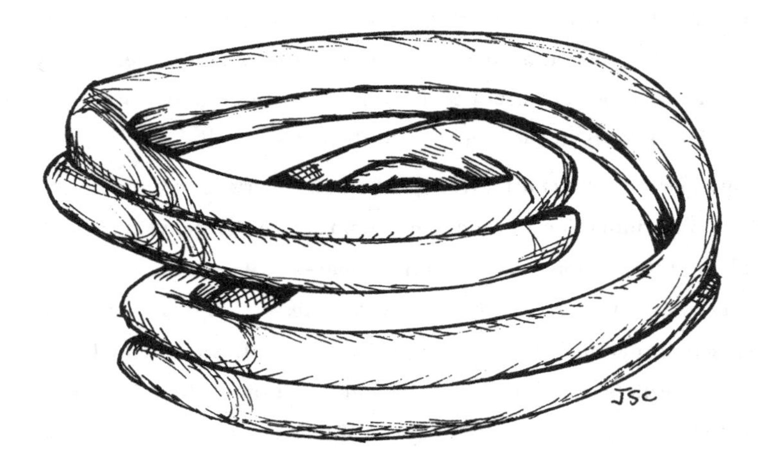

Figure 7.9: A twisted torus

The figure 8 does bound a disk in the ball, as we have seen in Chapter 1, but this disk has a branch point, and branch points are against the rules for immersions. In general, if an n-manifold had a single $(n+1)$-tuple point it couldn't be the boundary of another manifold because the $(n+1)$-tuple point would be on the end of an arc that has its other end on the inside of the ball. This interior end point would be a branch point. (The most amazing things can be proven from the simple fact that an arc has two ends.)

Thus when there is a single $(n+1)$-tuple point, the immersed n-manifold cannot be the boundary of an $(n+1)$-manifold mapped into the $(n+2)$-ball.

The other self intersection sets also provide ways of showing that an immersed n-manifold is not a boundary. Two data are needed. The first is the topological type of the manifold along which the intersection occurs. The second is twisting of the reference frames.

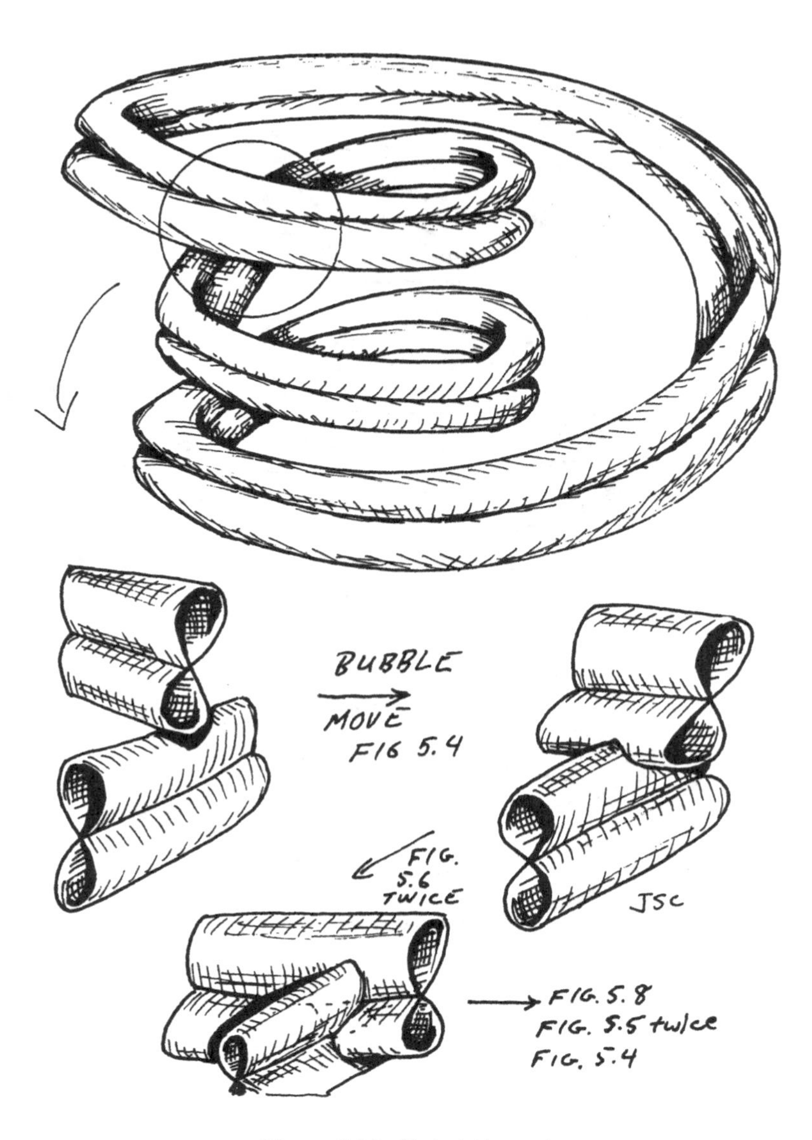

Figure 7.10: Untwisting a torus

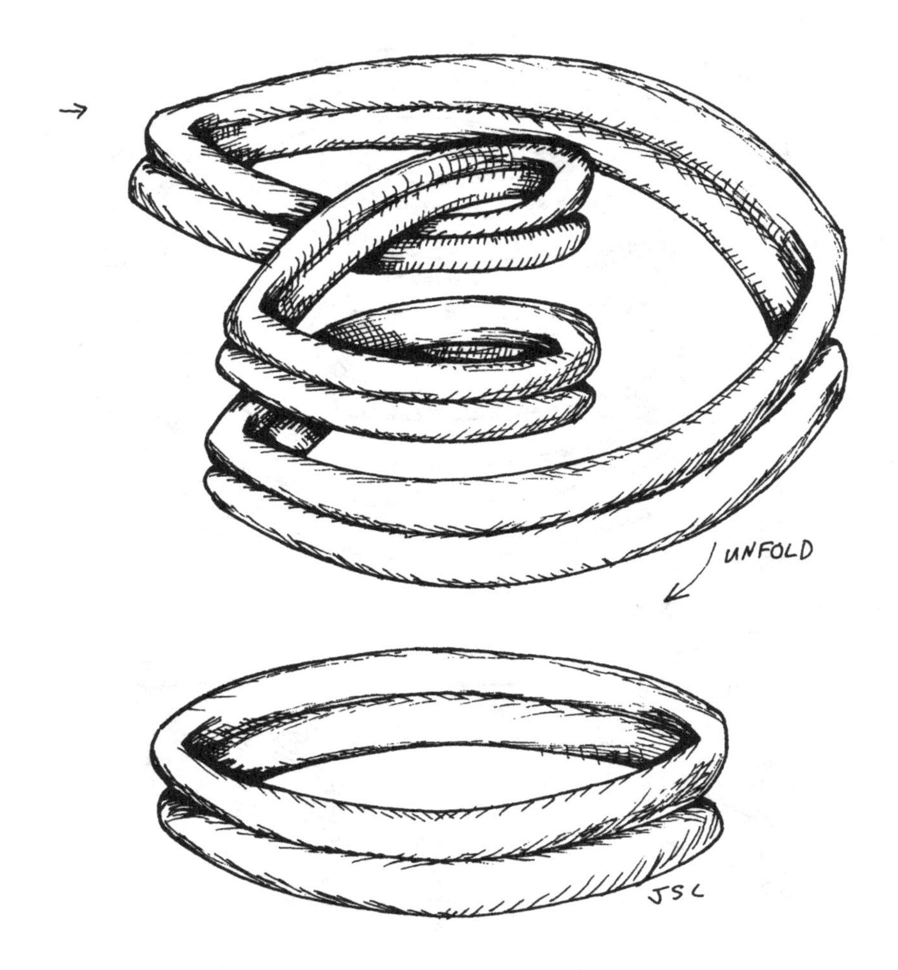
UNFOLD
JSC

The twisting can be a bit subtle, so let's do one last experiment with paper. Take a vertical strip and put two full twists in it, but don't tape the edges. Just hold them together with your fingers. Now keep the twists in tact but let the paper pass through itself once at the would-be taped edges. Hold on tightly, and tape the ends after the pass through. You can pull the twist out. This is again an illustration that there is a loop of rotations such that a journey twice around is trivial. ($\pi_1(SO(3)) = \mathbf{Z}/2$.)

To further illustrate this double twisting, a movie of undoing a doubly twisted figure 8 torus is given in Figure 7.10.

The Construction of Thom-Pontryagin

A fair question to ask now is, "Why does anyone care whether an immersed manifold is a boundary?" Well, the immersed manifold represents a very unintuitive idea that we will now explore. Suppose that there is a map from an $(n+k)$-dimensional sphere to an n-sphere. The set of points that map to a single point generically form an n-dimensional embedded manifold in the larger sphere.

There aren't many easy examples around, but I'll try to give one. The Hopf link represents the preimage of a pair of points under a map from the 3-sphere to the 2-sphere. In general, the preimage of a point is the set of points that solve a certain equation.

Given a map as above, the preimage is almost always a manifold. And if two different preimages are compared, they together bound a manifold of one higher dimension. So to determine whether or not the given unintuitive map is not trivial, we want to determine if the preimage manifold is not trivial.

The manifold contains some extra structure that is essentially the way a frame of reference that describes its neighborhood can twist around it. The framing is determined (once n is large enough) by the difference k in the dimensions of the bigger sphere and the little sphere. Finally, the framing really is determined by the way that one of the vectors twist. So the preimage manifold can be successively projected into codimension 1 where it intersects itself. (The codimension is the difference between the dimension of the bigger space and the smaller space.) All of these considerations imply that the manifold itself is orientable, but non-orientable manifolds can be fit into a similar scheme.

On the other hand, if we start with an n-manifold intersecting itself in $(n+1)$-space, we can gradually push it into a higher dimensional sphere until it is embedded. There is a way to use the framing to map the resulting $(n+k)$-dimensional sphere onto a k-dimensional sphere that is not so complicated. Everything outside a neighborhood is mapped to a point. The neighborhood looks like Cartesian product of a k-disk and the manifold, and the framing identifies it precisely as such. This k-disk is wrapped around the k-sphere as in the case of stereographic projection and the entire manifold is mapped to a point.

The construction from maps between spheres to manifolds and back is called the **Pontryagin-Thom construction.** It is a standard tool in higher dimensional topology.

To imagine the Pontryagin-Thom construction, I think of a cartoon in which a blob of ink grows first to an arc and then to an animated character. The character is a victim of cartoon violence and folds in on itself. It expands back to its original form and returns to the ink-well as a dot.

In summary, the Pontryagin-Thom construction gives a correspondence between maps between higher dimensional spheres on the one hand and immersed n-manifolds on the other. The self intersection sets of the immersed manifolds provide a way of measuring whether or not the maps between spheres are trivial. In essence, the self intersections measure the substance of the maps that they represent.

7.4 The General Notion of Space

This book opened with the question, "What is Space?" We have come to see that the correct questions are, "What are spaces?" and "How can spaces be distinguished from each other?" We have considered examples of 2- and 3-dimensional spaces, and we have mentioned the general notion of an n-dimensional manifold. Manifolds are important examples of spaces, but not every space that one would want to consider is a manifold.

The general notion of a space is extremely broad, so broad that the notion admits some spaces that some people consider pathological. Some sets should be endowed with a space-like structure, but in giving these sets the structure of spaces, we give other sets such a structure as well.

The notion of space includes a concept of openness and a concept of neighborhood. An arbitrary union of open subsets is an open subset, and a finite intersection of open subsets is an open subset. The whole space is open as is the empty set. A neighborhood of a point is a subset containing the point that also contains an open subset which contains the point. By specifying the types of open sets that occur, more structure is placed on the space.

We distinguish spaces only up to a continuously invertible map that has a continuous inverse. So we ask, "Given two spaces is there a continuous map from one to the other?" Continuous maps preserve connected sets. Therefore, we have analyzed spaces in terms of the subsets that do and do not separate them. That approach can distinguish some, but not all, spaces.

Dimension can distinguish spaces, but to show that dimension is an invariant took quite some doing. First, a rigorous notion of dimension had to be developed. Then homology was used to show that spaces of two different dimensions are different. The problem is difficult because the n-ball is the continuous image of an interval. The meaning of dimension is not at all obvious, and as sets of fractional dimension become more useful, the notion of dimension will, no doubt, undergo a major restructuring.

In algebraic topology, sequences of spaces that are related to each other are often considered as a single object. These spaces are related by certain continuous maps, and they form the back bones of some general algebraic machinery. Under suitable axioms these constitute a set of classifying spaces for a generalized homology theory in which the homology of a space is determined by a class of continuous functions from the space to the sequence. The lens spaces and the projective plane are spaces in some of these sequences.

The ways that surfaces intersect in ordinary space provides us with models for other intersections and other spaces. Closed surfaces separate space as simple closed curves separate the plane. There are surfaces that are not separated by closed curves, so there ought to be spaces that are not separated by surfaces, and there are. Non-separating arcs in surfaces lead to the general notion of non-separating sets. These define the substances that exist within a space. They are the geographical bearings for our journey into the unknown. Intersecting surfaces and spaces are the substance of maps between higher dimensional spheres. So higher dimensional spaces can be analyzed by means of their lower dimensional intersections.

Why do we want to know what space is? Because we want to understand our universe and our place within it. We must compare the space we see with the spaces we can imagine. Furthermore, we can use imagined spaces to model other aspects of our reality. Our economy is determined by numeric quantities (money), and these numbers must assemble themselves into some kind of multidimensional reality. We cannot uncover that reality as a single space unless we are familiar with many spaces. Ultimately, by exploring the spaces that we can imagine, we are exploring the deepest levels of our own minds, and to understand our place in the universe we must understand ourselves.

7.5 Notes

Milnor's notes [60] give backbone to the discussion on handles and general position. The book by Rourke and Sanderson [66] provides the rigorous framework in the piecewise linear setting. Both are excellent. Homology is developed in Greenburg's book [42] . And generalized homology is beautifully presented in Brayton Gray's book [41]. Several texts discuss the general notion of topological space. One outstanding reference is Kelley [53] because the information therein is encyclopaedic. For pedagogical reasons, I like Armstrong [4].

The material on equators in spheres is in my papers [11, 12, 13, 15, 16]. I don't know if any of these references is readable; occasionally the pictures are enlightening.

Bibliography

[1] Adams, Colin, "The Knot Book," W. H. Freeman and Company (1993).

[2] Apery, F., *La Surface du Boy*, Advances in Mathematics 61 (1986), 186-286.

[3] Apery, F., "Models of The Real Projective Plane," F. Vieweg & Sohn Braunschweig/Wiesbaden (1987).

[4] Armstrong, M. A., "Basic Topology," Springer-Verlag, New York, (1983).

[5] Banchoff, T.F., *Triple Points and Surgery of Immersed Surfaces*, Proc. AMS 46, No.3 (Dec. 1974), 403-413.

[6] Banchoff, T.F., "Beyond the Third Dimension," Scientific American Library, W. H. Freedman and Co., New York, (1990).

[7] Banchoff, T. F. *et al.*, *The Hypersphere: Foliation and Projections*, Video tape distributed by Thomas Banchoff Productions, Providence, RI. (Circa 1985).

[8] Boy, Werner, *Über die Curvature Integra und die Topologie Geshchlossener Flächen,* Math Ann. 57 (1903), 151-184.

[9] Bredon, G. E. and Wood, J. W., *Non-orientable Surfaces in Orientable 3-Manifolds*, Inventiones Math. 7 (1969), 83-110.

[10] Cairns, G., and Elton, D., *The Planarity Problem for Signed Gauss Words,* Journal of Knot Theory and Its Ramifications, Vol 2, No.4 (1993), 359-367.

[11] Carter, J. Scott, *Surgery on Codimension One Immersions in (n+1)-space: Removing n-tuple points*, Trans. AMS 298, No. 1, (Nov 1986), 83-102.

[12] Carter, J. Scott, *On Generalizing Boy's Surface: Constructing a Generator of the Third Stable Stem*, Trans. AMS 298, No. 1, (Nov 1986), 103-122.

[13] Carter, J. Scott, *A Further Generalization of Boy's Surface*, Houston Journal of Mathematics 12, No. 1 (1986), 11-31.

[14] Carter, J. Scott, *Simplifying the Self Intersection Sets of Codimension One Immersions in (n+1)-space*, Houston Journal of Mathematics, Vol. 13, No. 3, 1987, 353-365.

[15] Carter, J. Scott, *Surgery on the Equatorial Immersion I*, Illinois Journal of Mathematics 34, No. 4, (1988), 704-715.

[16] Carter, J. Scott, *Surgering the Equatorial Immersion in Low Dimensions*, Differential Topology Proceedings, Siegen 1987, ed. U.Koscorke, LNM 1350, 144-170.

[17] Carter, J. Scott, *Immersed Projective Planes in Lens Spaces*, Proc. AMS, 106, No. 1 (May 1989), 251-260.

[18] Carter, J. Scott, *Extending Immersed Circles in the Sphere to Immersed Disks in the Ball*, Comm. Math. Helv. Vol 67 (1992), 337- 348.

[19] Carter, J. Scott, *Classifying Immersed Curves*, Proc. of the AMS. 111, No. 1 (Jan. 1991), 281-287.

[20] Carter, J. Scott, *Extending Immersions of Curves to Properly Immersed Surfaces*, Topology and its Applications 40 (1991), 287-306.

[21] Carter, J. Scott, *Closed Curves That Never Extend to Proper Maps of Disks*, Proc. of the AMS 113, No. 3 (Nov 1992), 879-888.

[22] Carter, J. Scott and Saito, Masahico, *Syzygies among Elementary String Interactions in Dimension 2+1*, Letters in Math Physics 23 (1991), 287-300.

[23] Carter, J. Scott, and Saito, Masahico, *Planar Generalizations of the Yang-Baxter equation and Their Skeins*, Journal of Knot Theory and Its Ramifications 1, No 2 (1992), 207-217.

[24] Carter, J. Scott and Saito, Masahico, *Canceling Branch Points on Projections of Surfaces in 4-Space*, Proc. of the AMS. 116, No 1 (Sept 1992), 229-237.

[25] Carter, J. Scott, and Saito, Masahico, *Reidemeister Moves for Surface Isotopies and Their Interpretation As Moves to Movies*, Journal of Knot Theory and Its Ramifications Vol. 2, No 3 (1993), 251-284.

[26] Carter, J. Scott, and Saito, Masahico, *A Diagrammatic Theory of Knotted Surfaces*, in Baadhio, R. & Kauffman, L. "Quantum Topology," World Science Publishing (Series on Knots and Everything Vol. 3), (1993), 91-115.

[27] Carter, J. Scott, and Saito, Masahico, *A Seifert Algorithm for Knotted Surfaces*, Preprint.

[28] Carter, J. Scott, and Saito, Masahico, *Simplex Equations and Their Solutions*, Preprint.

[29] Carter, J. Scott, and Saito, Masahico, *Diagrammatic Invariants of Knotted Curves and Surfaces*, Preprint.

[30] Casson, A., *Lectures on New Infinite Constructions in 4-Dimensional Manifolds*, Notes by L. Guillou, Orsay.

[31] Chinn, W. G. and Steenrod, N. E., "First Concepts of Topology," Random House (1966).

[32] Crowell, P. and Marar, W. L. *Semi-regular Surfaces with a Single Triple Point,* Geometriae Dedicata 52 (1994), 142-153.

[33] Eccles, Peter J., *Codimension One Immersions and the Kervaire Invariant One Problem*, Math Proc. Cambr. Phil. Soc. 90 (1981), 483-493.

[34] Francis, G. K. "A Topological Picturebook," Springer-Verlag, New York (1987).

[35] Freedman, M. H., *Quadruple Points of 3-Manifolds in S^4*, Comment. Math. Helv. 53 (1978) 385-394.

[36] Freedman, M. H. and Quinn, F., "The Topology of 4- Manifolds," Princeton University Press (1990).

[37] Frohardt, Daniel, *On Equivalence Classes of Gauss Words*, Preprint.

[38] Gabai, David, *Foliations and the Topology of 3-manifolds III*, J. Differential Geometry, 26 (1987), 479-536.

[39] Gauss, C.F., Werke VIII, pages 271-286.

[40] Golubitsky, Martin, and Guillemin, Victor, "Stable Mappings and Their Singularities," Springer-Verlag, New York (1973).

[41] Gray, B., "Homotopy Theory: an Introduction to Algebraic Topology," Academic Press, New York (1975).

[42] Greenburg, M. J., "Lectures on Algebraic Topology," W. A. Bejamin (1967).

[43] Griffiths, H. B., "Surfaces," Cambridge University Press (1981).

[44] Hempel, J., "3-Manifolds," Prineton University Press (1976).

[45] Hilbert, D. and Cohn-Vassen, S., "Geometry and the Imagination," Chelsea, New York (1951).

[46] Homma, Tatsuo, and Nagase, Teruo, *On Elementary Deformations of the Maps of Surfaces into 3-manifolds I,* Yokohama Math. J. 33 (1985), 103-119.

[47] Hughes, J. F. *Boy's Surface,* a short computer animation, Circa (1989).

[48] Izumiya, H. and Marar, W. L. *The Euler Characteristic of a Generic Wave Fornt in a 3-Manifold*, Proc. AMS 118 No 4 (August 1993), 1347-1350.

[49] Jaco, Wm., "Lectures on Three Manifold Topology," Conference Board on Mathematical Sciences and The American Mathematical Society (1980).

[50] Jones, V. F. R., *Hecke Algebra Representations of Braid Groups and Link Polynomials,* Annals of Mathematics 126, (1987), 335-388, Reprinted in Kohno, T, "New Delopments in the Theory of Knots," World Science Publishing, (1989).

[51] Kauffman, L. H., "On Knots," Princeton Unviersity Press (1987).

[52] Kauffman, L. H., "Knots and Physics," World Scientific Press (1991).

[53] Kelley, J. L., "General Topology," Springer-Verlag, New York (1975).

[54] Lannes, *Sur les Immersion de Boy,* Preprint Circa (1982).

[55] Ki Hyoung and Carter, J. S., *Triple Points of Immersed Surfaces in Three Dimensional Manifolds,* Topology and Its Applications 32, (1989), 149-159.

[56] Koschorke, Ulrich *Multiple points of Immersions and the Kahn-Priddy Theorem,* Math Z. 169 (1979), 223-236.

[57] Massey, Wm. S., "Algebraic Topology: an Introduction," Springer Verlag, New York (1967).

[58] Max, Nelson, *Turining a Sphere Inside Out,* a 23 minute color sound 16 mm. film, International Film Bureau, Chicago, Illinois (1976).

[59] Milnor, J. W., "Topology from the Differentiable Viewpoint," University of Virginia Press (1965).

[60] Milnor, J. W., "Lectures on the h-Cobordism Theory," Princeton University Press (1965).

[61] Morin, B. *Equations du retournement de la sphere,* C. R. Acad. Sc. Paris 287 (1978), 876-882.

[62] Phillips, A., *Turning a Surface Inside Out,* Scientific American 214 (May 1966), 112-120.

[63] Pugh, C. *Chicken Wire Models* that were reported in Max's film to have been displayed at U. C. Berkeley.

[64] Reidemeister, Kurt, "Knotentheorie," Springer Verlag, Berlin (1932). "Knot Theory," BSC Associates, Moscow, Idaho (1983).

[65] Rochlin, V. A., *Proof of an Conjecture of Godkov,* Func. Anal. Appl. 6 (1972), 136-138.

[66] Rourke, C. P., and Sanderson, B. J., "Introduction to Piecewise Linear Topology," Springer-Verlag, New York (1982).

[67] Rolfsen, Dale, "Knots and Links," Publish or Perish Press, Berkley (1976).

[68] Roseman, Dennis, *Reidemeister-Type Moves for Surfaces in Four Dimensional Space,* Preprint.

[69] Rucker, R., "The Fourth Dimension: Toward a Geometry of Higher Relality," (1984).

[70] Seifert and Threfall, "Seifert and Threfall: A Textbook of Topology," Academic Press, (1980).

[71] Smale, S., *A Classification of Immersions of the two-sphere*, Transactions of the AMS 90 (1961), 391-406.

[72] Thurston, Wm. P. "The Geometry and Topology of 3-Manifolds," Princeton University Lecture Notes (1982).

[73] Vonnegut, K."Slaughterhouse Five, or the Childrens Crusade," Dell Publishing Co. New York (1978).

[74] Weeks, J. R., "The Shape of Space," Marcel Dekker, Inc. (1985).

[75] Whitney, H., *On the Topology of Differentiable Manifolds*, in Lectures in Topology, (Wilder and Ayres, eds.), University of Michigan Press (1941), 101-141.

[76] Wilson, G. "Everybody's Favorite Duck," Mysterious Press, New York (1988).

[77] Zelazny, R., "Doorways in the Sand," Harper Collins (1981).

Index